Lecture Notes in
Applied Differential
Equations of
Mathematical Physics

Lecture Notes in
Applied Differential Equations of Mathematical Physics

Luiz C L Botelho
Federal University, Brazil

W World Scientific

NEW JERSEY · LONDON · SINGAPORE · BEIJING · SHANGHAI · HONG KONG · TAIPEI · CHENNAI

Published by

World Scientific Publishing Co. Pte. Ltd.

5 Toh Tuck Link, Singapore 596224

USA office: 27 Warren Street, Suite 401-402, Hackensack, NJ 07601

UK office: 57 Shelton Street, Covent Garden, London WC2H 9HE

British Library Cataloguing-in-Publication Data
A catalogue record for this book is available from the British Library.

ISBN-13 978-981-281-457-9
ISBN-10 981-281-457-4

Printed in Singapore.

For all those who do not think that doing research
and writing is just to sign and be "co-author"
in other's papers.

Foreword

"...Both Mathematics and Physics are in a state of crisis in present days, from intellectual and social sides simultaneously. Both have much to contribute to each other's disciplines, and I believe that a renewal of their traditional liaison would go far to cure at least the intellectual side of the malaise felt by workers in both disciplines..." — Robert Hermann in *Lectures in Mathematical Physics* — Vol. I & Vol. II.

My aim in this set of informal lecture notes on applied mathematical analysis is to present mathematical author's original research material (Chaps. 3–9 and appendixes) revised and amplified of our two previous work on path-integrals in statistical and quantum physics and strongly focused at a pure and applied mathematical audience of readers.

We have a strong believe that mathematics students can find a source of inspiration to stimulate mutual interaction of their discipline and physics and thus, undertaking the reading of physics books, talking with physicists and try to read our previous two monographs in classical and quantum physics. The special emphasis in this supplementary volume 3 is on advanced mathematical analysis methods with its applications to solve partial differential equations in finite and infinite dimensions arising in applied settings. Another objective behind writing this set of author's mostly original results in the form of lectures is the hope it could serve to show that the "aridity" and "sterility" that one finds in much modern "estimate mathematical analysis" can be conterweighted side by side by mathematical analysis formulae exactly used in the modern applications.

A word on the "methodology" in our set of lecture notes as a set of complementary notes we feel that the students should read the chapter with a pencil and paper at hand, after which they should make additional studies in the specific topic in the literature and thus present their studies in seminars to others students with all lectures mathematical details worked out.

Luiz C.L. Botelho
Professor

Contents

Chapter 1

Elementary Aspects of Potential Theory in Mathematical Physics

1.1. Introduction

Many problems of elementary classical mathematical-physical analysis are connected with the solution through integral representations of the Dirichlet, Newmann, and Poisson problems for the Laplacian operator, including its generalization to the Cauchy problem for wave and diffusion equations. In this chapter, we present the basic and introductory mathematical methods analysis used in obtaining such above-mentioned integral representations.[1]

1.2. The Laplace Differential Operator and the Poisson–Dirichlet Potential Problem

Let Ω be an open-bounded set in R^N. We define $C^2(\Omega)$ to be the vectorial space composed by those functions $f(x) \equiv f(x^1, \ldots, x^n) \equiv f(\vec{r})$. (Here we adopt the physical notation $\vec{r} = \sum_{i=1}^{N} x^i e_i$ with $\vec{e}_i$ denoting the canonical vectorial basis of R^N), continuously differentiable until the second order in Ω, namely,

$$\frac{\partial^2 f(x)}{\partial x^k \partial x^m} \in C(\Omega), \quad \text{for any } (k, m) \in \{1, \ldots, n\}.$$

Let us consider the following linear transformation with domain $C^2(\Omega)$ and range $C(\Omega)$

$$\Delta_{(N)} \colon C^2(\Omega \to C(\Omega)$$

$$f(x) \to (\Delta f)(x) = \left(\sum_{i=1}^{n} \frac{\partial^2}{\partial x_i^2} \right) f(x^1, \ldots, x^n). \quad (1.1)$$

We now have the following simple result about the kernel of the above linear transformation in R^3 (holding true for any R^N, with $N \geq 2$).

1

Lemma 1.1. *The kernel of the Laplacean operator as defined by Eq. (1.1) has infinite dimension in R^3.*

Proof. Let $\bar{f} \in C^2(\mathbb{C} \times R)$ such that $\frac{\partial^2 \bar{f}}{\partial^2 w}(w, t)$ exists and the function defined below ($x^1 = x$, $x^2 = y$, $x^3 = z$ — the usual classical notation in potential theory in the physical space R^3)

$$U_t(x, y, z) = \bar{f}(ix \cos t + iy \sin t + z, t). \tag{1.2}$$

We wish to show that

$$\Delta_{(3)}\big(U_t(x, y, z)\big) = 0. \tag{1.3}$$

This result is a simple consequence of the following elementary identities:

$$\frac{\partial^2}{\partial^2 z^2} U_t(x, y, z) = \frac{\partial^2}{\partial^2 w} \bar{f}(w, t)|_{w = z + ix \cos t + iy \sin t}, \tag{1.4a}$$

$$\frac{\partial^2}{\partial^2 x^2} U_t(x, y, z) = \frac{\partial^2}{\partial^2 w} \bar{f}(w, t)|_{w = z + ix \cos t + iy \sin t}((i \cos t)^2), \tag{1.4b}$$

$$\frac{\partial^2}{\partial^2 y^2} U_t(x, y, z) = \frac{\partial^2}{\partial^2 w} \bar{f}(w, t)|_{w = z + ix \cos t + iy \sin t}(i \sin t)^2. \tag{1.4c}$$

As a consequence

$$\Delta_{(3)}\big(U_t(x, y, z)\big) = \frac{\partial^2}{\partial^2 w} f(w, t)\Big|_{w = z + ix \cos t + iy \sin t}(1 - \cos^2 t - \sin^2 t) \equiv 0. \tag{1.5}$$

Note that a whole class of harmonic functions $\tilde{H}(\Omega)$ in the class of $C^2(\Omega)$-functions in the kernel of the Laplacean operator Eq. (1.3) may be obtained from Eq. (1.3):

$$U(x, y, z) = \int_a^b dt\, g(t)(\bar{f}(w, t)|_{w = ix \cos t + i \sin ty + z}), \tag{1.6}$$

and for the special function below:

$$\bar{f}(w, t) = w^n\, e^{i\alpha t}, \quad (\alpha \in R, n \in \mathbb{Z}),$$

we have the harmonic legendre polynomials in the three-dimensional spherical ball

$$U(x,y,z) = \int_{-\pi}^{\pi} e^{i\alpha t} \left(z + ix\cos t + iy\sin t\right)^n dt$$

$$= \int_{-\pi}^{\pi} dt\, e^{i\alpha t} \left(r\cos\theta + ir\sin\theta\cos\varphi\cos t + ir\sin\theta\sin\varphi\sin t\right)^n$$

$$= 2r^n\, e^{i\alpha\varphi} \left(\int_{0}^{\pi} dt\, \cos(\alpha t)(\cos\theta + (\sin\theta\sin t)^n \right)$$

$$= r^n\, e^{i\alpha\varphi}\, P_{n,\alpha}(\cos\theta). \tag{1.7}$$

After these preliminary elementary remarks we pass on the discussion of the famous classical Poisson–Newton problem: Let $\Omega \subset R^3$ be a bounded open set of R^3 with a boundary $\partial\Omega$ being an orientable simply connected C^2-surface. We search for the solution of the nonhomogeneous linear problem in Ω through an integral representation (the Laplacean Green function).

$$-\Delta_3\, U(x,y,z) = f(x,y,z). \tag{1.8}$$

Here $f(x,y,z) \in C^1(\bar{\Omega})$ $(\bar{\Omega} = \Omega \cup \partial\Omega)$.

We have the basic result in answering the Poisson–Newton problem.

$\square$

Theorem 1.1. *The Laplacean operator*

$$-\Delta_3\colon C^2(\Omega)\big/ \tilde{H}(\Omega) \to C^1(\bar{\Omega}) \tag{1.9}$$

is inversible, and its inverse has the following integral representation:

$$U(x,y,z) = \frac{1}{4\pi}\left\{ \iiint_{\Omega} dx'dy'dz'\, \frac{f(x',y',z')}{\sqrt{(x-x')^2 + (y-y')^2 + (z-z')^2}} \right\}$$

$$\stackrel{\text{def}}{\equiv} \iiint_{\Omega} d^3\vec{r}'\, f(\vec{r}')\cdot G_{(0)}(|\vec{r}-\vec{r}'|). \tag{1.10}$$

Proof. Let us re-introduce the "classic mathematical-physicist" vectorial notation in our formulas

$$\vec{r} = (x, y, z); \quad \vec{r}' = (x', y', z').$$ (1.11)

Let us consider the δ-approximant solution for the problem

$$U^{(\delta)}(\vec{r}) = \frac{1}{4\pi} \iiint_{\Omega} d^3\vec{r}' \, f(\vec{r}') \, G^{(\delta)}_{(0)}(|\vec{r} - \vec{r}'|).$$ (1.12)

Here the "regularized" Green function is given by

$$G^{(\delta)}_{(0)}(|\vec{r} - \vec{r}'|) = \begin{cases} \dfrac{1}{|\vec{r} - \vec{r}'|}, & \text{for } |\vec{r} - \vec{r}'| > \delta, \\[2mm] \dfrac{1}{2\delta}\left(3 - \dfrac{|\vec{r} - \vec{r}'|^2}{\delta^2}\right), & \text{for } |\vec{r} - \vec{r}'| \le \delta. \end{cases}$$ (1.13)

Note that the above "regularized" Green function is a continuous function in both variables, i.e.

$$G^{(\delta)}_{(0)}(|\vec{r} - \vec{r}'|) \in C(\Omega \times \Omega)(\text{since } G^{(\delta)}_{(0)}(\delta^+) = G^{(\delta)}_{(0)}(\delta^-)).$$

We now have the estimate for $\delta \to 0$:

$$\sup_{\vec{r}\in\Omega}\left(|U^{(\delta)} - U|(\vec{r})\right) \le \|f\|^{\frac{1}{2}}_{L^2(\Omega)} \times \left\{ \iiint_{B_\delta(0)} d^3\vec{r} |G^{(\delta)}(|\vec{r} - \vec{r}'|)|^2 \right\}^{\frac{1}{2}}$$

$$\le \|f\|^{\frac{1}{2}}_{L^2(\Omega)}(\sqrt{2\pi}\delta),$$ (1.14)

which shows the uniform convergence of the sequence of approximate solutions to the (well-defined) function $U(\vec{r})$ by the Weierstrass uniform convergence criterium.

That the integral representation Eq. (1.10) defines a well-defined function comes straightforwardly from the estimate around our infinitesimal ball $B_\varepsilon(\vec{r}) = \{\vec{r}' \mid ||\vec{r}' - \vec{r}| \le \varepsilon\}$.

$$\left| \iiint_{B_\varepsilon(\vec{r})} d^3\vec{r}' \, \frac{f(\vec{r}')}{|\vec{r} - \vec{r}'|} \right| \le \|f\|^{\frac{1}{2}}_{L^2(\Omega)} \times \left\{ 4\pi \int_0^\varepsilon dv \cdot v^2 \cdot \frac{1}{v^2} \right\}$$

$$\le (4\pi\varepsilon)^{\frac{1}{2}} \|f\|^{\frac{1}{2}}_{L^2(\Omega)} < \infty.$$ (1.15)

As a second step to the proof of Theorem 1.1, we point out that the sequence of functions

$$\nabla \cdot U^{(\delta)} = \frac{1}{4\pi} \iiint_{\Omega} d^3\vec{r} \, f(\vec{r}')\nabla \cdot G^{(\delta)}(|\vec{r} - \vec{r}'|)$$ (1.16)

converges uniformly to the (well-defined) function in $C(\bar{\Omega})$

$$I = +\left\{ \iiint_{\Omega} d^3 \vec{r}' \, f(\vec{r}') \nabla_{\vec{r}} \left(\frac{1}{4\pi|\vec{r} - \vec{r}'|} \right) \right\}. \qquad (1.17)$$

The proof of the above made claim is a consequence of the following estimate:

$$\lim_{\delta \to 0} \left\{ \iiint_{B_\delta(\vec{r})} d^3 \vec{r}' \left| \left(\nabla_{\vec{r}_x} \left[-\frac{|\vec{r} - \vec{r}'|^2}{\delta^2} \right] + \frac{(x - x')}{|\vec{r} - \vec{r}'|^3} \right) \right|^2 \right\} = 0. \qquad (1.18)$$

Let us now pass on the problem of evaluating second derivatives.

In this case we have the Gauss theorem (integration by parts in R^3!)

$$\iiint_{\Omega} d^3 \vec{r}' \, f(\vec{r}') \, \nabla_{\vec{r}} \left(\frac{1}{|\vec{r} - \vec{r}'|} \right)$$

$$= -\iiint_{\Omega} d^3 \vec{r} \, f(\vec{r}') \, \nabla_{\vec{r}'} \left(\frac{1}{|\vec{r} - \vec{r}'|} \right)$$

$$= -\oiint_{\partial\Omega} \left(f(\vec{r}') \frac{1}{|\vec{r} - \vec{r}'|} \right) \left(\cos(\vec{N} \angle \vec{r}') \right) dS(\vec{r}')$$

$$+ \iiint_{\Omega} d^3 \vec{r}' \left(\nabla_{\vec{r}'} \cdot f(\vec{r}') \right) \frac{1}{|\vec{r} - \vec{r}'|}. \qquad (1.19)$$

Here $\cos(\vec{N} \angle \vec{r})$ is the cosine between the normal field of the globally oriented surface $\partial\Omega$ and the point $\vec{r}$, argument of the function $U(\vec{r})$. At this point we are using the fact that $f \in C^1(\bar{\Omega})$. However, Eq. (1.19) has a rigorous meaning for $f \in L^2(\bar{\Omega})$ (left as an exercise to our diligent reader!). It is important to remark that at this point of our exposition all these topological–geometrical constraints imposed on the surface $\partial\Omega$ by means of Gauss theorem must be imposed as complementary hypothesis on the geometrical nature of the Poisson problem.

By deriving a second time the left-hand side of Eq. (1.19) and taking into account that in the surface integral we always have $\vec{r} \neq \vec{r}'$ (we do not evaluate the solution $U(\vec{r})$ at the boundary points) and obviously $\nabla_{\vec{r}'} f(\vec{r}) \equiv 0$, we arrive at the following result (without bothering ourselves with the 4π-overall factor in Eq. (1.10)):

$$\Delta_{\vec{r}} U(\vec{r}) = -\oiint_{\partial\Omega} f(\vec{r}\,') \, \nabla_{\vec{r}}\left(\frac{1}{|\vec{r} - \vec{r}\,'|}\right) \cos(\vec{N} \angle \vec{r}\,') dS(\vec{r}\,')$$

$$+ \iiint_{\Omega} d^3\vec{r}\,' \left(\nabla_{\vec{r}\,'} (f(\vec{r}\,') - f(\vec{r})) \nabla_{\vec{r}}\left(\frac{1}{|\vec{r} - \vec{r}\,'|}\right)\right)$$

$$= \left(-\oiint_{\partial\Omega} f(\vec{r}\,') \left(\nabla_{\vec{r}} \frac{1}{|\vec{r} - \vec{r}\,'|}\right) \cos(\vec{N} \angle \vec{r}\,') dS(\vec{r}\,')\right)$$

$$+ \left(\oiint_{\partial\Omega} (f(\vec{r}\,') - f(\vec{r})) \nabla_{\vec{r}}\left(\frac{1}{|\vec{r} - \vec{r}\,'|}\right) \cos(\vec{N} \angle \vec{r}\,') dS(\vec{r}\,')\right)$$

$$- \left(\iiint_{\Omega} (f(\vec{r}\,') - f(\vec{r})) \Delta_{\vec{r}}\left(\frac{1}{|\vec{r} - \vec{r}\,'|}\right) d^3\vec{r}\,'\right). \qquad (1.20)$$

Now we have that

$$f(\vec{r}) \left\{ \oiint_{\partial\Omega} \nabla_{\vec{r}}\left(\frac{1}{|\vec{r} - \vec{r}\,'|}\right) \cos(\vec{N} \angle \vec{r}\,') dS(\vec{r}\,') \right\}$$

$$= f(\vec{r}) \left\{ \oiint_{\partial\Omega} \frac{\vec{N} \cdot \vec{r}\,'}{|\vec{r}\,'|^3} \, ds(\vec{r}) \right\} = 4\pi f(\vec{r}) \qquad (1.21)$$

and (with $x^a = (x, y, z)$ and $a = 1, 2, 3$):

$$\left| \iiint_{B_\varepsilon(\vec{r})} (f(\vec{r}\,') - f(\vec{r})) \left(-\frac{3}{|\vec{r} - \vec{r}\,'|^3} + \frac{(x_a - x_a')^2}{|\vec{r} - \vec{r}\,'|^5}\right) d^3\vec{r}\,' \right|$$

$$\leq \left(\sup_{\vec{r}\,' \in B_\varepsilon(\vec{r})} |Df|(\vec{r}\,')\right) \times \left\{ 3 \iiint_{B_\varepsilon(\vec{r})} d^3\vec{r}\,' \frac{|\vec{r} - \vec{r}\,'|}{|\vec{r} - \vec{r}\,'|^3} \right.$$

$$\left. + \iiint_{B_\varepsilon(\vec{r})} \frac{|x_a - x_a'|^2 |\vec{r} - \vec{r}\,'|}{|\vec{r} - \vec{r}\,'|^5} d^3\vec{r}\,' \right\} \underset{\to 0}{\varepsilon \to 0}. \qquad (1.22)$$

Precisely, to obtain this estimate, we need the condition that the data $f(\vec{r})$ must belong to the space $C^1(\bar{\Omega})$ (including the continuity of the derivative at the boundary $\partial\Omega$!).

As a general exercise in this section we left to our reader to prove Theorem 1.1 in the general case of the domain $\Omega \subset R^N$ ($N \geq 3$) with the Green function in R^N.

$$U(\vec{r}) = \frac{\Gamma\left(\frac{N}{2}\right)}{(N-2)2(\sqrt{\pi})^N} \times \left\{ \int_{\Omega} d^N\vec{r}\,' f(\vec{r}\,') |\vec{r} - \vec{r}\,'|^{2-N} \right\}. \qquad (1.23)$$

Note that the regularized Green function $G^{(\delta)}$ is now given by

$$
G^{(\delta)} \, |\vec{r}, \vec{r}'\,)
$$

$$
= \begin{cases}
\dfrac{1}{|\vec{r} - \vec{r}'|^{N-2}} & \text{if } |\vec{r} - \vec{r}'| > \delta, \\[4mm]
\dfrac{\delta^{2-N}}{(N-2)w_N} \left[2 - \dfrac{2}{N} - \left(\dfrac{N-2}{2}\right) \left(\dfrac{|\vec{r} - \vec{r}'|}{\delta}\right)^{N}\right] & \text{if } |\vec{r} - \vec{r}'| \leq \delta.
\end{cases}
$$

$$(1.24)$$

In R^2, we can show that (left to our readers)

$$
U(\vec{r}) = -\frac{1}{2\pi}\left\{ \int_{\Omega} d^2\vec{r}' \, \ell g\left(\frac{1}{|\vec{r} - \vec{r}'|}\right) f(\vec{r}')\right\}. \tag{1.25}
$$

This case can be seen as coming from the general case Eq. (1.23) after considering the (distributional) limit

$$
\lim_{N \to 2}\left\{ \frac{\Gamma(\tfrac{2}{2})}{2(\sqrt{\pi})^2}\frac{1}{N-2}\, e^{(2-N)\ell g|\vec{r} - \vec{r}'|}\right\} = -\frac{1}{2\pi}\, \ell g|\vec{r} - \vec{r}'|. \tag{1.26}
$$

We consider now the well-known Dirichlet problem: Determine a $U(\vec{r}) \in C^2(\Omega) \cap C^1(\bar{\Omega})$ such that

$$
\begin{cases}
\Delta U(\vec{r}) = 0, & \text{for } r \in \Omega, \\
U(\vec{r})\big|_{\partial\Omega} = g(\vec{r})\big|_{\partial\Omega},
\end{cases} \tag{1.27}
$$

a very useful problem in string path integrals for $\Omega \subset R^2$.

By a simple application of the first Green identity we obtain the uniqueness property for the Dirichlet problem

$$
\iiint_{\Omega} (U \cdot \overset{0}{\widehat{\Delta U}} + |\nabla U|^2)(\vec{r})d^3\vec{r} = \oiint_{\partial\Omega} (U \cdot \nabla_{\vec{N}} U)(\vec{r})d\Gamma(\vec{r}) = 0. \tag{1.28}
$$

As a consequence,

$$
\iiint_{\Omega} |\nabla U|^2(\vec{r})d^3\vec{r} = 0 \Leftrightarrow U(\vec{r}) \equiv 0 \quad \text{in } \Omega. \tag{1.29}
$$

An explicit solution for Eq. (1.27) is easily obtained by considering the Poincaré method of considering a Green function of the structural form below.

Let

$$G(\vec{r}, \vec{r}) = H(\vec{r}, \vec{r}') + \frac{1}{4\pi|\vec{r} - \vec{r}'|}, \tag{1.30}$$

where $H(\vec{r}, \vec{r}')$ is a harmonic function in Ω in relation to the variable $\vec{r}'$, satisfying the following condition:

$$H(\vec{r}, \vec{r}')|_{\vec{r}' \in \partial\Omega} = -\left. \frac{1}{4\pi|\vec{r} - \vec{r}'|} \right|_{\vec{r}' \in \partial\Omega}. \tag{1.31}$$

Then we have the following explicit representation: □

Theorem 1.2. *We have the integral representation for the Dirichlet problem*

$$U(\vec{r}) = \underbrace{-}_{(-1/2\pi \text{ if } \vec{r} \in \partial\Omega - \text{exercise})} \frac{1}{4\pi} \left\{ \oiint_{\partial\Omega} (g(\vec{r}')\nabla_{\vec{N}} G(\vec{r}, \vec{r}')) dS(\vec{r}') \right\}$$

$$= -\frac{1}{4\pi} \left\{ \oiint_{\partial\Omega} \left(g(\vec{r}')\nabla_{\vec{N}} \left(H(\vec{r}, \vec{r}') + \frac{1}{4\pi|\vec{r} - \vec{r}'|} \right) \right) dS(\vec{r}') \right\}. \tag{1.32}$$

Proof. Let us consider the regularized form for the Green function as proposed by the Poincaré ansatz

$$G^{(\delta)}(\vec{r}, \vec{r}') = H^{(\delta)}(\vec{r}, \vec{r}') + U^{(\delta)}(\vec{r} - \vec{r}'). \tag{1.33}$$

Here

$$\begin{cases} \Delta_{\vec{r}} H^{(\delta)}(\vec{r}, \vec{r}') = 0, \\ H^{(\delta)}(\vec{r}, \vec{r}')|_{\vec{r}' \in \partial\Omega} = -U^{(\delta)}(\vec{r} - \vec{r}'), \end{cases} \tag{1.34}$$

with

$$U^{(\delta)}(\vec{r} - \vec{r}') = \begin{cases} \dfrac{1}{4\pi|\vec{r} - \vec{r}'|}, & \text{if } |\vec{r} - \vec{r}'| > \delta, \\ \dfrac{1}{4\pi} \left[\dfrac{1}{2\delta} \left(3 - \dfrac{|\vec{r} - \vec{r}'|^2}{\delta^2} \right) \right], & \text{if } |\vec{r} - \vec{r}'| \le \delta. \end{cases} \tag{1.35}$$

By applying the second Green identity to the pair of functions $U(\vec{r})$ and $G^{(\delta)}(\vec{r}, \vec{r}')$, we obtain the relation

$$\iiint_{\Omega} [U(\vec{r}')(\Delta_{\vec{r}'} U^{(\delta)}(\vec{r}, \vec{r}'))] d^3\vec{r}'$$

$$= \oiint_{\partial\Omega} (U(\vec{r}') \nabla_{\vec{N}} G^{(\delta)}(\vec{r}, \vec{r}') dS(\vec{r}'). \tag{1.36}$$

By noting that we have explicitly the Laplacean action on the regularized function $U^{(\delta)}$:

$$\Delta_{\vec{r}'} U^{(\delta)}(\vec{r},\vec{r}') = \begin{cases} 0, & \text{if } |\vec{r}' - \vec{r}'| > \delta, \\ -\left(\dfrac{3}{4\pi\delta^3}\right), & \text{if } |\vec{r} - \vec{r}'| \leq \delta, \end{cases} \tag{1.37}$$

and by taking the limit of $\delta \to 0$, it yields the integral representation Eq. (1.30).

Our problem now is to obtain an explicit procedure to determine the Harmonic piece of the Green function, namely, $H(\vec{r},\vec{r}')$.

A solution was first given by H. Poincaré through the use of the so-called double-layer potential $\Phi(\vec{r},\vec{r}')$, satisfying the Poincaré integral equation in $L^2(\partial\Omega, dS)$:

$$\left(-\frac{1}{4\pi|\vec{r} - \vec{r}'|}\right) = \frac{\Phi(\vec{r},\vec{r}')}{2} - \oiint_{\partial\Omega} \nabla_{\vec{N}}\left(\frac{1}{4\pi}\frac{1}{|\vec{r} - \vec{r}'|}\right)\Phi(\vec{r},\vec{w}) \times dS(\vec{w}). \tag{1.38}$$

Note the somewhat geometrical complexity of the geometrical measure dS on the surface $\partial\Omega$, when the integral equation above is solved by the standard Fixed Point Theorems.[1,2]

After solving Eq. (1.38), Poincaré showed that one easily has a formula for the function $H(\vec{r},\vec{r}')$:

$$H(\vec{r},\vec{r}') = \oiint_{\partial\Omega} \Phi(\vec{r},\vec{r}')\nabla_{\vec{N}}\left(-\frac{1}{4\pi|\vec{r} - \vec{r}'|}\right)dS(\vec{r}'). \tag{1.39a}$$

Another point worth to call the reader's attention is that it appears more straightforward to consider the following integral equation for the surface derivative of the harmonic function $U(\vec{r})$. Since

$$U(\vec{r}) = \frac{1}{4\pi}\oiint_{\partial\Omega}\left\{\frac{1}{|\vec{r} - \vec{r}'|}\nabla_{\vec{N}}U(\vec{r}') - U(\vec{r}')\nabla_{\vec{N}}\left(\frac{1}{|\vec{r} - \vec{r}'|}\right)\right\}dS(\vec{r}), \tag{1.39b}$$

we have

$$(\vec{\nabla}_{\vec{r}}\vec{U})(\vec{r}) \cdot \vec{N}(\vec{r}) = \nabla_{\vec{N}}U(\vec{r}) = \rho(\vec{r})$$

$$= \frac{1}{4\pi}\oiint_{\partial\Omega}\left\{ -\left(\frac{(\vec{r} - \vec{r}')}{|\vec{r} - \vec{r}'|^3}\vec{N}(\vec{r})\right) \cdot \overbrace{\nabla_{\vec{N}}U(\vec{r}')}^{\rho(\vec{r}')} \right.$$

$$\left. -g(\vec{r}')\nabla_{\vec{N}}\left(\nabla_{\vec{N}}\frac{1}{|\vec{r} - \vec{r}'|}\right)\right\}dS(\vec{r}'). \tag{1.39c}$$

One can in principle solve the above-written integral equation in $L^2(\partial\Omega, dS(\vec{r}))$ by the Banach fixed point theorem at least for small volumes and for the normal derivative of the harmonic function $U(\vec{r})$ and then using again the Liouville integral representation Eq. (1.39b) to determine $U(\vec{r})$ in Ω.

A more useful approach for the explicit determination of the harmonic term $H(r, \vec{r}')$ in the Poincaré–Green function is by means of the eigenvalue problem for our Laplacean operator with a Dirichlet condition

$$-\Delta\,\varphi_n(\vec{r}) = \lambda_n\,\varphi_n(\vec{r}) \quad \text{for } \vec{r} \in \Omega,$$
$$\varphi_n(\vec{r})\big|_{\partial\Omega} = 0 \qquad \text{for } \vec{r} \in \partial\Omega, \tag{1.40}$$

or in the equivalent integral form (see Theorem 1.1)

$$\mu_n\,\varphi_n(\vec{r}) = \iiint_\Omega d^3\vec{r}'\,\frac{\varphi_n(\vec{r}')}{4\pi|\vec{r}-\vec{r}'|} \tag{1.41}$$

with

$$\mu_n = \frac{1}{\lambda_n} = \left(\left(\iiint_\Omega \varphi_n^2(\vec{r})d^3\vec{r}\right)\Big/\iiint_\Omega |\nabla\varphi_n|^2(\vec{r})d^3\vec{r}\right) > 0.$$

Now it is straightforward to see that

$$\iiint_\Omega d^3\vec{r}'\,G(\vec{r},\vec{r}')\varphi_n(\vec{r}') = \left(\frac{\varphi_n(\vec{r})}{\lambda_n} - \frac{1}{\lambda_n}\oiint_{\partial\Omega}\frac{(\nabla_{\vec{N}}\,\varphi_n(\vec{r}'))dS(\vec{r}')}{4\pi|\vec{r}-\vec{r}'|}\right)$$
$$+ \left(\iiint_\Omega H(\vec{r},\vec{r}')\frac{(-\Delta_{\vec{r}'}\,\varphi_n(\vec{r}'))}{\lambda_n}\,d^3\vec{r}'\right). \tag{1.42}$$

At the point we left as an exercise to our reader to show that the surface integral and the volume integral cancel out.

By the usual Mercer theorem (if $\sup_{\vec{r}\in\bar{\Omega}}|\varphi_n(\vec{r})| \leq M$, for $\forall n \in \mathbb{Z}^+$), we have the uniform convergence of the spectral series defining the Dirichlet Green Function as a result that $\lim_{n\to\infty}\left(\frac{\lambda_n}{|n|^2}\right) < \infty$,

$$|G(\vec{r},\vec{r}')| \leq \sum_{n\in\mathbb{Z}^3}\frac{|\varphi_n(\vec{r})\varphi_n(\vec{r}')|}{\lambda_n} \leq M^2\left\{\sum_{n\in\mathbb{Z}^3}\frac{1}{|n|^2}\right\} < \infty. \tag{1.43}$$

The same result holds true for the case of the Newmann problem written below:

$$\begin{cases}\Delta\,U(\vec{r}) = 0, & \text{for } \vec{r}\in\Omega, \\ \nabla_{\vec{N}}\,U(\vec{r}) = g(\vec{r}) & \left(\oiint_{\partial\Omega} g(\vec{r}')dS(\vec{r}') \equiv 0\right),\end{cases} \tag{1.44a}$$

with

$$U(\vec{r}) = + \oint_{\partial\Omega} g(\vec{r}) G(\vec{r}, \vec{r}') dS(\vec{r}'), \tag{1.44b}$$

where

$$G(\vec{r}, \vec{r}') = \sum_n \frac{\beta_n(\vec{r}) \beta_n(\vec{r}')}{\lambda_n}, \tag{1.44c}$$

and with the eigenfunction/eigenvalue problem

$$\begin{cases} -\Delta \beta_n(\vec{r}) = \lambda_n \beta_n(\vec{r}), & \text{for } \vec{r} \in \Omega, \\ \nabla_{\vec{N}} \cdot \beta_n(\vec{r}) = 0, & \text{for } \vec{r} \in \partial\Omega. \end{cases} \tag{1.45}$$

Finally, it is left to our readers as a highly nontrivial exercise to generalize all the above results (their proofs!) for the case of Ω possessing a Riemannian structure: The famous Laplace–Beltrami operator acting on scalar functions into the domain $(\Omega, g_{ab}(x))$, $\Omega \subset R^N$, $\partial_a = \frac{\partial}{\partial x^a}$ $(a = 1, \dots, n)$

$$-\Delta_g U = \left\{ -\frac{1}{\sqrt{g}} \partial_a (\sqrt{g}\, g^{ab}\, \partial_b) \right\} U(x^a) \tag{1.46}$$

mathematical object instrumental in quantum geometry of strings and quantum gravity.[4]

For instance, for two-dimensional Riemann manifolds, one can always locally parametrize the Riemann metric into the conformal form in the trivial topological Riemann surface sector $g_{ab}(x, y) = \rho(x, y)\delta_{ab}$.

An important application in potential theory is the use of the famous multipole expansion inside the Laplacean Green function $(r = \vec{r}|, r' = |\vec{r}'|)$:

$$\frac{1}{|\vec{r} - \vec{r}'|} = \frac{1}{r} + \frac{\vec{r} \cdot \vec{r}'}{r^3} + \left\{ \frac{3}{2\pi} \left(\frac{\vec{r} \cdot \vec{r}'}{r^2} \right) - \frac{(r')^2}{2r^3} \right\} + \cdots . \tag{1.47a}$$

A useful explicit expression for the Laplace Green function can be obtained for the Ball $B_a(0) = \{ \vec{r} \in R^3 \mid |\vec{r}| = r \leq a \}$ in spherical coordinates

$$U(r, \theta, \varphi) = \frac{a(r^2 - a^2)}{4\pi} \left\{ \int_0^{2\pi} d\varphi' \int_0^{\pi} \sin\theta' \, d\theta' \left(\frac{f(\theta', \varphi')}{(a^2 + r^2 - 2ar\cos\Omega)} \right) \right\}, \tag{1.47b}$$

where the composed angle in the above-written formula is defined as

$$\Omega\big[(\theta, \theta'); (\varphi, \varphi')\big] = \cos\theta \cos\theta' + \sin\theta \sin\theta' \cos(\varphi - \varphi'). \tag{1.47c}$$

Another point worth remarking is that Eqs. (1.38), (1.39a) and (1.39c) have potentialities for problems involving weak perturbations of the domain boundary $\partial\Omega^{(0)} \rightarrow \partial\Omega^{(0)} + \varepsilon\,\partial\,\Omega^{(1)}$, since we have for instance the explicit perturbation result for the integrand of Eq. (1.38) for a given parametrization of the perturbed surface

$$
\begin{cases}
\vec{W} = W_1(u,v)\hat{e}_1 + W_2(u,v)\hat{e}_2 + W_3(u,v)\hat{e}_3 = \big(\vec{W}^{(0)} + \varepsilon\vec{W}^{(1)}\big)(u,v) \\[2mm]
\vec{N}(u,v) = \dfrac{(\vec{W}^{(0)} + \varepsilon\vec{W}^{(1)}) \times (\vec{W}^{(0)} + \varepsilon\vec{W}^{(1)})}{|(\vec{W}^{(0)} + \varepsilon\vec{W}^{(1)}) \times (\vec{W}^{(0)} + \varepsilon\vec{W}^{(1)})|_{R^3}} = \vec{N}^{(0)} + \varepsilon\vec{N}^{(1)} + \cdots \\[2mm]
\dfrac{1}{|\vec{r} - \vec{W}^{(0)} - \varepsilon\vec{W}^{(1)}|} = \dfrac{1}{|\vec{r} - \vec{W}^{(0)}|} + \dfrac{(\vec{r} - \vec{W}.^{(0)}\varepsilon\vec{W}^{(1)})}{|\vec{r} - \vec{W}^{(0)}|^3} + \cdots \\[2mm]
\Phi(\vec{r},\vec{W}) = \Phi^{(0)}(\vec{r}, W^{(0)}) + \varepsilon\nabla\Phi^{(1)}(\vec{r}, W^{(0)}) \cdot \nabla\vec{W}^{(1)} + \cdots.
\end{cases}
$$

$$(1.47\text{d})$$

Rigorous mathematical analysis of such ε-perturbative expansion is left to our mathematically oriented readers, including the open problem in advanced calculus to find explicitly the Green function for a toroidal surface and the Möebius surface band and its weakly random perturbations. $\square$

An important mathematical question is the one about the well-posedness properties of the Dirichlet problem in relation to the uniform convergence of the datum. We have the following result in this direction:

Lemma 1.2. *If in Eq. (1.27) $g_n(\vec{r}) \in C^0(\partial\Omega)$ converges uniformly for a function $g(\vec{r}) \in C_0(\partial\Omega)$, then we have the uniform convergence of the solutions (1.32) associated with the sequence data $\{g_n\}$ to the solution associated with $g(\vec{r})$.*

Proof. Let $\bar{G}$ be a compact in Ω such that $\text{dist}(\bar{G}, \partial\Omega) = \delta > 0$. Since we have the following estimate for the Green function

$$
\sup_{\vec{r}\in\bar{G};\,\vec{r}\,'\in\partial\Omega} |G(\vec{r},\vec{r}\,')| \leq \sup_{\vec{r}\in\bar{G};\,\vec{r}\,'\in\partial\Omega} \left| \frac{1}{4\pi|\vec{r} - \vec{r}\,'|} \right| \leq \frac{1}{4\pi\delta}
\tag{1.47e}
$$

and so on,

$$
\sup_{\vec{r}\in\bar{G}} |U_n(\vec{r}\,') - U(\vec{r})| \leq \frac{\text{Area}(\partial\Omega)}{4\pi\delta}\,\|g_n - g\|_{C^0(\partial\Omega)} \to 0.
\tag{1.47f}
$$

$\square$

Finally, let us announce the combined problem of Poisson–Dirichlet in R^3 and its associated integral representation.

Theorem 1.3. *Let $\Omega \subset R^3$ be a region satisfying the hypothesis of the Gauss theorem. Let $g(\vec{r})$ and $f(\vec{r})$ be elements in $C^1(\bar{\Omega})$; the solution of this problem in $C^2(\Omega) \cap C^1(\bar{\Omega})$*

$$\begin{cases} \Delta U(\vec{r}) = f(\vec{r}); \quad \vec{r} \in \Omega, \\ U(\vec{r})\big|_{\partial\Omega} = g(\vec{r})\big|_{\partial\Omega}, \end{cases} \tag{1.48}$$

is (uniquely) given by the integral representation written below:

$$U(\vec{r}) = \frac{1}{4\pi}\left(\iiint_\Omega d^3\vec{r}\,' \frac{f(\vec{r}\,')}{|\vec{r} - \vec{r}\,'|} \right) + v(\vec{r}\,'). \tag{1.49}$$

Here the function $v(\vec{r})$ is the harmonic function related to the boundary data

$$\Delta\, v(\vec{r}) = 0,$$

$$v(\vec{r})\big|_{\partial\Omega} = \left\{ \left(-\iiint_\Omega d^3\vec{r}\,' \frac{f(\vec{r}\,')}{|\vec{r} - \vec{r}\,'|} + g(\vec{r}) \right) \right\}\bigg|_{\vec{r}\in\partial\Omega}. \tag{1.50}$$

1.3. The Dirichlet Problem in Connected Planar Regions: A Conformal Transformation Method for Green Functions in String Theory

Let Ω be a simply connected region in R^2 and $W \subset R^2$ another region with similar topological properties. A coordinate transformation (diffeomorphism) of W into Ω

$$\begin{array}{ccc} W \subset R^2 & \xrightarrow{F} & \Omega \subset R^2 \\ (\xi, \eta) & & (x(\xi,\eta), y(\xi,\eta)) \end{array} \tag{1.51}$$

is called conformal when one has the validity of the conditions in W:

$$\begin{cases} \dfrac{\partial x(\xi,\eta)}{\partial \xi} = \dfrac{\partial y(\xi,\eta)}{\partial \eta}, \\ \dfrac{\partial x(\xi,\eta)}{\partial \eta} = -\dfrac{\partial y(\xi,\eta)}{\partial \xi}. \end{cases} \tag{1.52}$$

In this case one can see that $z = x + iy = f^{-1}(\xi + i\eta)$ and $f(z)$ is an analytic function in Ω. We have the following relation among the functional expressions of the operator Laplaceans among the two given domains W and Ω:

$$\Delta_{\xi,\eta} U(\xi,\eta) = \left(\Delta_{x,y}\, \bar{U}(x,y)\right)\left(\left(\frac{\partial x}{\partial \xi}\right)^2 + \left(\frac{\partial y}{\partial \eta}\right)^2\right)\Bigg|_{\varepsilon+i\eta=f(x+iy)}, \quad (1.53)$$

where $U(\xi(x,y),\eta(x,y)) \equiv \bar{U}(x,y)$.

In the case of the given domain transformation F possessing a unique extension as a diffeomorphism between the domain boundaries $\Gamma_1 = \partial W$ and $\Gamma_2 = \partial\Omega$, we have the following relationships among the Dirichlet problems in each region:

(a)
$$\begin{cases} \Delta_{\xi,\eta} U(\xi,\eta) = 0 \quad (\xi,\eta) \in W, \\ \\ U(\xi,\eta)\big|_{\Gamma_1} = f(\xi,\eta)\big|_{\Gamma_1}, \end{cases} \qquad (1.54)$$

(b)
$$\begin{cases} \Delta_{x,y}\, \bar{U}(x,y) = 0 \quad (x,y) \in \Omega, \\ \\ \bar{U}(x,y)\big|_{\Gamma_2} = \bar{f}(x,y)\big|_{\Gamma_2}. \end{cases} \qquad (1.55)$$

Then,

$$U(\xi,\eta) = \bar{U}(x,y). \qquad (1.56)$$

It is a deep theorem of B. Riemann in the theory of one-complex variables that given two simply connected *bounded* domains where the boundaries are curved with continuous curvature there is a conformal transformation of the given domains realizing a coordinate (sense preserving) of the boundaries.[3] A proof of such a result can be envisaged along the following arguments: Firstly, one considers the universal domain $B_1(0) = \{(x,y) \mid x^2 + y^2 < 1\}$ as the domain Ω. Secondly, one considers a triangularization $T_{(n)}$ of the region W. It is possible to write explicitly a conformal transformation of $T_{(n)}$ into $B_1(0)$ by means of the well-known Schwartz–Christoffell transformation of a polygon into the disc $B_1(0)$ denoted by $f_n(z)$. One can show now that the family of functions $\{f_n(z)\}$ is a uniformly bounded (compact) set in the space $H(\Omega)$ (Holomorphic function in Ω). By choosing the family of triangulations T_n in such a way that $T_n \subset T_{n+1}$, one can expect that $f_n(z)$ converges into $H(\Omega)$ to a function $f(z) \in H(\Omega)$ carrying the conformal transformation of W into $B_1(0)$. (Details of our suggested proof are left to our mathematically oriented readers again.)

At this point we intend to present a Hilbert space approach to construct explicitly such canonical conformal mapping of the domain W into the disc $B_1(0)$.

Let $H = \{F(W, \mathbb{C}), f: W \to \mathbb{C}, f \text{ is a holomorphic function in } W\}$. We introduce the following Hilbert space inner product in this space of holomorphic (analytical) functions in W:

$$\langle f, g \rangle_c = \frac{1}{2i} \iint_W dz \wedge d\bar{z} \, f(z)\overline{g(z)}. \tag{1.57}$$

We have the following straightforward result for two elements of the Hilbert space $(H, \langle\,,\,\rangle_c)$ obtained by a simple application of the Green theorem in the plane for any domain $\bar{W} \subset W$:

$$\int_{\partial \bar{W} = \Gamma_1} g(z)\bar{h}(z)\,dz = \frac{1}{2i} \iint_{\bar{W}} (g(z)\,\overline{h'(z)})dz \wedge d\bar{z}. \tag{1.58}$$

If one considers $h(z) = z - z_0$ and $\bar{W} = B_r(z_0)$ in Eq. (1.58), one obtains that

$$\left\{ \int_{|z-z_0|=r} g(z)(\overline{z - z_0})dz \right\} = r^2 \int_{|z-z_0|=r} \frac{g(z)}{z - z_0} = \frac{r^2}{2\pi i} g(z_0). \tag{1.59}$$

In other words,

$$|g(z_0)| \leq \frac{2\pi}{r^2} \left| \frac{1}{2} \iint_{|z-z_0|\leq r} g(z)\,dxdy \right|$$

$$\leq \frac{2\pi}{r^2} \frac{1}{2} \left(\iint_{|z-z_0|\leq r} |g^2(z)|\,dxdy \right)^{\frac{1}{2}} \left(\iint_{|z-z_0|\leq r} dxdy \right)^{\frac{1}{2}}$$

$$\leq \left(\frac{\pi}{r^2} \sqrt{\pi} \cdot r \right) \|g\|^2_{H(W)}$$

$$\leq \frac{\pi^{3/2}}{r} \|g\|^2_{H(W)} \leq c\|g\|^2_{W(D)}. \tag{1.60}$$

By the Riesz theorem for representations of linear functionals in Hilbert spaces, the bounded (so continuous) linear evaluation functional on $(H(D), \langle\,,\,\rangle_c) = H_D$ has the following representation:

$$g(z_0) = \frac{1}{2i} \iint_D dz \wedge d\bar{z}\, R(z, z_0)g(z) \tag{1.61}$$

and

$$R(z, z_0) \overset{H}{=} \sum_{k=1}^{\infty} e_k(z)\overline{e_k(z_0)}, \tag{1.62}$$

where $\{e_k(z)\}$ denotes any orthonormal set of holomorphic functions in H_D.

Let $f \colon W \to B_1(0)$ be the aforementioned Riemann function mapping holomorphically the given bounded simple connected-smooth boundary domain in $B_1(0)$. Since $f(z)$ is univalent, we have that for $f(z_0) = 0$ and $f'(z_0) > 0$ the following identity holds true:

$$\frac{1}{2\pi i} \int_{\partial W = \Gamma_1} \frac{g(z)}{f(z)} = \frac{g(z_0)}{f'(z_0)} \quad \text{for any } g \in H_D. \tag{1.63}$$

Let $C_r = \{(f(z))(\overline{f(z)}) = r < 1\}$ be the curve in the complex plane. For this special contour, we have that (see Eq. (1.58))

$$\frac{1}{2\pi i} \int_{C_r} \frac{g(z)}{f(z)} \, dz = \frac{1}{2\pi i r^2} \int_{C_r} g(z)\bar{f}(z) \, dz$$

$$= \frac{1}{\pi r^2} \iint_{(W \setminus f^{-1}(B_r(0)))} \left(g(z)\overline{f'(z)}\right) dx dy. \tag{1.64}$$

Note that we have implicitly assumed that the domain W is a homotopical deformation of its boundary Γ_1!

As a consequence (by considering the limit of $r \to 1$)

$$g(z_0) = \frac{1}{\pi} f'(z_0) \left\{ \iint_W g(z) \, \overline{f'(z)} \, dx dy \right\}. \tag{1.65}$$

In view of the uniqueness of the Reproducing Kernel Eq. (1.57), we have the explicit representation for the conformal transformation ($f(z_0) = 0$ and $f'(z_0) > 0$)

$$f(z) = \left(\left(\frac{\pi}{R(z_0, z_0)} \right)^{\frac{1}{2}} \left\{ \int_{z_0}^{z} R(\zeta, z_0) d\zeta \right\} \right). \tag{1.66}$$

For instance, one could consider the Graham–Schmidt orthogonalization process applied to the functions $\{z^n, \bar{z}^n\}_{n \in \mathbb{Z}^+}$ and obtain Eq. (1.66) as an infinite sum of holomorphic functions, or use triangulations of the domain D in order to evaluate the functions $e_n(z)$ in Eq. (1.62).

Finally let us devise some useful formulae for the Dirichlet problem in planar "smooth" regions, useful in string path integral theory.

Let $U_0 f^{-1} \colon B_1(0) \to R$ be the associated harmonic function on the disc $B_1(0)$ related to the Dirichlet problem in the region W. Note that by construction $f(z_0) = 0$ and $f'(z_0) > 0$ for a given $z_0 \in W$. As a consequence of the mean value for harmonic function, we have that

$$U(z_0) = (U_0 f^{-1})(0) = \frac{1}{2\pi} \int_0^{2\pi} (U_0 f^{-1})(\cos\theta, \sin\theta)d\theta$$

$$= \frac{1}{2\pi i} \oint_{\partial W} U(z) \frac{f'_{z_0}(z)}{f_{z_0}(z)} \, dz. \qquad (1.67)$$

From Eq. (1.67) we can write explicitly the Green functions $G(z, z_0) = f'_{z_0}(z)/f_{z_0}(z)$ for the given canonical regions. For instance, $f_{z_0}(z) = R(z - z'_0)/(R^2 - \bar{z}'_0 z)$ applies $B_R(0)$ conformally into $B_1(0)$, leading to the Poisson formulae in the circle of radius R

$$U(r, \theta) = \frac{1}{2\pi} \int_0^{2\pi} d\phi \, U(Re^{i\phi}) \left(\frac{R^2 - r^2}{R^2 + r^2 - 2Rr\cos(\phi - \theta)} \right). \qquad (1.68)$$

The function $f_{z_0}(z) = (z - z_0)/(z + \bar{z}_0)$ applies conformally the right-half-plane $\prod^+ = \{ z = x + iy, x \geq 0, -\infty < y < \infty \}$ into $B_1(0)$ and $f'_{z_0}(z_0) = 0$. As a consequence, one has the integral representation for this region

$$U(x, y) = \frac{x}{\pi} \int_{-\infty}^{+\infty} \frac{U(it) \, dt}{(t - y)^2 + x^2}. \qquad (1.69)$$

The function $f_{z_0}(z) = (z^2 - z_0^2)/(z^2 - (\bar{z}_0)^2)$ solves the Dirichlet problem in the first quadrant plane, leading to the integral representation below:

$$U(x, y) = \frac{4xy}{\pi} \left[\int_0^\infty \frac{t \, U(t) \, dt}{t^4 - 2t^2(x^2 - y^2) + (x^2 - y')^2} \right]$$

$$+ \left[\int_0^\infty \frac{t \, U(it) \, dt}{t^4 + 2t^2(x^2 - y^2) + (x^2 - y^2)'} \right]. \qquad (1.70a)$$

In general

$$G(z, z') = \frac{f_{z_0}(z') - f_{z_0}(z)}{1 - \overline{f_{z_0}(z)} \, f_{z_0}(z')} \qquad (1.70b)$$

with $f(z_0) = 0$ is the explicit Green function associated with the Dirichlet problem in the planar region W.

Just for completeness let us write the Green function of the Dirichlet problem in the S_R^2 in polar coordinates

$$G((\rho,\theta),(\rho',\theta')) = \frac{1}{4\pi}\Big\{ -\log[\rho^2 + \rho'^2 - 2\rho\rho'\cos(\theta-\theta')]$$

$$+ \log\Big[R^2 + \frac{\rho^2\rho'^2}{R^2} - 2\rho\rho'\cos(\theta-\theta')\Big]\Big\}. \qquad (1.70c)$$

Here

$$\begin{cases} \Delta_{\rho,\theta}\, G((\rho,\theta),(\rho',\theta') = 0, \\ G((R,\theta),(R,\theta')) = +\Big(\dfrac{1}{2\pi}\ell g|\vec{r}-\vec{r}'|\Big)\Big|_{\substack{\vec{r}=Re^{i\theta} \\ \vec{r}'=Re^{i\theta'}}}. \end{cases} \qquad (1.70d)$$

1.4. Hilbert Spaces Methods in the Poisson Problem

Let us pass now to the formulation of the Poisson and Dirichlet problems in the Hilbert spaces $L^2(\Omega)$, with $\bar{\Omega}$ a compact set of R^N. This formulation is known in the mathematical literature as the Hilbert space weak formulation of the problem.

Let $H_0^1(\Omega)$ be the completion of the vector space $C_c^1(\Omega)$ in relation to the inner product $\langle f,g\rangle_{H^1} = \displaystyle\int_\Omega d^4x\, \nabla f\, \overline{\nabla}g$; the weak formulation of Poisson problem reads as of

$$\langle U,\varphi\rangle_{H^1} = \int_\Omega d^N x\, \nabla U(x)\cdot(\overline{\nabla\varphi})(x)$$

$$= \int_\Omega d^N x(f\cdot\bar\varphi)(x) = \langle f,\varphi\rangle_{L^2} \quad \forall\,\varphi\in H_0^1(\Omega), \qquad (1.71)$$

where one must determine a $U(x)\in H_0^1(\Omega)$ for a given $f\in L^2(\Omega)$. That such (unique) $U(x)$ exists is readily seen from the fact that the functional linear below is bounded in $H_0^1(\Omega)$:

$$L_f\colon H_0^1(\Omega\to\mathbb{C})$$

$$\varphi\to\int_\Omega d^N x(f\cdot\bar\varphi)(x) = L_f(\varphi) \qquad (1.72)$$

as a consequence of the Friedrichs inequality

$$\underbrace{\int_\Omega d^N x|U|^2(x)}_{\|U\|^2_{L^2(\Omega)}} \leq (\mathrm{diam}(\bar\Omega))^{N-1}\underbrace{\Big(\int_\Omega d^N x|\nabla U|^2(x)\Big)}_{\leq C(\Omega)\,\|U\|^2_{H_0^1(\Omega)}}. \qquad \begin{array}{c}(1.73a)\\[20pt](1.73b)\end{array}$$

By the Riesz representation theorem, there is a unique element $U(x) \in H_0^1(\Omega)$, such that it satisfies Eq. (1.72) and has the explicit expansion

$$U(x) \stackrel{H_0^1(\Omega)}{=} \sum_n \overline{L_f(e_n)} e_n(x)$$

$$= \sum_n \left(\int_\Omega d^N y \, \bar{f}(y) e_n(y) \right) \bar{e}_n(x)$$

$$= \int_\Omega d^N y \left\{ \sum_n \bar{e}_n(x) e_n(y) \right\} \bar{f}(y) = \int_\Omega d^N y \, G(x,y) \bar{f}(y). \quad (1.74)$$

Here, $e_n(x)$ is an arbitrary orthonormal basis in $H_0^1(\Omega)$, which can be constructed from the application of the Grahm–Schmidt orthogonalization process to any linear independent set of $C_c^\infty(\Omega)$, for instance, the set of the polynomials $\{x^{|n|}, x^{|m|}\}$ with $x \in R^N$.

This method of determining weak solutions is easily generalized to the Poisson problem with variable coefficients — the Lax–Milgram theorem, a useful result in covariant euclidean path integrals in string theory.

Let $a_{ij}(x)$ and $U(x)$ be functions in $C(\Omega)$, with $\bar{\Omega}$ a compact set of R^N. Let us consider the following Hilbert space:

$$L_{\{a\}}^2(\Omega) = \text{topological closure of}$$

$$\cdot \left\{ f \in C_0(\Omega) \left(\int_\Omega d^N x \, a_{ij}(x) \frac{\partial}{\partial x^i} f \cdot \frac{\partial}{\partial x^j} \bar{f} + \int_\Omega d^N x \, V(x)(f\bar{f})(x) < \infty \right) \right.$$

$$\text{with } a_{ij}(x) = a_{ji}(x), \, a_{ij}(x)\lambda^i\lambda^j \geq a_0|\lambda|^2$$

$$\left. \text{and } V(x) > V_0 > 0 \right\}. \quad (1.75)$$

We left as an exercise to our readers to mimic with some improvements the previous study to prove the existence and uniqueness of the Poisson problem in $L_{\{a\}}^2(\Omega)$ for $f \in L_{\{a\}}^2(\Omega) \supseteq H_0^1(\Omega)$ and $\varphi(x) \in H_0^1(\Omega)$

$$\int_\Omega d^N x \left(a_{ij}(x) \frac{\partial}{\partial x^i} U(x) \overline{\frac{\partial}{\partial x^j} \varphi(x)} \right) + \int_\Omega d^N x \, V(x) U(x) \bar{\varphi}(x)$$

$$= \int_\Omega d^N x \, f(x) \bar{\varphi}(x). \quad (1.76)$$

It is another nontrivial problem to find conditions on the data function $f(x)$ (see Theorem 1.1) in order to have the weak solution $U(x)$ to belong to the space $C^2(\Omega) \cap C^1(\bar{\Omega})$, which will lead to the strong solution of the Poisson problem in Ω

$$-\sum_{i,j=1}^{N} \left\{ \frac{\partial}{\partial x^j} \left(a_{ij}(x) \frac{\partial}{\partial x^i} U(x) \right) \right\} + V(x)U(x) = f(x) \quad x \in \Omega. \quad (1.77)$$

Another important generalization of the Poisson problem is the class of semilinear Poisson problems defined by Lipschitzian nonlinearities as written below for $\Omega \subset R^3$:

$$(-\Delta U)(x) + F(U(x)) + V(x)U(x) = f(x). \quad (1.78)$$

Here $f(x) \in L^2(\Omega)$ and $F(z)$ are real Lipschitzian functions (i.e. $\forall (z_1, z_2) \in R \times R \Rightarrow |F(z_1) - F(z_2)| \leq C|z_1 - z_2|$ for a uniform bound c).

In order to show the existence and uniqueness through a construction technique, we rewrite Eq. (1.78) into the form of an integral equation (see Theorem 1.1)

$$U(x) = -\frac{1}{4\pi} \iiint \left(\frac{F(U(x')) + V(x')U(x') - f(x')}{|x - x'|} \right) d^3x' = (TU)(x). \quad (1.79)$$

One can see easily that T defines an application with domain $L^2(\Omega)$ and range in $L^2(\Omega)$. Besides, T is a contraction application in $L^2(\Omega)$ for small domains, as we can see from the following estimate:

$$\|Tu - Tv\|_{L^2(\Omega)}^2$$

$$\leq \frac{1}{16\pi^2} \|u - v\|_{L^2(\Omega)}^2 \left\{ (c^2 + \|V\|_{L^2(\Omega)}^2) \times \iiint_\Omega d^3x \frac{1}{|x - x'|^2} \right\}$$

$$\leq \left(\frac{C^2 + \|V\|_{L^2(\Omega)}^2}{16\pi^2} \right) (4\pi \operatorname{diam}(\Omega)) \|u - v\|_{L^2(\Omega)}^2. \quad (1.80)$$

As a consequence of the Banach Fixed Point Theorem, we have that the sequence

$$U_n = T(U_{n+1}) \quad (1.81)$$

converges strongly to the solution in $L^2(\Omega)$ of Eq. (1.79).

The reader should repeat the above-made analysis in the general case $\Omega \subset R^N$ (see Eq. (1.23)).

1.5. The Abstract Formulation of the Poisson Problem

Let A be a linear operator defined in a Hilbert space $H = (H, \langle , \rangle)$ $(A \colon H \to H$, $A(\alpha U + \beta v) = \alpha A(u) + \beta A(v))$ such that its domain $D(A)$ forms a (non-closed) vectorial subspace dense in H. Note that $D(A)$ has the explicit analytical representation

$$D(A) = \big\{ u \in H \mid Au \in H \Leftrightarrow \langle Au, Au \rangle_H < \infty \big\}. \qquad (1.82)$$

Let $\lambda \in \mathbb{C}$ such that there is a vector $U_\lambda \in H$, such that $AU_\lambda = \lambda U_\lambda$. Such vector is called the eigenvector of A associated with the eigenvalue λ. The set of all these eigenvectors $\{U_\lambda\}$ makes an invariant subspace for the operator A.

We have the following results in operator theory in separable Hilbert spaces, useful to the Poisson problem solution.

Theorem 1.4. *If the operator A satisfies the symmetric condition (so A is called a symmetric operator) $\langle Au, v \rangle_H = (u, Av)_H (u, v \in D(A))$, then all eigenvalues λ are real numbers.*

Proof. Let $AU_\lambda = \lambda U_\lambda$. As a consequence of the symmetry property of the operator A, we have that

$$\langle AU_\lambda, U_\lambda \rangle_H = \langle \lambda U_\lambda, U_\lambda \rangle = \lambda \|U_\lambda\|^2 = \langle U_\lambda, AU_\lambda \rangle = \bar{\lambda} \|U_\lambda\|^2. \qquad (1.83)$$

$\square$

Theorem 1.5. *If A is a symmetric operator, then different eigenvectors corresponding to different eigenvalues are orthogonals. So A has at most a countable number of eigenvalues in the separable H.*

Proof. Let $AU_{\lambda_1} = \lambda_1 U_{\lambda_1}$ and $AU_{\lambda_2} = \lambda_2 U_{\lambda_2}$ $(\lambda_1 \neq \lambda_2)$. Thus we have the identity

$$0 = \langle AU_{\lambda_1}, U_{\lambda_2} \rangle_H - \langle U_{\lambda_1}, AU_{\lambda_2} \rangle_H = (\lambda_1 - \lambda_2)\langle U_{\lambda_1}, U_{\lambda_2} \rangle_H, \qquad (1.84)$$

which means the result searched

$$\langle U_{\lambda_1}, U_{\lambda_2} \rangle_H = 0. \qquad (1.85)$$

$\square$

We have our basic result in the theory of the Poisson problem in Hilbert spaces.

Theorem 1.6. (*Dirichlet–Riemann Variational formulation of Poisson problems.*) *Let A be a positive-definite symmetric operator in a given separable Hilbert space $H = (H, \langle\,,\,\rangle)$. (Exists $c \in R^+$ such that $\langle AU, U \rangle_H \geq c \langle U, U \rangle_H$ for any $U \in D(A)$.) Let the functional given below for any $F \in \text{Range}(A)$ be*

$$F_{(A)}(\varphi) = \langle A\varphi, \varphi \rangle_H - \langle f, \varphi \rangle_H . \tag{1.86}$$

Then, the functional equation (1.86) has a minimum value at $U \in D(A)$ such that

$$A U = f, \tag{1.87}$$

the converse still holds true.

Let us exemplify the above result for the Poisson problem in bounded open sets in R^N with a Riemannian structure $\{g_{ab}(x^a); a, b = 1, \ldots, n\}$ (manifolds charts).

Thus we consider the following covariant Hilbert space associated with the given metric structures

$$L_g^2(\Omega) = \text{ closure of } \left\{ f \in C_c(\Omega) \mid \langle\,,\,\rangle \equiv \int_{\bar{W}} d^N x \sqrt{g(x)} f(x) \overline{f(x)} \right\},$$
$$\tag{1.88a}$$

$$H_{0,g}^1(\Omega) = \text{ closure of } \left\{ f \in C_0^1(\Omega) \mid \langle\,,\,\rangle = \int_{\bar{W}} d^N x \sqrt{g}\, g^{ab} \partial_a f\, \overline{\partial_b f} \right\},$$
$$\tag{1.88b}$$

$$H_{0,g}^p(\Omega) = \text{ closure of } \left\{ f \in C_0^p(\Omega) \mid \langle\,,\,\rangle_{H^0} \right.$$
$$= \int_{\bar{W}} d^N x \sqrt{g} (\nabla_{a_1} \ldots \nabla_{a_{(p/2)}} f)(x)$$
$$\left. \times \{ g^{a_1, a'_{(p/2+1)}} \ldots g^{a_{(p/2)} a'_p} \}(x) \overline{(\nabla_{a'_{(p/2+1)}} \ldots \nabla_{a'_p} f)}(x). \right. \tag{1.88c}$$

In the covariant Hilbert space $H_{0,g}^1(\Omega)$, we consider the functional (positive-definite) associated to the Laplace–Beltrami operator $\Delta_g = -\frac{1}{\sqrt{g}} \partial_a(\sqrt{g}\, \delta^{ab} \partial_b)$ acting on real functions

$$F_{\Delta_q}(\varphi) = \int_{\bar{W}} d^N x \sqrt{g} (\varphi(-\Delta_g)\varphi)(x) - \int_{\bar{W}} d^N x \sqrt{g}\, f(x)\varphi(x). \tag{1.89}$$

The minimizing $\bar{U} \in H^1_{0,g}(\Omega)$ of the above-written functional is the solution of the covariant Poisson problem in the open domain W:

$$(-\Delta_g \bar{U})(x) = f(x) \quad \text{for } x \in W. \tag{1.90}$$

Let us now analyze the eigenvalue problem in the Dirichlet problem in the Hilbert space $L^2(\Omega)$.

We can accomplish such studies by considering the problem rewritten into the integral operator form

$$\mu \, U_\lambda(\vec{r}) = \frac{1}{4\pi} \int_\Omega d^3\vec{r}' \, \frac{U_\lambda(\vec{r}')}{|\vec{r} - \vec{r}'|} = (T \, U_\lambda)(F). \tag{1.91}$$

First, let us see that the integral operator T has as its domain $C^1(\Omega) \subset L^2(\Omega)$ and as its range the functional space $C^2(\Omega)$. However, it can be extended to the whole space $L^2(\Omega)$ by the Hahn–Banach theorem

$$T \colon L^2(\Omega) \to L^2(\Omega) \tag{1.92}$$

since we have the estimate

$$|(TU)(\vec{r})| \le \left(\frac{1}{4\pi} \int_\Omega d^3\vec{r}' \, \frac{1}{|\vec{r} - \vec{r}'|^2} \right)^{\frac{1}{2}} \times \left(\int_\Omega d^3\vec{r}' \, U^2(\vec{r}') \right)^{\frac{1}{2}}$$

$$\le \left(\frac{1}{3} \operatorname{diam}(\Omega) \right)^{\frac{1}{2}} \|U\|_{L^2(\Omega)}. \tag{1.93}$$

We can see that T is a symmetric bounded compact operator with a set of eigenvalues $|\mu_r| \le \frac{4\pi}{3} \operatorname{diam}(\Omega)$ and $\mu_n \to 0$ for $n \to \infty$.

1.6. Potential Theory for the Wave Equation in R^3 — Kirchhoff Potentials (Spherical Means)

Let us start this section by defining the Poisson–Kirchhoff potential associated with the given $C^2(\Omega)$ functions. The first potential associated with a certain $f(\vec{r}) \in C^2(\Omega)$ is defined by the spherical means $(\vec{r} = (x, y, z) \in \Omega)$

$$U^{(1)}(\vec{r}, t, [f]) = \frac{t}{4\pi} \int_{S^3_{ct}(\vec{r})} f(x' + \alpha ct, y' + \beta ct, z' + \gamma ct) d\Omega. \tag{1.94}$$

Here, $S^3_{ct}(\vec{r}) = \partial\{B^3_{ct}(\vec{r})\}$ is the spherical surface centered at the point $\vec{r} = (x, y, z)$ with radius $|\vec{r}' - \vec{r}| = ct$. The spherical parameter α, β, and γ are defined as usual as

$$\begin{aligned}
\alpha &= \cos\varphi\sin\theta, \\
\beta &= \sin\varphi\sin\theta, \\
\gamma &= \cos\theta, \\
d\Omega &= \sin\theta d\theta \cdot d\varphi.
\end{aligned} \tag{1.95}$$

The second potential is defined in an analogous way as

$$U^{(2)}(\vec{r}, t, [g]) = \frac{1}{4\pi}\frac{\partial}{\partial t}\left\{\frac{1}{2}U^{(1)}(\vec{r}, t, [g])\right\}. \tag{1.96}$$

We have the following elementary initial-value properties for the above wave potentials, which are the functions defined in the infinite domain $R^3 \times R^+$:

$$U^{(1)}(\vec{r}, 0, [f]) = 0,$$
$$\frac{\partial}{\partial t}U^{(1)}(\vec{r}, 0, [f]) = f(x, y, z) \times \lim_{t\to 0^+}\left\{\frac{t}{4\pi}\int_{S^3_{ct}(\vec{r})}d\Omega\right\} = f(\vec{r}),$$
$$U^{(2)}(\vec{r}, 0, [g]) = g(\vec{r}), \tag{1.97}$$
$$\frac{\partial}{\partial t}U_2(\vec{r}, 0, [g]) = 0 \qquad \text{(exercise)}.$$

The main differentiability property of the above-written wave potentials is the following:

$$\Delta_{\vec{r}}U^{(1)}(\vec{r}, t, [f]) = \frac{t}{4\pi}\left\{\int_0^{2\pi}d\varphi\int_0^\pi \sin\theta d\theta\Delta_{\vec{r}}f(x + \alpha ct, y + \beta ct, z + \gamma ct)\right\}. \tag{1.98}$$

We should now evaluate the second-time derivative of the potential $(\bar{x} = x + \alpha ct, \bar{y} = y + \beta ct, \bar{z} = z + \gamma ct)$

$$\frac{\partial U^{(1)}(\vec{r}, t, [f])}{\partial t}$$

$$= \frac{U^{(1)}(\vec{r}, t, [f])}{t} + \frac{t}{4\pi}\left\{\int_{S^3_{ct}(\vec{r})}\left(c\alpha\frac{\partial}{\partial\bar{x}} + c\beta\frac{\partial}{\partial\bar{y}} + c\gamma\frac{\partial}{\partial\bar{z}}\right)f(\bar{x}, \bar{y}, \bar{z})d^2\Omega\right\}$$

$$= \frac{U^{(1)}(\vec{r},t,[f])}{t} + \frac{1}{4\pi ct}\left\{ \int_{S^3_{ct}(\vec{r})} \overbrace{\left(\alpha\frac{\partial}{\partial\bar{x}} + \beta\frac{\partial}{\partial\bar{y}} + \gamma\frac{\partial}{\partial\bar{z}}\right)}^{\vec{n}\cdot\vec{\nabla}} f(\bar{x},\bar{y},\bar{z})dA \right\}$$

(Gauss theorem)

$$= \frac{U^{(1)}(\vec{r},t,[f])}{t} + \frac{1}{4\pi ct}\left\{ \int_{B^3_{ct}(\vec{r})} (\Delta f)(\bar{x},\bar{y},\bar{z})d\bar{x}\,d\bar{y}\,d\bar{z} \right\}$$

$$= \frac{U^{(1)}(\vec{r},t,[f])}{t} + \frac{1}{4\pi ct}\left\{ \int_0^{ct} d\rho \int_{S^3_{ct}(\vec{r})} d^2A(\Delta f)(\bar{x},\bar{y},\bar{z}) \right\} \qquad (1.99)$$

since

$$\begin{cases} d\bar{x}\,d\bar{y}\,d\bar{z} = d\rho\,d^2A, \\ 0 \le \rho \le ct. \end{cases} \qquad (1.100)$$

Since we have the usual Leibnitz rule for derivatives inside integrals

$$\frac{d}{dt}\left\{ \int_0^{ct} f(x)\,dx \right\} = f(ct)\cdot c. \qquad (1.101)$$

We have the result for evaluation of the second-time derivative of Eq. (1.94):

$$\frac{\partial^2}{\partial t^2} U^{(1)}(\vec{r},t,[f]) = \frac{\partial U^{(1)}(\vec{r},t,[f])}{\partial t}\cdot\frac{1}{t} - \frac{1}{t^2} U^{(1)}(\vec{r},t,[f])$$

$$+ \frac{1}{4\pi ct} \times c \times \left[\int_{S^3_{ct}(\vec{r})} d^2A(\Delta f)(\bar{x},\bar{y},\bar{z}) \right]$$

$$- \frac{1}{4\pi ct^2}\left[\int_0^{ct} d\rho \int_{S^3_\rho(\vec{r})} d^2A(\Delta f)(\bar{x},\bar{y},\bar{z}) \right]. \qquad (1.102)$$

Using Eq. (1.99) again to simplify Eq. (1.102), yields

$$\frac{\partial^2 U^{(1)}(\vec{r},t,[f])}{\partial^2 t} = \frac{1}{4\pi t}\left[\int_{S^3_\rho(\vec{r})} d^2A(\Delta f)(\bar{x},\bar{y},\bar{z}) \right]$$

$$= c^2 t\left[\int_{S^3_{ct}(\vec{r})} d^2\Omega(\Delta f)(\bar{x},\bar{y},\bar{z}) \right] = \frac{1}{c^2}\Delta_{\vec{r}} U^{(1)}(\vec{r},t,[f]). \qquad (1.103)$$

As a result of the above differential calculus evaluations we have the analogous of Theorem 1.1 (the Cauchy problem for the wave equation).

Theorem 1.7. *The solution for the Cauchy (initial values) problem for the wave equation with "smooth" $C^2(\Omega)$ data (including an external forcing $F(\vec{r}, t)$, continuous in the t-variable), is given by the wave potentials*

$$
\begin{cases}
\dfrac{\partial^2 U(\vec{r}, t)}{\partial^2 t} = \dfrac{1}{c^2} \Delta_{\vec{r}} U(\vec{r}, t) + F(\vec{r}, t), \\[2mm]
U(\vec{r}, 0) = f(\vec{r}), \quad \vec{r} \in \Omega, \\[2mm]
U_t(\vec{r}, 0) = g(\vec{r}), \quad \vec{r} \in \Omega,
\end{cases}
\tag{1.104}
$$

$$
U(\vec{r}, t) = U^{(1)}(\vec{r}, t, [f]) + U^{(2)}(\vec{r}, t, [g])
$$

$$
+ \frac{1}{4\pi} \left\{ \int_{|\vec{r} - \vec{r}'| \leq ct} \frac{F(\vec{r}', ct - |\vec{r} - \vec{r}'|)}{|\vec{r} - \vec{r}'|} d^3\vec{r}' \right\}. \tag{1.105}
$$

In the case of $\Omega \subset R^2$ we have the Poisson wave potentials:

$$
U(\vec{r}, t) = \left[\frac{\partial}{\partial t} \left(\frac{1}{2\pi} \int_{|\vec{r} - \vec{r}'| < ct} \frac{f(\vec{r}')}{\sqrt{c^2 t^2 - |\vec{r} - \vec{r}'|^2}} d^2\vec{r}' \right) \right]
$$

$$
+ \left[\frac{1}{2\pi} \int_{|\vec{r} - \vec{r}'| < ct} \frac{g(\vec{r}')}{\sqrt{c^2 t^2 - |\vec{r} - \vec{r}'|^2}} d^2\vec{r}' \right]
$$

$$
+ \left[\frac{1}{2\pi} \int_0^t dt' \int_{|\vec{r} - \vec{r}'| < ct} \frac{F(\vec{r}', c(t - t'))}{\sqrt{(t')^2 - |\vec{r} - \vec{r}'|^2}} d^2\vec{r}' \right]. \tag{1.106}
$$

Let us give a proof for the problem uniqueness for the Cauchy initial value Eq. (1.104).

In order to show such a result, let us consider the energy wave (functional) inside a spherical ball of radius R:

$$
E_R(t) = \iiint_{B_R(0)} [U_t^2 + c^2 |\nabla U|^2](\vec{r}, t) d^3\vec{r}. \tag{1.107}
$$

Let us analyze its time derivative

$$
\lim_{R \to \infty} \frac{dE_R(t)}{dt} = \iiint_{B_R(0)} (2 U_t U_{tt} + 2c^2 \nabla \partial_t U \cdot \nabla U)(\vec{r}, t)
$$

$$
= 2 \left\{ \iiint_{B_R^3(0)} U_t (U_{tt} - c^2 \Delta U) + \oiint_{S_R^3(0)} (U_t \cdot \nabla_{\vec{N}} U) dS \right\}
$$

$$
= 2 \lim_{R \to \infty} \left\{ \oiint_{S_R^3(0)} (U_t \cdot \nabla_{\vec{N}} U) dS \right\} = 0, \tag{1.108}
$$

if one imposes a sort of Helmholtz–Sommerfeld radiation condition at infinity (the vanishing of the surfaces term in Eq. (1.108). As a consequence, if $f(r) = g(r) \equiv 0$ (vanishing of the initial conditions), we have that

$$E(t) = E(0)$$

$$= \lim_{R \to \infty} \left\{ -c^2 \oiint_{B_R(0)} (U \cdot \Delta U)(\vec{r}, 0) + \lim \oiint_{S_R^3(0)} (U_t \cdot \nabla_{\vec{N}} U) dS \right\} = 0.$$

$$(1.109)$$

1.7. The Dirichlet Problem for the Diffusion Equation — Seminar Exercises

Let us consider the following Dirichlet problem for the diffusion equation in a given domain $\Omega \subset R^3$ with a nontrivial time-dependence on the boundary (an interior problem):

$$\begin{cases} \Delta U(\vec{r}, t) = \dfrac{1}{a^2} \dfrac{\partial}{\partial t} U_{(\vec{r}, t)}, \\[2mm] U(\vec{r}, 0) = f(\vec{r}) \in C^\infty(\Omega), \\[2mm] U(\vec{r}, t)\big|_{\partial\Omega} = g(\vec{r}, t)\big|_{\partial\Omega} \in C^\infty(\partial\Omega). \end{cases} \qquad (1.110)$$

Let us introduce the double-layer like Poisson potential for a not-yet determined density $\Phi(\vec{r}, t)$

$$U^{(1)}(\vec{r}, t) = \frac{1}{4\pi} \int_0^t d\zeta \left\{ \oiint_{\partial\Omega} \frac{\Phi(\vec{r}', \zeta)}{(t - \zeta)} \nabla_{\vec{N}} \left(e^{-\frac{r}{4a^2(t-\zeta)}} \right) dS(\vec{r}') \right\}$$

$$= \frac{1}{8\pi a^2} \int_0^t d\zeta \left\{ \oiint_{\partial\Omega} \frac{\Phi(\vec{r}', \zeta)}{(t - \zeta)^2} e^{-\frac{r}{4a^2(t-\zeta)}} \cdot r \cdot \cos(\vec{N}(\vec{r}) \angle dS(\vec{r}') \right\}.$$

$$(1.111)$$

We observe that $U^{(1)}(\vec{r}, t)$ satisfies Eq. (1.110) and vanishes at $t = 0$ (exercise).

The solution for the Dirichlet-like problem Eq. (1.110) will be determined from the ansatz

$$U(\vec{r}, t) = U^{(1)}(\vec{r}, t) + \overbrace{\frac{1}{(2\pi a^2 t)^{\frac{3}{2}}} \left\{ \int_\Omega d^3 \vec{r}' \, f(\vec{r}') e^{-\frac{(\vec{r} - \vec{r}')^2}{4a^2 t}} \right\}}^{U^{(2)}(\vec{r}, t)}. \qquad (1.112)$$

It is clear that the formal solution written above satisfies the boundary condition in Eq. (1.110) if one can determine the density function $\Phi(\vec{r}, t)$ from the integral equation coming from the imposition of the boundary condition as given by Eq. (1.110) (exercise):

$$U(\vec{r}, t)\big|_{\partial\Omega} = g(\vec{r}, t)\big|_{\partial\Omega} = U^{(2)}(\vec{r}, t)\big|_{\partial\Omega}$$

$$+ \left[-\Phi(\vec{r}, t)\big|_{\partial\Omega} + \frac{1}{8\pi a^2} \int_0^t d\zeta \oiint_{\partial\Omega} \frac{\Pi(\vec{r}', t)}{(t-\zeta)^2} \, e^{-\frac{r'^2}{4a^2(t-\zeta)}} \cdot r' \right.$$

$$\left. \times \cos(N\angle\vec{r}')dS(\vec{r}') \right]. \tag{1.113}$$

One can show that for small densities $\big(\Phi(\vec{r}, t) \to \lambda\Phi(\vec{r}, t)$ with $\lambda \ll 1\big)$, one can solve Eq. (1.113) by means of the fixed point theorem of Banach (exercise) and yielding the solution as a power series in λ (the size of the area of $\partial\Omega$).

1.8. The Potential Theory in Distributional Spaces — The Gelfand–Chilov Method

Seminar Exercises

In this brief section, we intend to show the general method of Kirchhoff–Poisson potentials in the context of Schwartz distribution theory in $S'(R^N)$.

Let us start this section by considering the problem of determining the fundamental solution of an arbitrary elliptic differential operator of order m in the whole space R^N with constant coefficients

$$\left(\sum_{|\alpha| \leq m} a_\alpha D_x^\alpha \right) U(x) = \delta^{(N)}(x). \tag{1.114}$$

Here $U(x)$ shall belong to $S'(R^N)$. In the distributional sense[2]

$$\langle U(x), \varphi(x) \rangle = \left[\left(\sum_{|\alpha| \leq m} a_\alpha D_x^\alpha \right)^t \varphi(x) \right]_{x=0}. \tag{1.115}$$

Let us consider the analytical regularization of the Dirac distribution in the right-hand side of Eq. (1.114) (the well-known analytical regularization scheme in quantum field theory) leading to a usual function-theoretic partial differential equation expressed below:

$$\left(\sum_{|\alpha| \leq m} a_\alpha D_x^\alpha U^{(\lambda)}(x) \right) = \frac{2|x|^\lambda}{\omega_n \, \Gamma\left(\frac{\lambda+n}{2}\right)}$$

$$= \frac{1}{\omega_n \pi^{\frac{n-1}{2}} \, \Gamma\left(\frac{\lambda+n}{2}\right)} \left\{ \int_{S_1^{(n)}(0)} |\vec{x} \cdot \vec{r}'|^\lambda dS(\vec{r}') \right\}. \tag{1.116}$$

At this point one can consider the following analytical regularized associated problem:

$$(LU)(\vec{r}) = \left(\sum_{|\alpha| \leq m} a_\alpha D_{\vec{r}}^\alpha v^{(\lambda)}(\vec{r}) \right) = \frac{|\vec{r} \cdot \vec{r}'|^\lambda}{\omega_n \, \pi^{\frac{n-1}{2}} \, \Gamma\left(\frac{\lambda+n}{2}\right)}. \tag{1.117}$$

Due to the linearity of Eq. (1.116), we should search for a solution of this analytical regularized equation in the linear-superposing form of a spherical mean (Radon transforms)

$$U^{(\lambda)}(\vec{r}) = \int_{S_1^{(n)}(0)} d\Omega(\vec{r}') \, v^{(\lambda)}(\vec{r}, \vec{r}'). \tag{1.118}$$

The solution of Eq. (1.117) reduces to the solution of an ordinary non-homogeneous differential equation in R after introducing the variable change $\frac{\partial}{\partial x^a} = \omega_a \frac{d}{d\xi}$, where $\vec{r} = \{x^a\}_{a=1,\ldots,n}$ and $\vec{r}' \in S_1^n(0) \Leftrightarrow \vec{r}' = \{\omega^a\}_{a=1,\ldots,n}$ $[(\omega^1)^2 + \cdots + (\omega^n)^2 = 1)]$ with $-\infty < \xi < +\infty$.
It yields the following simple result:

$$L\left(\omega^a \frac{d}{d\xi} \right) v^\lambda(\xi) = \frac{|\xi|^\lambda}{\omega_n \, \pi^{\frac{n-1}{2}} \, \Gamma\left(\frac{\lambda+n}{2}\right)}. \tag{1.119}$$

The solution of Eq. (1.119) is easily written down in the distributional sense in $S'(R)$:

$$v^{(\lambda)}(\xi) = \frac{1}{\omega_n \pi^{\frac{n-1}{2}} \, \Gamma\left(\frac{\lambda+n}{2}\right)} \left\{ \int_{-\infty}^{+\infty} G(\xi, \xi')|\xi'|^\lambda \, d\xi' \right\}. \tag{1.120}$$

Here, $G(\xi, \xi')$ is the Green function of the ordinary differential operator in Eq. (1.119), obtained through Fourier transforms in $S'(R)$, namely,

$$L\left(\omega^a \frac{d}{d\xi} \right) G(\xi, \xi') = \delta^{(1)}(\xi - \xi'). \tag{1.121}$$

Now it can be shown case by case that the limit of $\lambda \to -n$ in our already-obtained distribution-regularized solutions converges in the distributional sense to the expected $S'(R^N)$-distributional solution of Eq. (1.114).

As an exercise, the reader can find the distributional solution of the α-power Laplacean through the analytical regularized nonhomogeneous ordinary differential equation. (Here $\alpha \in \mathbb{C}$, and $0 < \mathrm{Real}(\alpha) < n$).

$$L\left(\omega^a \frac{d}{d\omega}\right) = -\left[((\omega_1)^2 + (\omega_2)^2 + (\omega_3)^2)\frac{d^2}{d^2\xi}\right\}^{2\alpha} v^\lambda(\xi) = \frac{|\xi|^\lambda}{\omega_3\,\pi\,\Gamma\left(\frac{\lambda+n}{2}\right)}.$$

$$(1.122)$$

The full solution is given by the famous Riesz–Poisson potential in R^N (compare with Eq. (1.41):

$$\begin{cases} (-\Delta)^{\frac{\alpha}{2}}\, U(x) = f(x), \\ U(x) = \dfrac{\Gamma\left(\frac{(n-\alpha)}{2}\right)}{\Gamma\left(\frac{\alpha}{2}\right)2^\alpha \cdot \pi^{(n/2)}}\left[\displaystyle\int_{R^N} d^N x\, \frac{f(x')}{|x-x'|^{n-\alpha}}\right]. \end{cases} \qquad (1.123)$$

References

1. C. Hilbert, *Methods of Mathematics Physics*, Vol. I, II (Interscience Publishers, Inc., New York, 1978).
2. B. Simon, *J. Math. Phys.* **12**, 140 (1971).
3. M. Lavrentiev and B. Chabat, *Méthodes de la théorie des fonctions d'une variable complexe*, Édition MIR (Moscou, 1976).
4. L. C. L. Botelho, *Methods of Bosonic and Fermionic Path Integrals Representations: Continuum Random Geometry in Quantum Field Theory* (Nova Science Publisher, NY, USA, 2008).
5. C. Luiz and L. Botelho, *Il Nuovo Cimento* **112A**, 1615 (1999).
6. S. Weinberg, *Gravitation and Cosmology* (John Wiley & Sons Inc., 1992).
7. P. G. Bergmann, *Introduction to the Theory of Relativity* (Prentice-Hall, Inc., England, Cliffs, NJ, USA).
8. V. Fock, *The Theory on Space, Time and Gravitation*, 2nd revised edn. (Pergamon Press, 1964).
9. S. Sternberg, *Lectures on Differential Geometry* (Chelsea Publishing Company, NY, 1983), Theorem 4.4, p. 63.
10. S. W. Hawking, *General Relativity: An Einstein Centenary, Survey* (Cambridge University Press, 1979).
11. H. Flanders, *Differential Forms* (Academic Press, 1963).
12. J. Milnor, *Topology from the Differentiable Viewpoint* (University Press of Virginia, 1965).

Appendix A. Light Deflection on de-Sitter Space

A.1. The Light Deflection

The most general covariant second-order equation for the gravitation field generated by a given (covariant) enegy–matter distribution on the space–time is given by the famous Einstein field equation with a cosmological constant Λ with dimension (length)$^{-2}$ (Ref. 5), namely,

$$\left(R_{\mu\nu}(g) - \frac{1}{2}g_{\mu\nu}R + \Lambda g_{\mu\nu} \right)(x) = 8\pi G T_{\mu\nu}(x), \tag{A.1}$$

where x belongs to a space–time local chart.

It is well known that studies on the light deflection by a gravitational field generated by a massive point-particle with a pure time like geodesic trajectory (a rest particle "sun" for a three-dimensional spatial space–time section observer!) is always carried out by considering $\Lambda \equiv 0$.[6,7]

Our purpose in this appendix is to understand the light deflection phenomena in the presence of a nonvanishing cosmological term in Einstein equation (A.1), at least on a formal mathematical level of solving trajectory motion equations.

Let us, thus, look for a static spherically symmetric solution of Eq. (A.1) in the standard isotropic form[6,7]

$$(ds)^2 = B(r)(dt)^2 - A(r)(dr)^2 - r^2[(d\theta)^2 + \sin^2\theta(d\phi)^2]. \tag{A.2}$$

In the space–time region $r = + \mid \vec{x} \mid^2 > 0$, where the matter–energy tensor vanishes identically, we have that the Einstein equation takes the following form:

$$R_{\mu\nu}(g)(x) = -\Lambda(g_{\mu\nu}(x)). \tag{A.3}$$

In the above cited region, the Ricci tensor is given by

$$-\Lambda g_{\mu\nu} = \begin{pmatrix} -\Lambda B(r) & 0 & 0 & 0 \\ 0 & \Lambda A(r) & 0 & 0 \\ 0 & 0 & \Lambda r^2 & 0 \\ 0 & 0 & 0 & \Lambda r^2 \sin^2\theta \end{pmatrix}$$

$$= \begin{pmatrix} R_{tt} & 0 & 0 & 0 \\ 0 & R_{rr} & 0 & 0 \\ 0 & 0 & R_{\theta\theta} & 0 \\ 0 & 0 & 0 & R_{\phi\phi} \end{pmatrix}. \tag{A.4}$$

We have, thus, the following set of ordinary differential equations in place of Einstein partial differential Eq. (A.1)

$$R_{tt} = -\frac{B''}{2A} + \frac{1}{4}\left(\frac{B'}{A}\right)\left(\frac{A'}{A}+\frac{B'}{B}\right) - \frac{1}{r}\left(\frac{B'}{A}\right) = -\Lambda B, \qquad (A.5)$$

$$R_{rr} \equiv \frac{B''}{2B} - \frac{1}{4}\left(\frac{B'}{B}\right)\left(\frac{A'}{A}+\frac{B'}{B}\right) - \frac{1}{r}\left(\frac{A'}{A}\right) = \Lambda A, \qquad (A.6)$$

$$R_{\theta\theta} = -1 + \frac{r}{2A}\left(-\frac{A'}{A}+\frac{B'}{B}\right) + \frac{1}{A} = \Lambda r^2, \qquad (A.7)$$

$$R_{\phi\phi} = \sin^2\theta R_{\theta\theta} = \Lambda(\sin^2\theta)r^2. \qquad (A.8)$$

At this point we note that

$$\frac{R_{tt}}{B(r)} + \frac{Rrr}{A(r)} = 0, \qquad (A.9)$$

or equivalently

$$A(r) = \frac{\alpha}{B(r)}, \qquad (A.10)$$

where α is an integration constant.

Since $R_{\theta\theta} = -1 + \frac{r}{\alpha}B' + \frac{B}{\alpha} = \Lambda r^2$, we get the following expression for the $B(r)$ function:

$$B(r) = \frac{\alpha\Lambda r^2}{3} + \alpha + \frac{\beta}{r}, \qquad (A.11)$$

with β denoting another integration constant.

In the literature situation,[6,7] one always considers the case $\Lambda \neq 0$ in a pure classical mathematical vacuum situation context, the so-called de-Sitter vacuum pure gravity. However, in our case it becomes physical to consider that our solution depends analytically on the cosmological constant. In other words, if the parameter $\Lambda \to 0$ in our solution, it must converge to the usual Schwarzschild solution with a mass singularity at the origin $r = 0$, that is our boundary condition hypothesis imposed on our solution.

As a consequence, one gets our proposed Schwarzschild–de-Sitter solution

$$(ds)^2 = \left(\frac{\Lambda r^2}{3} + 1 - \frac{2MG}{r} \right)(dt^2)$$

$$- \left(\frac{\Lambda r^2}{3} + 1 - \frac{2MG}{r} \right)^{-1}(dr)^2 - r^2[(d\theta)^2 + (\sin^2 \theta)(d\phi)^2].$$

$$(A.12)$$

At this point let us comment that for the space–time region exterior to the spatial sphere $r > (\frac{3mG}{\Lambda})^{(1/3)}$, the field gravitation approximation leads to the antigravity (a repulsion gravity force) if $\Lambda < 0$; So, explain from this Einstein Gravitation theory of ours the famous "Huble accelerating Universe expansion".

In what follows we are going to consider a nonvanishing $\Lambda < 0$ and study the path of a light ray on such negative cosmological constant Einstein manifold Eq. (A.12).

We have the following null-geodesic equation for light propagating in $\theta = \pi/2$ plane (Einstein hypothesis) for light propagation in the presence of the sun (Sec. A.2 for the related formulae):

$$0 = B(r) - \frac{1}{B(r)} \left(\frac{dr}{dt} \right)^2 - r^2 \left(\frac{d\phi}{dt} \right)^2. \qquad (A.13)$$

At this point we note that

$$\left(\frac{d\phi}{dt} \right) = \left(\frac{B(r)J}{r^2} \right), \qquad (A.14)$$

where J is an integration constant.

After substituting Eq. (A.14) into Eq. (A.13) we have the following differential equation for the light trajectory as a function of the deflection angle ϕ

$$\left(\frac{dr}{d\phi} \frac{d\phi}{dt} \right)^2 + \frac{J^2 B^3(r)}{r^2} - B^2(r) = 0, \qquad (A.15)$$

which is exactly integrable:

$$d\phi = \frac{dr}{r^2 \sqrt{\frac{1}{J^2} - \frac{B(r)}{r^2}}}. \qquad (A.16)$$

By supposing a deflection point r_m where $\frac{dr}{dt} = 0$ and, thus, $J = r_m/\sqrt{B(r_m)}$, we get the deflection angle

$$\Delta_1\phi = \int_\infty^{r_m} \frac{dr}{r^2[\frac{B(r_m)}{r_m^2} - \frac{B(r)}{r^2}]^{(1/2)}} = \int_0^{\frac{1}{r_m}} \frac{dU}{[(U_m^2 - U^2) - 2MG(U_m^3 - U^3)]^{(1/2)}},$$

(A.17)

which is exactly the one given in the pure ($\Lambda = 0$) Schwarzschild famous case. However, if one supposes that there is no deflection (a continuous monotone trajectory $r = r(\phi)$!), the total deflection angle now depends on the cosmological constant and is given formally by the expression below:

$$\Delta_2\phi = \int_\infty^{r_m} dr \left\{ \frac{1}{r^2\sqrt{-\frac{1}{r^2} + \frac{2mG}{r^3}}} \left[\frac{1}{\sqrt{1 + \left[\frac{r^3(\frac{3-\Lambda J^2}{3J^2})}{2MG-r} \right]}} \right] \right\} \neq \Delta_1\phi. \quad (A.18)$$

As a general conclusion of our note we claim that the usual light-deflection experimented test does not make any difference between the usual noncosmological Schwarzschild case and our case Eq. (A.12), and, thus, it should not be considered as a definitive physical support for Einstein General Relativity without cosmological constant.

A.2. The Trajectory Motion Equations

The body trajectory $(t(p), r(p), \theta(p), \varphi(p))$ in the presence of the gravitational field generated by the metric Eqs. (A.2)–(A.10) is described by the following geodesic equations:

$$\frac{d^2t}{d^2p} + \frac{B'}{B}\left(\frac{dr}{dp}\right)\left(\frac{dt}{dp}\right) = 0, \quad (A.19)$$

$$\frac{d^2r}{d^2p} + \frac{A'}{2A}\left(\frac{dr}{dp}\right)^2 - \frac{r}{A}\left(\frac{d\theta}{dp}\right)^2 - \frac{r\sin^2\theta}{A}\left(\frac{d\phi}{dp}\right)^2 + \frac{B'}{2A}\left(\frac{dt}{dp}\right)^2 = 0,$$

(A.20)

$$\frac{d^2\theta}{d^2p} + \frac{2}{r}\frac{d\theta}{dr}\frac{dr}{dp} - \sin\theta \cdot \cos\theta \left(\frac{d\phi}{dp}\right)^2 = 0, \quad (A.21)$$

$$\frac{d^2\phi}{d^2p} + \frac{2}{r}\frac{d\phi}{dp}\frac{dr}{dp} + 2cotg(\theta)\frac{d\phi}{dp}\frac{d\theta}{dp} = 0. \quad (A.22)$$

At this point we remark that by multiplying Eq. (A.19) by $B(r(p))$, it reduces to the exact integral form relating the Einstein proper-time (physical evolution parameter) p with the geometrical-dependent coordinate Newtonian time t:

$$\frac{dt}{dp} = \frac{1}{B(r)}. \tag{A.23}$$

We remark either that Eq. (A.22) can be rewritten in the form

$$\frac{d}{dp}\left(\ell n \frac{d\phi}{dp} + \ell n r^2 + 2\ell n \sin\theta \right) = 0, \tag{A.24}$$

which reduces to the following form:

$$\left(\frac{d\phi}{dp} r^2(p) \sin^2(\theta(p)) \right) = J, \tag{A.25}$$

where J is an integration constant.

By substituting Eqs. (A.23) and (A.25) into Eqs. (A.20) and (A.21) we obtain the full set of equations describing the body trajectory in relation to the (r, θ) variables

$$\frac{d^2 r}{d^2 p} - \frac{B'}{2B}\left(\frac{dr}{dp} \right)^2 - rB\left(\frac{d\theta}{dp} \right)^2 - \frac{J^2 B}{r^3} + \frac{B'}{2B} = 0, \tag{A.26}$$

$$\frac{d^2\theta}{d^2 p} + \frac{2}{r}\frac{d\theta}{dr}\frac{dr}{dp} - \frac{\cos\theta}{\sin^3\theta}\frac{J^2}{r^4} = 0. \tag{A.27}$$

For Einstein hypothesis of light propagation on the plane $\theta = \pi/2$, Eq. (A.27) vanishes and Eq. (A.26) takes the form

$$\frac{d^2 r}{d^2 p} - \frac{B'}{2B}\left(\frac{dr}{dp} \right)^2 - \frac{J^2 B}{r^3} + \frac{B'}{2B} = 0 \tag{A.28}$$

or in a more manageable alternative form after multiplying Eq. (A.28) by $\frac{2}{B}\left(\frac{dr}{dp} \right)$ and by using Eq. (A.23) for exchanging the geometrical parameter p by the time manifold coordinate

$$\left(\frac{dr}{dt} \right)^2 \frac{1}{B^3} + \frac{J^2}{r^2} - \frac{1}{B} + E = 0, \tag{A.29}$$

where E denotes another integration constant.

By writing r as a function of ϕ and using Eq. (A.25) $\left(\frac{d\phi}{dt}\frac{r^2}{B} = J!\right)$, we get our final trajectory equation

$$\frac{dr}{d\phi} = \pm r^2 \left[\frac{1}{J^2} - \frac{B}{r^2} - \frac{BE}{J^2}\right]^{(1/2)}, \qquad (A.30)$$

which leads to the body trajectory geometric form

$$\phi = \pm \int \frac{dr}{r^2 B^{(1/2)}\left[\frac{1}{J^2 B} - \frac{E}{J^2} - \frac{1}{r^2}\right]^{(1/2)}}. \qquad (A.31)$$

Note that for light propagation the integration constant E always vanishes, a result used in the text by means of Eq. (A.16).

A.3. On the Topology of the Euclidean Space–Time

One of the most interesting aspects of Einstein gravitation theory is the question of the nonexistence of "holes" in the space–time C^2-manifold from the view point of a mathematical observer situated on the Euclidean space R^9 associated with the "minimal" Whitney imbedding theorem of M on Euclidean spaces.[9]

In order to conjecture the validity of such a topological space–time property, let us suppose that M is a C^2-manifold, and the analytically continued (Euclidean) matter distribution tensor generating the (Euclidean) gravitation field on M allows a well-defined Euclidean metric tensor (solution of Euclidean Einstein equation).[10]

At this point we note that M must be always orientable in order to have a well-defined theory of integration on M and, thus, the validity of the rule of integration by parts: Stokes' theorem is always needed in order to construct matter tensor energy momentum. Since Euclidean Einstein's equations say simply that the sum of sectional curvatures is a measure of the (classical) matter energy density generating gravity, which must be always considered positive, it will be natural to expect the positivity of the Euclidean. Energy–Momentum of the matter content leads to the result that the associated sectional curvatures are positive individually. Since M is even-dimensional (four), the famous Synge's theorem[8] leads to the result that M is simply connected (note that this topological property is obviously independent of the metric structure being Lorentzian or Euclidean!) and as a direct consequence of this result, any physical geodesic (particles trajectory) on M can be topologically deformed to a point, and, thus, M does not possess "holes" from the point of view of the Whitney imbedding extrinsic minimal space R^9.

Finally, let us argue that the existence of a (symmetric) energy–momentum tensor on M is associated with the "General Relativity" description of the space–time manifold M by means of charts (the Physics is invariant under the action of the diffeomorphism group of M), which by its turn leads to the existence of the matter energy–momentum tensor by means of Noether theorem (a metric-independent result) applied to the matter distribution Lagrangean (a scalar function defined on the tangent bundle of M).

As a consequence, let us conjecture again that the introduction of a cosmological term on Einstein equation spoils the physical results presented on the whole topology and the physical requirement of positivity of the matter–energy universe moments tensor, given, thus, a plausible topological argument for the vanishing of the cosmological constant at the level of the global–topological aspects of the space–time manifold.

Finally, let us show the mathematical formulae associated with our ideas and conjectures written above.

Let $e_0, \ldots, e_3$ be an orthonormal frame at a point of M (Euclidean). It is well known that the Ricci quadratic form can be expressed in terms of sectional curvatures

$$\mathrm{Ric}\,(e_i, e_i) = \sum_{j \neq i} K(e_i \wedge e_j), \qquad (A.32)$$

and the Einstein tensor is defined by

$$G_{ij} = R_{ij} - \frac{1}{2} g_{ij} R. \qquad (A.33)$$

Since the Einstein equation reads in terms of quadratic forms associated with the sectional curvatures as

$$G(e_p, e_p) = + \sum K(e_p^\perp) = T(e_p, e_p), \qquad (A.34)$$

with T_{ij} being the matter energy tensor and $e_p^\perp$ the basis two-plane orthogonal to e_p, one can in principle write the sectional curvatures $K(e_p \wedge e_q)$ in terms of the quadratic energy–momentum sectional curvatures $T_{ij}(e_p, e_q)$ at least for "short-time" cylindrical geometrodynamical space–time configurations as expected in a Quantum theory of gravitation (see

Ref. 5). For the two-dimensional case this assertive is straightforward as one can see from the relations below:

$$G(e_0, e_0) = K(e_1 \wedge e_1), \qquad (A.35)$$

$$G(e_1, e_1) = K(e_0 \wedge e_0). \qquad (A.36)$$

As a consequence, one should conjecture that the positivity of the energy–momentum tensor $T(e_p, e_q)$ leads to the individual positivity of the sectional curvature set $K(e_r, e_s)$ on the basis of Eq. (A.34), namely,

$$G(e_p, e_q) = T(e_p, e_q) = \text{Ric}\,(e_p, e_q) - \delta_{pq} \left[\sum_{i,j,i \neq j} K(e_i \wedge e_j) \right]. \qquad (A.37)$$

Chapter 2

Scattering Theory in Non-Relativistic One-Body Short-Range Quantum Mechanics: Möller Wave Operators and Asymptotic Completeness

2.1. The Wave Operators in One-Body Quantum Mechanics

We know that the quantum dynamics in R^N is postulated to be given (see the previous chapter) by unitary group generated by a self-adjoint operator of the form

$$H = H_0 + V, \tag{2.1}$$

where $H_0 = -\frac{1}{2}\Delta$ is the Laplacean operator defined in the Sobolev space $H^2(R^N)$ and $U(x)$ is a given multiplication operator in $H^2(R^N)$ (for instance: $V \in L^2_{\mathrm{loc}}(R^N)$ and with a lower bound in R^N).

The problem in scattering theory is to prove that for a class of special quantum initial states $\phi \in \mathcal{H}_c(H)$ (the subspace of absolute continuity of H), there are states $\phi_\pm \in \mathrm{Dom}(H_0)$ such that the interacting dynamics is asymptotically equivalent to the free dynamics. Namely,

$$\lim_{t \to \pm\infty} \|e^{-iHt}\phi - e^{-itH_0}\phi_\pm\|_{L^2(R^N)} = 0 \tag{2.2}$$

or equivalently

$$\lim_{t \to \pm\infty} \|\phi - e^{iHt}e^{-iH_0t}\phi_\pm\|_{L^2(R^N)} = 0. \tag{2.3}$$

Let us thus make a general formulation of the (formal) suggestion as given in Eqs. (2.2) and (2.3).

Definition 2.1. Let H and H_0 be self-adjoint operators in a given Hilbert space $(\mathcal{H}, \langle, \rangle)$ and $P_{\mathrm{AC}}(H_0)$ the projection onto the subspace of absolute continuity of the "free" hamiltonian H_0. Let us consider the operator family,[1-4] where $P_{\mathrm{AC}}(H_0)$ denotes the operator projection on the absolute continuous spectral subspace of H_0

$$W(t) = \exp(iHt)\exp(-iH_0 t)P_{AC}(H_0). \tag{2.4}$$

We define the wave operators $W^{\pm}(H, H_0)$ associated with the operator family equation (2.4) by the strong limit (if it exists)

$$W^{\pm}(H, H_0) = S - \lim_{t\to\pm\infty} W(t). \tag{2.5}$$

We now show the existence of the wave operators equation (2.5) for a certain class of self-adjoint operators (H, H_0).

First we need the following lemma[1]:

Lemma 2.1. *Let us suppose that $W(t)$ is strongly continuous differentiable and there are numbers $\delta_{\pm} > 0$, such that*

$$\int_{\delta_+}^{\infty} \left\| \frac{d}{dt}W(t)f \right\|_{\mathcal{H}} < \infty, \qquad \int_{-\infty}^{-\delta^-} \left\| \frac{d}{dt}W(t)f \right\|_{\mathcal{H}} < \infty \tag{2.6}$$

for $f \in \mathcal{H}$. Then $W(t)f$ converges for $t \to \pm\infty$.

Proof. It is a single consequence of the estimate

$$\|W_{t_1}f - W_{t_2}f\|_{\mathcal{H}} \le \int_{t_2}^{t_1} ds \left\| \frac{d}{ds}W(t)f \right\|_{\mathcal{H}} \xrightarrow{(t_1,t_2)\to+\infty} 0. \tag{2.7}$$

$\square$

We now have the basic result on the existence of wave operators (Cook–Kuroda–Jauch).[1,3,4]

Theorem 2.1. *Let M be a dense subset of $\mathrm{Dom}(H) \cap \mathrm{Dom}(H_0)$ with the following properties:*

(a) $\|(H - H_0)\exp(-itH_0)f\|_H$ *is a continuous function in R.*
(b) $\|(H - H_0)\exp(-itH_0)f\|_H$ *is integrable in $[\delta, \infty)$ (or in $(-\infty, -\delta]$).*

Then the wave operators are well defined.

Proof. By the Stone theorem

$$\frac{d}{dt}(W(t)f) = (e^{itH}(H - H_0)e^{-itH_0})f. \tag{2.8}$$

As a straightforward consequence of hypothesis (b), we have

$$\left\|\frac{d}{dt}W(t)f\right\|_H = \|(H - H_0)e^{-itH_0}f\|_H \in L^1([\delta_1, \infty)). \qquad (2.9)$$

Since M is dense in $P_{AC}(H)\mathcal{H}$ and $W(t)$ are uniformly bounded ($\|W(t)\|_{\text{operator}} = 1$) we have the validity of the lemma of Cock–Kuroda–Jauch. $\qquad \square$

Let us apply the above result to concrete examples.

First, let the quantum mechanical potential $V(x) \in L^2_{\text{loc}}(R^N)$ be such that there exists $\varepsilon > 0$ with

$$\|V(x)(1 + |x|)^{-\left(\frac{N}{2}-1\right)+\varepsilon}\|_{L^2(R^N)} = N_v < \infty. \qquad (2.10)$$

(The so-called short-range quantum mechanical scattering potential.)

Additionally, let us suppose that $H = -\frac{1}{2}\Delta + V(x)$ is a self-adjoint operator in $L^2(R^N)$. Then there are wave operators associated with the pair (H, H_0) in $L^2(R^N)$.

In order to show this result, let us remark that the set of functions

$$M = \left\{ \psi_a(x) \in L^2(R^N) \mid \psi_a(x) = \phi(x - a), \text{ for } a \in R^N \right.$$

$$\left. \text{and } \phi(x) = 2^{-\frac{N}{2}}\left(\prod_{j=1}^{N} x_j e^{-|x|^2/4}\right) \right\} \qquad (2.11)$$

is dense in $L^2(R^N)$ (Wiener theorem[2]) and that (exercise!)

$$(\exp(-it(-\Delta))\psi_a)(x) = \frac{2^{\frac{N}{2}}}{(1+it)^{\frac{3n}{2}}}\left(\prod_{j=1}^{N}(x_j - a_j)e^{-\left[\frac{(x-a)^2}{4(1+it)}\right]}\right). \qquad (2.12)$$

We have thus the estimate for any δ

$$|(\exp(-it(-\Delta))\psi_a)(x)|$$

$$\leq 2^{-\frac{N}{2}}\frac{|x-a|^{-\left(\frac{N}{2}-1\right)+\delta}}{|1+it|^{1+\delta}}\left[\left|\frac{x-a}{1+it}\right|^{\frac{3N}{2}-1-\delta}\exp\left(-\frac{|(x-a)|^2}{4(1+t^2|)}\right)\right].$$

$$(2.13)$$

As a consequence and by choosing $0 < \delta < \frac{1}{2}$, we have the combined estimate involving the scattering potential

$$|(V \exp(-it(-\Delta))\psi_a)(x)| \leq C_a 2^{-\frac{N}{2}} \frac{|V(x)|x-a|^{-(\frac{N}{2}-1)+\delta|}}{|1+it|^{1+\delta}}, \qquad (2.14)$$

where $t > 0$ and $C_a = \sup_{\Omega \in R} \|(2\Omega)^{\frac{3N}{2}-1-\delta} e^{-\Omega^2}\|_{L^\infty(R^N)} < \infty$, here $\Omega = |(x-a)|/(1+t^2)^{\frac{1}{2}}$.

If we choose $\delta < \min(\frac{1}{2}, \epsilon)$, we have

$$\|(Ve^{-it(-\Delta)}\psi_a)(x)\|_{L^2(R^N)} \leq \frac{N_V}{|1+it|^{1+\delta}}, \qquad (2.15)$$

which obviously leads to the searched integrability result

$$\int_{\delta+}^{\infty} \|Ve^{-it(-\Delta)}\psi_a\|_{L^2(R^N)} \leq N_V \left(\int_{\delta+}^{\infty} \frac{dt}{|(1+it)|^{1+\delta}}\right) < \infty \qquad (2.16)$$

and to the existence of the wave operators.

Another important class of potentials well-behaved in quantum mechanics is those which belongs to the Enss' class formed by those functions in $L^\infty(R^N) \cap L^2(R^N)$ (in such a way that $V((-\Delta)+i1)^{-1}$ is a compact operator) and satisfying the x-localizibility Enss property[5]

$$\|V(-\Delta + z1)^{-M}(1 - \chi_{B_R(0)}(x)\|_{\text{operator}} = h_z(R) \in L^1([0,\infty)). \qquad (2.17)$$

Here, $\chi_{B_R(0)}(x) \equiv F(|x| < R)$ is the characteristic function of the Ball of radius R in R^N and M a positive real number, $M \geq 1$.

Let us consider $f \in S(R^N)$, such that for a given $a > 0$

$$|\exp(-it(-\Delta))(-\Delta + 1)^N f|(x) \leq \frac{c}{(1+|x|+|t|)^{N+1}} F\left(|x| < \frac{1}{2}a|t|\right). \qquad (2.18)$$

This set is dense in $L^2(R^N)$ since $\operatorname{supp} \tilde{f}(x) \subset (B_a(0))^c$.

We have the identity

$$\|V \exp(-it(-\Delta))f\|_{L^2(R^N)}$$

$$\leq \left\|V(H_0 + i)^{-M} F\left(|x| \leq \frac{1}{2}a|t|\right) e^{-iH_0 t}(H_0 + i)^M f\right\|_{L^2(R^N)}$$

$$+ \left\|V(H_0 + i)^{-M} F\left(|x| \geq \frac{1}{2}a|t|\right) e^{-iH_0 t}(H_0 + i)^M f\right\|_{L^2(R^N)}. \qquad (2.19)$$

Since every compact operator is a bounded operator and the estimate equation (2.18) added to the Enss localizibility property equation (2.17), one can easily obtain the Cook–Kuroda–Jauch lemma validity for the existence of wave operator for Enss potentials.

At this point of our exposition we call the readers' attention to the following theorem which is easy to proof and is left to the reader's completion.[3]

Theorem 2.2. *Let us suppose that the wave operators* $W^\pm(H, H_0)$ *exist in* $\mathcal{H}$, *then we have*

(a) $W^\pm$ *are partial isometries between* $H_{AC}(H_0)$ *(the subspace of absolute continuity of* H_0*) and the final range spaces* $H^+ = \mathrm{Range}(W^+(H, H_0))$ *(* $H^- = \mathrm{Range}(W^-(H, H_0))$ *).*

(b) $HW^\pm = W^\pm H_0$.

(c) $\mathcal{H}^\pm \subset \mathcal{H}_{AC}(H)$.

(d) *The S-matrix operator* $S = (W^+)^*(W^-)\colon \mathcal{H} \to \mathcal{H}$.

An important physical requirement on quantum mechanics is the requirement that $W^\pm(H, H^0)$ should be unitary applications between the subspaces $\mathcal{H}(H_0)$ and $\mathcal{H}(H)$, with the condition of $H^+ = H^- = \mathcal{H}_{AC}(H_0)$, and in such a way that the operator $\bar{S} = (W^+)^*W^-$ must be a unitary operator between $\mathcal{H}^-$ and $\mathcal{H}^+$, and thus, insuring that the physical S-matrix operator $\bar{S}$ is a unitary operator, the backbone of the Copenhagen School interpetration of the quantum phenomena.

We intend to show the confirmation of this result in two situations. First, we show that it is possible to always have asymptotic completeness by reducing the physical Hilbert space of the scattering states. Second, we present mathematical details in next sections, the beautiful original Enss's geometrical work on the subject.[5]

Our first claim is a simple result of ours obtained from a result of Halmos (exercise).

Halmos Lemma. *Let T be an isometry in $\mathcal{H}$. Let us consider the closed subspace $N = (\mathrm{Range}\ T)^\perp$. Then we have the following subspace decomposition of the original Hilbert space in terms of invariant subspace of the isometry T*

$$\mathcal{H} = M \oplus \left(\sum_{k=0}^{\infty} T^k(N) \right), \tag{2.20}$$

with T^k denoting the kth power of T and most importantly, the composite operator $T_M = TP_M$ (when $P\colon \mathcal{H} \to M$ is the orthogonal projection onto the closed subspace M) is a unitary operator $T_M\colon M \to M$.

In our case $M \subset \mathcal{H}^+ \cap \mathcal{H}^-$ and $S_M = SP_M$ is automatically a unitary operator in M.

2.2. Asymptotic Properties of States in the Continuous Spectra of the Enss Hamiltonian

Let $H = \overbrace{(-\frac{1}{2}\Delta)}^{H_0} + V$ be a self-adjoint operator defined by an Enss potential with $M = 1$ (see Eq. (2.13)). Then we have the result[5]:

Enss Lemma. *Let* $f \in \mathcal{H}_c(H)$ *be a state belonging to the continuity subspace of the Enss Hamiltonian. Then there is a sequence of times* t_n *such that* $\lim_{n \to \infty} t_n = +\infty$ $(t_n \le t_{n+1}, \forall n \in \mathbb{Z})$ *and*

$$\lim_{n \to \infty} \|F(|x| < 13n)e^{iHt_n}f\|_{L^2(R^N)} = 0, \qquad (2.21a)$$

$$\lim_{n \to \infty} \int_{-n}^{n} dt \||F(|x| < n)e^{-iH(t+t_n)}(H+i\mathbf{1})f\|_{L^2} = 0. \qquad (2.21b)$$

In order to show such result let us remark that if a continuous function has an ergodic mean zero, i.e. $\lim_{T \to \infty} \frac{1}{T}(\int_0^T g(z)dz) = 0$, then there is a monotone sequence (t_n) such that $\lim_{n \to \infty} g(t_n) = 0$ with $t_n \le t_{n+1}$ and $t_n \to \infty$. Now it is a deep theorem that (see Chap. 8)

$$\lim_{T \to \infty} \frac{1}{T} \int_0^T dt \|f(|x| < n)e^{-iH(t)}(H+i\mathbf{1})f\|_{L^2(R^N)} = 0. \qquad (2.22)$$

So, the result of lemma is a simple consequence of the above-stated RAGE theorem applied to the z-functions below written together with Fubbini theorem in order to interchange an integration order

$$W_{2n}(z) = \int_{-n}^{n} \|F(|x| < n)e^{-iH(t+z)}(H+i\mathbf{1})f\|_{L^2} dt, \qquad (2.23)$$

$$W_{2n+1}(z) = \|F(|x| < n)e^{-iHz}f\|_{L^2}. \qquad (2.24)$$

Let us now define the Enss' scattering states for a given state $f \in \mathcal{H}_c(H)$

$$f_n = \exp(-iHt_n)f. \qquad (2.25)$$

We now have the following operatorial limit:

$$\lim_{n \to \infty} \|((H+z\mathbf{1})^{-1} - (H_0+z\mathbf{1})^{-1})F(|x| > (13n)\|_{\text{operator}} = 0. \qquad (2.26)$$

The validity of Eq. (2.26) is a simple consequence of the continuity and integrability of the Enss function (see Eq. (2.13))

$$\|((H + z\mathbf{1})^{-1} - (H_0 + z\mathbf{1})^{-1})F(|x| > (13n)\|_{\text{operator}}$$

$$\leq \|V(H_0 + z\mathbf{1})^{-1}F(|x| \geq (13n)\|_{\text{operator}} = h_z(R) \qquad (2.27)$$

since

$$0 = \lim_{n\to\infty} \left(\int_{13n}^{14n} dR h_z(R) \right) \geq \lim_{n\to\infty} n h_z(13n). \qquad (2.28)$$

Similar result is obtained through derivatives in relation to the z-parameter. As the polynomials in the variables $(x + i)^{-1}$ and $(x - i)^{-1}$ are dense in the space of the continuous functions vanishing at ∞ (the Stone–Weierstrass theorem — Chap. 1), by a simple density argument we arrive at the following result for any function $\phi(x) \in C_0(R)$:

$$\lim_{n\to\infty} \|(\phi(H) - \phi(H_0))f_n\|_{L^2(R^N)} = 0. \qquad (2.29)$$

At this point we introduce the Enss scattering states $f_n \in \mathcal{H}_c(H)$, the physical requirement that they are really physical by possessing finite (interacting) energy. This last condition is achieved by considering the energy cut-off new-states $(0 \leq a < b < \infty)$

$$\bar{f}_n = E_H \left(73a^2, \frac{|b - a|^2}{4} \right) f_n, \qquad (2.30)$$

where $E_H([a, b])$ denotes the spectral projection of the full Enss Hamiltonian H, associated with the spectral interval $[a, b]$. (Note that $\mathcal{H}_c(H) \subset R^+$ since $(H + z\mathbf{1})^{-1}$ is a compact operator — see Appendix C.)

Let us consider $\tilde{\phi}(x) \in C^\infty(R)$ such that $\phi(x) = 1$ for $x \in [73a^2, (b - a)^2/4]$ and $\phi(x) \equiv 0$ for $r \in [72a^2, (b - a)^2/4]$. We now consider the "physical free-energy packet" of scattering states

$$\phi_n = \tilde{\varphi} \left(-\frac{1}{2}\Delta \right) \bar{f}_n. \qquad (2.31)$$

Let us point out that the Fourier transforms of the states $\phi_n(x)$ all have support in the region $\sqrt{2 \cdot 72a^2} \leq |P| \leq \sqrt{2\frac{(b-a)^2}{2}} \Leftrightarrow$

$$\text{supp}(\mathcal{F}[\phi_n](p)) \subseteq \{p \in R^N \mid 12a \leq |p| \leq b - a\}. \qquad (2.32)$$

At this point we remark that we need to further localize our scattering states by taking into account fully its quantum nature as given by the Heisenberg Principle.[5] In order to achieve such a result, we introduce a function $s(x) \in S(R^N)$ such that its Fourier transform has support in the ball $B_a(0)$ in R^N with $\tilde{S}(0) = \mathcal{F}[S](0) = 1$.

Let us define the quantum mechanical classical localization operators for a given Borelian A

$$F_0(A)(x) = \int_{R^N} dy S(x - y)\chi_A(y). \qquad (2.33)$$

At this point we can see that the family of functions $\{F_0(A_j)\}_{j \in I}$ is a unity decomposition if $\{A_j\}_{j \in J}$ is a disjoint decomposition of R^N

$$\sum_{j \in I} F_0(A_j) = \int S(y)d^N y = \tilde{S}(0) = 1. \qquad (2.34)$$

Another important point to call the readers' attention is the fact that the quantum localized states $F_0(A)\phi_n$ all have their momenta in the Ball with radius b. This result is easily obtained through the result that for $m(A) < \infty$, we have

$$\text{supp}\,\mathcal{F}[F_0(A)](p) \subseteq \text{supp}\,\tilde{S}(p) \subseteq B_a(0). \qquad (2.35)$$

In the case of $m(A) = +\infty$ where $\chi_A(x) \notin L^1(R^N)$, we left it for our readers by just considering the equivalent expression

$$F_0(A) = \int d^N y \cdot S(y)(|(x - y)^{2N} + 1|)\frac{\chi_A(x - y)}{(|x - y|^{2N} + 1)}. \qquad (2.36)$$

Let us now consider the best quantum mechanical spatial classical localization decomposition of R^N in Balls and truncated cones $\{B_{12n}(0), C_{12n}^{(j)}\}_{j \in I}$,[6] where the cones $C_{12n}^{(j)} = \{x \in R^N \mid |x| \geq 12n, x\vec{e}_j \geq |x|; 2\}$ have axis along the vector $\{\hat{e}_j\}$, set of unity vector to be suitably choosen.

Then we have the following results:

$$\lim_{n \to \infty} \|F_0(B_{12n})\phi_n\|_{L^2} = 0, \qquad (2.37a)$$

$$\lim_{n \to \infty} \left\| f_n - \left(\sum_{j \in I} F_0(C_{12n}^{(j)})\phi_n \right) \right\|_{L^2} = 0. \qquad (2.37b)$$

The proof is achieved by using the estimates

$$\|F_0(B_{12n})\phi_n - F(|x| \leq 13n)\bar{f}_n\|_{L^2}$$

$$\leq \|\tilde{\varphi}(H_0)\bar{f}_n - \tilde{\varphi}(H)\bar{f}_n\|_{L^2} \overbrace{\|F_0(B_{12n})\|_{L^2}}^{\leq \tilde{S}(0)=1)}$$

$$+ \|[(F_0(B_{12n}) - f(|x| \leq 13n)]\tilde{\varphi}(H)\bar{f}_n\|_{L^2} \qquad (2.38a)$$

and

$$\lim_{n \to \infty} \left\| \left(\sum_{j \in I} F_0(B_{12n}(f_n)) \right) \right\|_{L^2} = 0. \qquad (2.38b)$$

Let us finally consider the physical Enss localized scattering states by its classical motion direction. Let $h_j(p) \in C_0^\infty(R^N)$ be the functions satisfying the conditions

$$h_j(p) = 0 \qquad \text{for } p \cdot \hat{e}_j \leq -2a,$$

$$h_j(p) + h_j(-p) = 1 \quad \text{for } |p| < b \qquad (2.39)$$

with $\hat{e}_j$ denoting the vector in the direction of the cone's axis $C_{12n}^{(j)}$ as previously stated.

The Enss OUT and IN states are explicitly given by[5]

$$\phi_n(j)^{\text{OUT}} = F_0(C_{12n}^{(j)})h_j(p)\phi_n, \qquad (2.40a)$$

$$\phi_n(j)^{\text{IN}} = F_0(C_{12n}^{(j)})h_j(-p)\phi_n. \qquad (2.40b)$$

We now have the important technical result of free evolution of the above considered directional scattering states (for a detailed proof see Appendix A).

Enss Theorem (Technical). *We have the following result:*

$$\lim_{n \to \infty} \int_0^\infty dt \|F(|x| \leq n + at)e^{-iH_0 t}(H_0 + i\mathbf{1})\phi_n(j)^{\text{OUT}}\| = 0, \quad (2.41a)$$

$$\lim_{n \to \infty} \int_{-\infty}^0 dt \|F(|x| \leq n - at)e^{-iH_0 t}(H_0 + i\mathbf{1})\phi_n(j)^{\text{IN}}\| = 0. \quad (2.41b)$$

After displaying their geometrical properties of the Enss scattering states, we now pass to the proof of the asymptotic completeness of the Enss Hamiltonian in Sec. 2.3.

2.3. The Enss Proof of the Non-Relativistic One-Body Quantum Mechanical Scattering

Enss Theorem (Asymptotic completeness).[5] *Let H be the Enss Hamiltonian and its associated wave operators $W^{\pm}(H, H_0)$. Then, the subspaces $\mathcal{H}^{\pm} \equiv Range(W^{\pm})$ coincide with the subspace of continuity $\mathcal{H}_c(H)$ of the Enss Hamiltonian.*

Proof. Let $\psi \in M$ be one of the quantum mechanical localized state of the previous section. Note that M is dense in $\mathcal{H}_c(H)$. We now observe that (Eq. (2.37b))

$$\lim_{n \to \infty} \left\| \psi_n - \left(\sum_{j \in I} \phi_n(j)^{\text{IN}} + \sum_{j \in I} \phi_n(j)^{\text{OUT}} \right) \right\|_{L^2}$$

$$\lim_{n \to \infty} \left\| \psi_n - \left(\sum_{j \in I} F_0(C_{12n}^{(j)}) \phi_n \right) \right\|_{L^2} = 0. \tag{2.42}$$

Let us suppose that there is a state $\psi \in \mathcal{H}_c(H)$ such that it is orthogonal to $\mathcal{H}^+ = \text{Range}(W^+)$.

We observe now that the estimate below holds true

$$\lim_{n \to \infty} |\langle \psi_n, \psi_n \rangle|$$

$$= \lim_{n \to \infty} \left| \left\langle \psi_n, \psi_n - \left\{ \left(\sum_{j \in I} \phi_n(j)^{\text{IN}} + \sum_{j \in I} \phi_n(j)^{\text{OUT}} \right) \right. \right. \right.$$

$$\left. \left. \left. - \left(\sum_{j \in I} \phi_n(j)^{\text{IN}} + \sum_{j \in I} \phi_n(j)^{\text{OUT}} \right) \right\} \right\rangle \right|$$

$$\leq \lim_{n \to \infty} \left| \left\langle \psi_n, \psi_n - \left(\sum_{j \in I} \phi_n(j)^{\text{IN}} + \sum_{j \in I} \phi_n(j)^{\text{OUT}} \right) \right\rangle \right|$$

$$+ \lim_{n \to \infty} \left| \left\langle \psi_n, \sum_{j \in I} \phi_n(j)^{\text{IN}} \right\rangle \right| + \lim_{n \to \infty} \left| \left\langle \psi_n, \sum_{j \in I} \phi_n(j)^{\text{OUT}} \right\rangle \right|$$

$$\tag{2.43a}$$

$$\leq \lim_{n \to \infty} \left\{ \left[\|\psi_n\| \left\| \psi_n - \left(\sum_{j \in I} \phi_n(j)^{\text{IN}} + \sum_{j \in I} \phi_n(j)^{\text{OUT}} \right) \right\| \right] \right\}$$

$$\tag{2.43b}$$

$$+ \lim_{n \to \infty} \left\{ \left\langle \left| \psi_n, \sum_{j \in I} \phi_n(j)^{\text{IN}} \right| \right\rangle \right\} \tag{2.43c}$$

$$+ \lim_{n \to \infty} \left\{ \left| \left\langle \psi_n, \sum_{j \in I} W^+ \phi_n(j)^{\text{OUT}} \right\rangle \right| \right\}. \tag{2.43d}$$

Here, we have used the technical lemma (see Appendix B)

$$\lim_{n \to \infty} \| (W^+ - 1)\phi_n(j)^{\text{OUT}} \| = 0, \tag{2.44a}$$

$$\lim_{n \to \infty} |\langle \phi_n, \phi_n(j)^{\text{IN}} \| = 0. \tag{2.44b}$$

As a consequence that $\langle \psi, W^+ w \rangle = 0$, $\forall w \in \text{Dom}(W^+)$, we have that the only remaining term in the estimate considered above is the following:

$$\lim_{n \to \infty} \left| \left\langle \psi_n, \psi_n - \left(\sum_{j \in E} \phi_n(j)^{\text{IN}} + \phi_n(j)^{\text{OUT}} \right) \right\rangle \right| = 0. \tag{2.45}$$

In view of Eq. (2.42), we have the contradiction

$$0 \neq \|\psi\|^2 = \lim_{n \to \infty} |\langle \psi_n, \psi_n \rangle|$$

$$\leq \lim_{n \to \infty} \left| \left\langle \psi_n, \psi_n - \left(\sum_{j \in I} \phi_n(j)^{\text{IN}} + \phi_n(j)^{\text{OUT}} \right) \right\rangle \right| = 0 \tag{2.46}$$

The proof of $\mathcal{H}^- = \mathcal{H}_c(H)$ is similar and only need to consider $\phi_n(j)^{\text{IN}}$ in Eq. (2.44a). Details are left as an exercise. Note that this result implies that the singularly continuous spectrum of the Enss Hamiltonian is empty and all bound states are orthogonal to $\mathcal{H}_{\text{AC}}(H)$. $\square$

References

1. S. T. Kuroda, On the existence and the unitary property of the scattering operator, *Nuovo Cimento* **12**, 431–454 (1959).
2. N. Wiener, *The Fourier Integral and Certain of its Applications* (Cambridge University Press, NY, 1933).
3. W. O. Amrein, M. M. Jauch and B. L. Sinha, *Scattering Theory in Quantum Mechanics* (W.A. Benjamin, Ins., Massachussetts, 1977).
4. J. B. Dollard, Adiabatic switching in the Schrödinger theory of scattering, *J. Math. Phys.* 802–810 (1966).
5. V. Enss, Asymptotic completeness for quantum mechanical potential scattering, *Commun. Math. Phys.* **61**, 285–291 (1979).

6. E. B. Davies, *Quantum Theorem of Open Systems* (Academic Press, New York, 1976).
7. S. Steinberg, Meromorphic families of compact operators, *Arch. Rational Mech. Anal.* **31** (1969).

Appendix A

Theorem A.1. *Let us consider the Enss scatterbug states* $\phi_n(j)^{\text{OUT}}_{\text{IN}}$. *Then we have the result*

$$\lim_{n\to\infty} \int_0^\infty dt \| F(|x| \leq n + at) e^{-iH_0 t}(H_0 + i\mathbf{1})\phi_n(j)^{\text{OUT}} \| = 0, \quad \text{(A.1)}$$

$$\lim_{n\to\infty} \int_0^\infty dt \| F(|x| \leq n - at) e^{-iH_0 t}(H_0 + i\mathbf{1})\phi_n(j)^{\text{IN}} \| = 0. \quad \text{(A.2)}$$

Proof. Our aim is to reduce the above estimates to the one-dimensional case. Thus, let us analyze Eq. (A.1) by considering a new decomposition of the Enss scattering states $\phi_n(j)^{\text{IN}}$ in the x-space R^N. First, let us observe that the half-spaces $x \cdot F \leq n + at$ with $f = \{f_n\}_{1 \leq k \leq N} \in R^N$ and $\|f\| = 1$ contains the Ball $B_{n+at}(0) = \{x \in R^N \mid |x| \leq n + at\}$. We let the function $h_j(p)$ with $\{p \in R^N \mid 12a \leq |p| \leq b-a\}$ be decomposed into a finite number of summands $\xi_k^{(j)}(p) \in C_0^\infty(R^N)$ in such a way that support $\{\xi_k^{(j)}(p)\} \subseteq \{p \in R^N \mid P \cdot f_n \geq 2a\}$.

Let us choose $f_k^{(j)}$ unity vectors with an angle of $\pi/9$ rads with the cones vectors axis $\hat{e}_j$ of $C_{12n}^{(j)}$.

As a consequence $x \cdot f_k^{(j)} > 2n$ for $x \in C_{12n}^{(j)}$. We remark that Eq. (A.1) will be a result of the validity of the following limit:

$$\lim_{n\to\infty} \int_0^\infty dt \| F(x \cdot f_k^{(j)} \leq n + at) e^{-iH_0 t}(H_0 + i\mathbf{1})F_0(C_{12n}^{(j)})\xi_k^{(j)}(p)\phi_n \| = 0$$

$$\text{(A.3)}$$

since any state ϕ_n has its "momenta" support in the region $12a \leq |p| \leq b-a$. Within this region one has decomposition of the Enss scattering state

$$\phi_n(j)^{\text{OUT}} = F_0(C_{12n}^{(j)})h_0(p)\phi_n = \sum_k F_0(C_{12n}^{(j)})\xi_k^{(j)}(p)\phi_n. \quad \text{(A.4)}$$

Just in order to simplify the estimates which follows and choose the axis x_1 parallel to $f_j^{(k)}$ under consideration. Let us thus consider

$$F_m = (H_0 + i\mathbf{1})F_0(C_{12n}^{(j)} \cap \{2n + m \leq x_1 \leq 2n + m + 1\}). \quad \text{(A.5)}$$

Obviously

$$(H_0 + i1)F_0(C_{12n}^{(j)}) = \sum_{m=0}^{\infty} F_m. \tag{A.6}$$

It will be necessary to implement the action of $(H_0 + i1)$ by means of a multiplication operator in p-space by means of a function with compact support. This can be implemented by noting that all the states $F_0(A)\phi_n$, with A denoting a Borelian of R^N, have p-momentum support in the region $|p| \leq b$. Another point worth remarking is that the momentum component p_1 (related to the x_1-axis) of the states

$$(H_0 + i1)F_0(C_{12n}^{(j)} \cap \{2n + m \leq x_1 \leq 2n + m + 1\})\xi_k^{(j)}(p)\phi_n \tag{A.7}$$

is in the interval $a \leq p_1 \leq b$.

Let us now pass to the estimates. First, the estimate equation (A.3) reduces to the one-dimensional case as a consequence of the dependence of solely the coordinate x_1 into our sliced up states equation (A.7).

We have thus reduced our analysis to the one-dimensional case

$$\int_0^{\infty} dt \left\| F(x_1 \leq n + at)e^{-i\left(\frac{\partial^2}{\partial^2 x_1}\right)t}\phi(x_1, \hat{x}) \right\|_{L^2} \tag{A.8}$$

with $\phi(x, \hat{x})$ denoting our new sliced up states.

Let us define the following operator family of operators in $L^2(R^N)$:

$$(W_\zeta \phi) = \frac{e^{ix_1^2/2\zeta}}{(2\pi i\zeta)^{1/2}} \left\{ \int_{-\infty}^{+\infty} dy_1 \left[1 + \frac{iy_1^2}{2\zeta} + \left(\frac{iy_1^2}{2\zeta}\right)^2 \Big/ 2 \right] e^{-i\frac{x_1}{\zeta}y_1}\phi(y_1, \hat{x}) \right\}. \tag{A.9}$$

It is a consequence of the support of the state $\phi(y_1, \hat{x})$ that

$$(W_t \phi)(x) = 0 \quad \text{for } x_1 < at \tag{A.10}$$

by just considering Eq. (A.9) by means of the Fourier transform

$$(W_t \phi)(x) = \frac{e^{ix_1^2/2t}}{(2\pi i t)^{1/2}} \left[1 + \left(\frac{i}{2t}\frac{\partial^2}{\partial^2(\frac{x_1}{t})}\right) \right.$$
$$\left. + \left(\left(\frac{i}{2t}\right)^2 \frac{\partial^3}{\partial^3(\frac{x_1}{t})}\right) \mathcal{F}_{y_1}[\phi]\left(\frac{x_1}{t}, \hat{x}\right) \right] \tag{A.11}$$

and $\mathcal{F}_{y_1}[\phi](p_1, \hat{x}) \equiv 0$ for $p \leq \frac{x_1}{\zeta} < a$.

We now note the validity of the result written as

$$
\left\| (1+t)(-x_1+2at)^2 \left[\left(\left(e^{-i\frac{\partial^2}{\partial^2 x_1}\frac{t}{2}} - W_t \right)\phi \right)(x) \right] \right\|_{L^2}
$$

$$
= \left\| (1+t)t^2 \left(-i\frac{\partial}{\partial y_1} + 2a \right)^2 \left[e^{ib_1^2/2t} - 1 + \frac{iy_1^2}{2t} + \frac{\frac{iy_1^2}{2t}}{2} \right] \phi(y_1,\hat{x}) \right\|_{L^2}
$$

$$
\leq \left\| \frac{(1+t)t^2}{t^3} |P_1^*(|y_1|)|(\phi(y_1,\hat{x})) + |P_2^*(|y_1|)| \frac{\partial}{\partial y_1}\phi(y_1,\hat{x}) \right.
$$

$$
\left. + |P_3^*(|y_1|)| \left| \frac{\partial^2}{\partial y_1^2}\phi(y_1,\hat{x}) \right| \right\|_{L^2}, \tag{A.12}
$$

where $P_\ell^*(|y_1|)$, $\ell = 1,2,3$ are polynomials satisfying the relations ($x_1 \geq \tilde{a}$)

$$
P_3^*(|y_1|) = |y_1|^6/48,
$$

$$
P_2^*(|y_1|) = \frac{4a|y_1|^6}{48} + \frac{2|y_1|^5}{8},
$$

$$
P_1^*(|y_1|) \geq P\left(|y_1|, \frac{1}{\tilde{a}} \right) \geq \left| P\left(|y_1|, \left|\frac{1}{t}\right| \right) \right|. \tag{A.13}
$$

We have this for $x_1 \leq at$ the estimate

$$
\left\| (1+t)t^2 \left(-i\frac{\partial}{\partial y_1} + 2a \right)^2 \left[e^{iy_1^2/2t} - \left(1 + \frac{iy_1^2}{2t} + \frac{(iy_1^2)}{(2t)}\cdot\frac{1}{2} \right) \right]\phi(y_1,\hat{x}) \right\|
$$

$$
\leq \left(\frac{a}{\tilde{a}}+1\right)\left\| P_1^*(|y_1|)\phi(y_1,\hat{x}) + P_2^*(|y_1|)\frac{\partial}{\partial y_1}\phi(y_1,\hat{x}) \right.
$$

$$
\left. + P_3^*(|y_1|)\frac{\partial^2\phi(y_1,\hat{x})}{\partial^2 y_1} \right\| = \left(\frac{a}{\tilde{a}}\right)\left\| \left\{ P_1^*\left(\frac{\partial}{\partial p_1}\right) + P_2^*\left(\frac{\partial}{\partial p_1}\right)ip_1 \right. \right.
$$

$$
\left. \left. + P_3^*\left(\frac{\partial}{\partial p_1}\right)(-p_1)^2 \right\}\mathcal{F}[\phi](p_1,\hat{p}) \right\|_{L^2} \leq A \tag{A.14}
$$

since supp $\tilde{\phi}(p_1,\hat{p}) \subset \{p \in R^N \mid |p| \leq b\}$.

As a consequence (see Eqs. (A.10) and (A.14))

$$
\left\| (1+t)(-x_1+2at)^2 F(x_1 \leq at)e^{-\left(i\frac{\partial^2}{\partial x_1^2}\frac{t}{2}\right)}\phi(x_1,\hat{x}) \right\| \leq A. \tag{A.15}
$$

The result of Eq. (A.15) is the following:

$$
\left\| F(x_1 \leq -v+at)\left(e^{-i\frac{\partial^2}{\partial x_1^2}\frac{t}{2}}\phi \right)(x) \right\| \leq \frac{A}{(1+t)(u+at)^2} \tag{A.16}
$$

since for $v > 0$, $(x_1 \le -v + at)$

$$\frac{1}{(-x_1 + 2at)^2} \le \frac{1}{(v + at)^2}. \tag{A.17}$$

As a consequence we have the following estimate for $v = m + n$ and $\phi(x) = F_m \zeta_k^j(p)\phi_n(x_1 + 2m + n, \hat{x})$ and $x_1 \in C_{12n}^{(j)} \cap \{2n + m \le x_1 \le 2n + m + 1\}$ (Ref. 5) by using translational invariance $x_1 \to x_1 + v$:

$$\|F(x_1 - 2n - m < -(n + m) + at)e^{-iH_0 t} F_m \zeta_k^{(j)}(p)\phi_n(x_1, \hat{x})\|_{L^2}$$

$$\le \frac{A}{|(1 + t)|m + n + at|^2|}. \tag{A.18}$$

By noting again that

$$\sum_{m=0}^{\infty} \|F(x_1 - 2n - m < -n - m + at)e^{-iH_0 t} F_n \xi_k^{(j)}(p)\phi_n(x)\|$$

$$= \sum_{m=0}^{\infty} \|F(x_1 < n + at)e^{-iH_0 t} F_n \xi_k^{(j)}(p)\phi_n(x)\|.$$

We have

$$\|F(x_1 < n + at)e^{-iH_0 t}(H_0 + i1)\xi_k^{(j)}(p)\phi_n(x)\|$$

$$= \left\| F(x_1 < n + at)e^{-iH_0 t}\left(\sum_{m=0}^{\infty} F_n \xi_k^{(j)}(p)\phi_n(x) \right) \right\|$$

$$\le \sum_{m=0}^{\infty} \|F(x_1 + n + at)e^{-iH_0 t} F_n \xi_k^{(j)}(p)\phi_n(x)\|$$

$$\le A\left\{ \sum_{m=0}^{\infty} \frac{1}{|(1 + t)(m + n + at)^2|} \right\}.$$

We now left as an exercise to our diligent reader to establish the result[5]

$$\lim_{n \to \infty} \int_0^\infty dt \|F(x \cdot f_k^{(j)} \le n + at)e^{-iH_0 t}(H_0 + i1)F_0(C_{12n}^{(j)})\xi_k^{(j)}(p)\phi_n\|$$

$$\le A \lim_{n \to \infty} \left\{ \sum_{m=0}^{\infty} \int_0^\infty dt \left[\frac{1}{|(1 + t)(m + n + at)^2|} \right] \right\} = 0.$$

$\square$

Appendix B

Let us show the following technical result used in the proof of the Enss theorem.

Lemma B.1. *Let $W_\pm$ be the Enss wave operators and $\phi_n(j)^{\mathrm{OUT}}_{\mathrm{IN}}$ considered in Sec. 2.2, then we have the result ($\|\ \| = \|\ \|_{L^2}$).*

$$\lim_{n\to\infty} \|(W^+ - 1)\phi_n(j)^{\mathrm{OUT}}\| = 0, \tag{B.1}$$

$$\lim_{n\to\infty} \|(W^- - 1)\phi_n(j)^{\mathrm{IN}}\| = 0. \tag{B.2}$$

Proof. Since

$$\|(W(t) - W(0)f\| \le \left\| \int_0^t d\zeta (V e^{-iH_0\zeta} f) \right\| \tag{B.3}$$

we have that

$$\|(W^+ - 1)\phi_n(j)^{\mathrm{OUT}}\| \le \left\| \int_0^\infty d\zeta V e^{-iH_0\zeta} \phi_n(j)^{\mathrm{OUT}} \right\|$$

$$\le \|V(H_0 + i1)^{-1}\|_{\mathrm{op}} \left(\int_0^\infty d\zeta \|F(|x| \le n + a\zeta) e^{-iH_0\zeta}(H_0 + i1)\phi_n(j)^{\mathrm{OUT}}\| \right)$$

$$+ \left(\int_0^\infty d\zeta \|V(H_0 + i1)^{-1} F(|x| \ge n + a\zeta)\|_{\mathrm{op}} \right)$$

$$\times \|e^{-iH_0\zeta}(H_0 + i1)\phi_n(j)^{\mathrm{OUT}}\|_{L^2}. \tag{B.4}$$

The first term in Eq. (B.4) goes to zero by the long estimate given in Appendix A.

The second term can be estimated as follows ($\|e^{-iH_0\zeta}(\cdots)\| \le \|(\cdots)\|$!)

$$\|(H_0 + i1)\phi_n(j)^{\mathrm{OUT}}\| \int_0^\infty h(n + a\zeta)d\zeta$$

$$\le \|(H_0 + i1)\phi_n(j)^{\mathrm{OUT}}\| \left(\int_n^\infty h(R)dR \right). \tag{B.5}$$

Since

$$\|(H_0 + i\mathbf{1})\phi_n(j)^{\mathrm{OUT}}\|_{L^2} = \|(p^2 + i\mathbf{1})\mathcal{F}(\phi_n(j)^{\mathrm{OUT}})(p)\|_{L^2}$$
$$\leq (b^2 + 1)\|\mathcal{F}(\phi_n(j)^{\mathrm{OUT}})\|_{L^2} < \infty \tag{B.6}$$

since $\mathcal{F}[\phi_n(j)^{\mathrm{OUT}}](p)$ has support in $|p| \leq b$ and obviously due to the fact that $h(R) \in L^1(R)$, we have

$$\lim_{n \to \infty} \left(\int_n^\infty dR\, h(R) \right) = 0. \tag{B.7}$$

We now give a proof of the following additional result in the Enss proof of asymptotic completeness presented in Sec. 2.3

$$\lim_{n \to \infty} |\langle \phi_n(j)^{\mathrm{IN}}, \psi_n \rangle| = 0. \tag{B.8}$$

First, we remark that

$$|\langle \phi_n(j)^{\mathrm{IN}}, \psi_n \rangle| \leq |\langle (W^- \phi_n(j)^{\mathrm{IN}}, e^{-iHt_n}\psi \rangle + \langle (I - W^-)\phi_n(j)^{\mathrm{IN}}, e^{-iHt_n}\psi \rangle| \tag{B.9}$$
$$\leq |\langle W_- e^{iH_0 t_n}\phi_n(j)^{\mathrm{IN}}, \psi \rangle| + \|(I - W^-)\phi_n(j)^{\mathrm{IN}}\|\|\psi\|. \tag{B.10}$$

Here we have used the result (exercise)

$$e^{iHt}W_- = W_- e^{iH_0 t}. \tag{B.11}$$

As a consequence

$$|\langle \phi_n(j)^{\mathrm{IN}}, \psi_n \rangle| \leq \|(I - W^-)\phi_n(j)^{\mathrm{IN}}\| + |\langle e^{iH_0 t_n}\phi_n(j)^{\mathrm{IN}}, (W^-)^*\psi \rangle| \tag{B.12}$$

since

$$\lim_{n \to \infty} \|(1 - W^-)\phi_n(j)^{\mathrm{IN}}\| = 0 \tag{B.13}$$

we only need to prove that

$$\lim_{n \to \infty} |\langle e^{iH_0 t_n}\phi_n(j)^{\mathrm{IN}}, (W^-)^*\psi \rangle| = 0 \tag{B.14}$$

in order to show the validity of Eq. (B.8).

Now, we have the set of estimates

$$|\langle e^{iH_0 t_n} \phi_n(j)^{\text{IN}}, (W^-)^* \psi \rangle|$$
$$= |\langle (F(|x| \le n + at_n) + F(|x| \ge n + at_n)) e^{iH_0 t_n} \phi_n(j)^{\text{IN}}, (W^-)^* \psi \rangle|$$
$$\le \{ \|F(|x| \le n + at_n)(W^-)^* \psi\| \|F(|x| \le n + at_n) e^{iH_0 t_n} \phi_n(j)^{\text{IN}}\|$$
$$+ \|F(|x| \ge n + at_n)(W^-)^* \psi\| \|F(|x| \ge n + at_n) e^{iH_0 t_n} \phi_n(j)^{\text{IN}}\|$$
$$\le \|(W^-)^* \psi\| \|F(|x| \le n + at_n) e^{iH_0 t_n} \phi_n(j)^{\text{IN}}\|$$
$$+ \|F(|x| \ge n + at_n)(W^-)^* \psi\| \|\phi_n(j)^{\text{IN}}\|. \tag{B.15}$$

At this point we observe that

$$\lim_{n \to \infty} \|F(|x| \le n + at_n) e^{iH_0 t_n} \phi_n(j)^{\text{IN}}\| = 0 \tag{B.16}$$

by a similar analysis made in Appendix A.[5]   □

Appendix C

In the appendix, we intend to show that compact resolvent perturbations of the free quantum dynamics $H_0 = -\frac{1}{2}\Delta$ does not modify the essential operator spectrum: defined as the union of the continuous spectrum and the accumulation points of the bound-states (point spectrum). Let us announce this deep result in its general terms.

Theorem C.1. *Let $H = H_0 + U$ and H_0 self-adjoint operators acting on a separable Hilbert Space $\mathcal{H}$. Let us suppose that all H_0 spectrum is contained in positive real axis and such that $(H_0 + U + \lambda 1)^{-1} - (H_0 + \lambda 1)^{-1}$ is a compact operator for some complex λ_0, with $\text{Im}(\lambda_0) \ne 0$. Then the spectral part of the operator H contained in $\text{Real}(\lambda) \le 0$ is formed entirely by discrete points possesing zero as the only possible accumulation point for them.*

Proof. This result is obtained through an application of the Holomorphic Fredholm theorem to the following operatorial-valued holomorphic function:

$$f(\lambda) = 1 - (\lambda - \lambda_0)[1 - (\lambda - \lambda_0)(H_0 + \lambda 1)^{-1}]^{-1}$$
$$\times \underbrace{[(H_0 + U + \lambda 1)^{-1} - (H_0 + \lambda 1)^{-1}]}_{C(\lambda)\text{-compact operator}} \tag{C.1}$$

Since there is $\lambda_0 \in \mathbb{C}\backslash R^+$ such that $f(x)$ is inversible ($f(\lambda_0) = 1$!), we have that $C(\lambda)$ has inverse in all $\mathbb{C}\backslash R^+$, with the exception of a discret set $S \subset \mathbb{C}\backslash R^+$. Since for λ with non-zero imaginary part $C(\lambda)$ is always inversible, we have that $S \subset R^-$. It is straightforward to see that S coincides with the negtive part of the spectrum of H. $\qquad\square$

Just for completeness let us announce the holomorphic Fredholm theorem.[7]

Holomorphic Fredholm Theorem. *Let Ω be a connected open set of $\mathbb{C}$ (a domain). Let*

$$f \colon \Omega \to (\mathbb{L}(\mathcal{H}, \mathcal{H}), \|\,\|_{\text{uniform}}),$$

a holomorphic bounded-operator valued function such that for each $z \in \Omega$, $f(z)$ is a compact operator. Then we have the Fredholm alternatives:

(a) *$(1 - f(z))^{-1}$ does not exists for any $z \in \Omega$*

or

(b) *$(1 - f(z))^{-1}$ exists for $z \in \Omega\backslash S$, where S is a discret subset of Ω. In this case $(1 - f(z))^{-1}$ is a meromorphic function in Ω, holomorphic in $\Omega\backslash S$ and in $z \in S$, the equation $f(z)\psi = \psi$, has solely a finite number of linearly independent solutions in $\mathcal{H}$.*

Chapter 3

On the Hilbert Space Integration Method for the Wave Equation and Some Applications to Wave Physics*

3.1. Introduction

One important topic of mathematical methods for wave propagation physics is to devise functional analytic techniques useful to implement approximate-numerical schemes.[1] One important framework to implement such time-evolution studies is the famous Trotter–Kato–Feynman short-time method to approximate rigorously the full-wave evolution by means of a finite number of suitable short-time propagations — the well-known Feynman idea of a path-integral representation for the Cauchy problem.[2]

In this chapter we propose a pure Hilbert space analytical framework — the content of Secs. 3.2 and 3.3 — and suitable to generalize rigorously to the hyperbolic case, the powerful Trotter–Kato–Feynman formulae in Sec. 3.4. In Sec. 3.5, we present as a further example of usefulness of our abstract Hilbert space Techniques, a simple proof (by means of the use of the integral Cook–Kuroda criterium) of the existence of reduced (finite-time) scattering wave operators $W_{\pm}$ for elastic wave scattering by a compact supported inhomogenous medium (a "small" space-dependent Young medium elasticity modulus $E(x) \in C_c^{\infty}(R^3)$, i.e. $\sup |\frac{E(x)}{E_0}| \ll 1$ and $\sup |\frac{\nabla E(x)}{E_0}| \ll 1$).[3-7]

In Sec. 3.6 and Appendix A, we address the very important problem of giving a complete proof for the asymptotic stability for systems in thermoelasticity in the relevant linearizing approximation for the nonlinear magnetoelasticity wave equations,[8,9] i.e. a coupling between the divergences Lamé system of elastic vibrations and the convective parabolic equation for a constant external magnetic field.

We give a (new) proof of stability for this system by producing an exponential uniform bound in time for the total energy of the linearized

---

*Author's original results.

magnetoelastic model through the systematic use of simple estimates on the Trotter formula representations for our magnetoelastic wave propagators in our proposed Hilbert space abstract scheme.

The resulting exponential bound solve an important long-standing problem on the structural form of the decay behavior of magnetoelastic systems in the presence of external constant magnetic fields.

Finally, in Sec. 3.7, we use again Hilbert space techniques to produce a proof of existence and uniqueness of a semilinear problem of Abstract Klein Gordon propagation[11] as another example of usefulness of our proposed abstract approach.

3.2. The Abstract Spectral Method — The Nondissipative Case

Let A be a positive (essential) self-adjoint operator with domain $D(A)$ acting on a separable Hilbert space $H^{(0)}$ with the inner product denoted by $(\,,\,)^{(0)}$.

Associated to the above abstract operator A, one can consider the wave equation (Cauchy problem) on $H^{(0)} \times [0, T]$

$$\frac{\partial^2 U}{\partial^2 t} = -AU + f, \tag{3.1}$$

$$U(0) = U_0; \quad U_t(0) = v_0 \tag{3.2}$$

with initial dates given in suitable subspaces of $H^{(0)}$ to be specified later and $f \in L^2([0, T], H^{(0)})$.

It is an important problem in the mathematical physics associated with Eqs. (3.1) and (3.2) to determine exactly the subspaces where the initial date should belong in order to insure the existence of a global solution of Eqs. (3.1) and (3.2) in $C^\infty((0, \infty), H^{(0)})$ when the time-interval $[0, T]$ of the external source $f(x, t)$ is infinite $(T = +\infty)$ and besides, when there exist a natural extension of this global solution to the whole time interval $(-\infty, \infty)$, namely,

$$U(x, t) \in C^\infty((-\infty, \infty); H^{(0)}).$$

Let us present in this section a solution for this problem by means of a convenient generalization method exposed in Ref. 1 and based uniquely on the Stone spectral theorem for one-parameter unitary groups in Hilbert space.

In order to implement this Hilbert space abstract method,[1] we introduce the following Hilbert space:

$$H^{(1)} = \overline{\{(\varphi, \pi) \mid \varphi \in D(A); \ \pi \in H^{(0)}, \langle \ ; \ \rangle\}}, \qquad (3.3a)$$

where the closure is taken in relation to the inner product below:

$$\langle (\varphi_1, \pi_1); (\varphi_2, \pi_2) \rangle = (A\varphi_1, \varphi_2)^{(0)} + (\pi_1, \pi_2)^{(0)}. \qquad (3.3b)$$

Let us introduce the following operator L_A on $H^{(1)}$ and defined by the rule

$$L_A(\varphi, \pi) = (\pi, A\varphi). \qquad (3.4)$$

Note that Eq. (3.4) is well defined since we assume that the range of A is dense on $H^{(0)}$. L_A is a symmetric operator on $H^{(1)}$ since

$$\begin{aligned} \langle (\varphi_1, \pi_1); L_A(\varphi_2, \pi_2) \rangle &= (A\phi_1, \pi_2)^{(0)} + (\pi_1, A\phi_2)^{(0)} \\ &= \langle L_A(\varphi_1, \pi_1), (\varphi_2, \pi_2) \rangle. \end{aligned} \qquad (3.5)$$

Let us consider the smallest closed extension of L_A, denoted by $\bar{L}_A$. We claim that its deficiency indices $n_+(\bar{L}_A)$ and $n_-(\bar{L}_A)$ are identically zero and as a consequence leading to its self-adjointness on its domain $D(\bar{L}_A) \subset H^{(1)}$ which is explicitly given by the vectors (φ, π) such that $(A\varphi, \varphi)^{(0)} < \infty$ and $\pi \in H^{(0)}$.

Let (φ_1, π_1) be a nonzero vector such that $(\varphi_1, \pi_1) \in (\text{Range}(i + L_A))^\perp$, or equivalently for any $\varphi_2 \in D(A)$ and $\pi_2 \in H^{(0)}$:

$$\langle (\varphi_1, \pi_1); (i + L_A)(\varphi_2, \pi_2) \rangle^{(1)} = 0, \qquad (3.6)$$

which is the same content of the equation below:

$$-i(A\varphi_1, \varphi_2)^{(0)} + (\pi_1, A\varphi_2)^{(0)} + i(\pi_1, \pi_2)^{(0)} + (A\varphi_1, \pi_2) = 0, \qquad (3.7)$$

and due to the fact that $\text{Range}(A \pm i)$ is dense on $H^{(0)}$ and since A is self-adjoint it leads to $\varphi_1 = 0$ and $\pi_1 = 0$. In the same way one can prove that $n_-(\bar{L}_A) = 0$.

Collecting all the above exposed results, one has the following theorem:

Theorem 3.1. *Let the spectral theorem be applied to the one-parameter group $U(t) = \exp(it\bar{L}_A)$ acting on $D(\bar{L}_A)$; one has that it satisfies the abstract wave equation (3.1) on the strong sense for dates $U_0 \in D(A)$ and $v_0 \in H^{(0)}$ and leading to a global (time-reversing) solution on $C^\infty((-\infty, \infty), H^{(0)})$.*

Proof. It is straightforward that in components the wave equation associated with the one-parameter semigroup $U(t)(\varphi_0, \pi_0) = (\varphi(t), \pi(t))$ is given by

$$\frac{\partial}{\partial t}\varphi(t) = i\pi(t), \tag{3.8}$$

$$\frac{\partial}{\partial t}\pi(t) = i(A\phi)(t), \tag{3.9}$$

which is the same as Eqs. (3.1) and (3.2) without the external term $f(x, t)$:

$$\frac{\partial^2 \varphi(t)}{\partial^2 t} = -(A\varphi)(t) \tag{3.10}$$

with the initial date

$$\varphi(0) = U_0, \tag{3.11}$$

$$\varphi_t(0) = v_0. \tag{3.12}$$

It is worth noting that one has always a weak solution of Eqs. (3.1) and (3.2) *with* a nonzero external term

$$\begin{pmatrix} \varphi \\ \pi \end{pmatrix}(t) = e^{it\bar{L}_A}\begin{pmatrix} U_0 \\ -i\varphi_0 \end{pmatrix} + \int_0^t ds \left\{ e^{i(t-s)\bar{L}_A}\begin{pmatrix} 0 \\ -if(s) \end{pmatrix} \right\}, \tag{3.13}$$

namely, $\varphi(\cdot, t) \in C((-\infty, \infty); H^{(0)})$, unless $f(\cdot, t)$ belongs to $D(A)(f: (-\infty, \infty) \to D(A))$, in which case it produces a global C^∞ time-differentiable solution.

At this point, we make one of the most important remark that comes from such a theorem applied to the subject of random wave propagation.[2] It is important to know the correct mathematical meaning of considering the Cauchy problem for the wave equation with both initial conditions now being stochastic process with realizations on $L^2(\Omega)$ space of functions (see Appendix B). In this case, Eq. (3.13) in its weak-integral form should be considered the correct mathematical solution for this problem since one can take safely the initial dates (U_0, v_0) *both* in the Hilbert space $H^{(0)}$, which in the usual case of wave propagation is the $L^2(\Omega)$ functional Hilbert space $(H^{(0)} = L^2(\Omega))$.

Let us point out an abstract version of the wave total energy conservation during the time evolution.

First, we observe that for any $-\infty < t < \infty$ we have the identity

$$\text{Im}\{(\pi(t), A\varphi(t))^{(0)}\} \equiv 0, \tag{3.14}$$

which is obtained as an immediate consequence of the unitary property of the semigroup $U(t)$ on the two-component Hilbert space $H^{(1)}$ given by Eqs. (3.3a) and (3.3b):

$$\langle \exp(it\bar{L}_A)(\varphi_0, \pi_0); \exp(it\bar{L}_A)(\varphi_0, \pi_0) \rangle$$
$$= \langle (\varphi_0, \pi_0); (\varphi_0, \pi_0) \rangle, \tag{3.15}$$

and by deriving the above equation in relation to time, it yields

$$\langle i\bar{L}_A(\varphi(t), \pi(t)), (\varphi(t), \pi(t)) \rangle$$
$$+ \langle (\varphi(t), \pi(t)); i\bar{L}_A(\varphi(t), \pi(t)) \rangle = 0, \tag{3.16}$$

which is Eq. (3.14) when it is written in terms of components.

At this point one can easily see that the energy functional given by

$$E(t) = \frac{1}{2}(\pi(t), \pi(t))^{(0)} + \frac{1}{2}(\phi(t), (A\phi)(t)) \tag{3.17}$$

is such that

$$\frac{dE(t)}{dt} = \frac{1}{2}(i(A\phi)(t), \pi(t)) + \frac{1}{2}(\pi(t), i(A\phi)(t)) \Big\} 0$$
$$+ \frac{1}{2}(i\pi(t), (A\phi)(t)) + \frac{1}{2}((A\phi)(t), i\pi(t)) \Big\} 0$$
$$= 0. \tag{3.18}$$

As a consequence, we have the energy abstract conservation law

$$E(t) = \frac{1}{2}(v_0, v_0)^{(0)} + \frac{1}{2}(U_0, AU_0)^{(0)}. \tag{3.19}$$

Finally, we remark that by a straightforward application of the Banach fixed-point theorem within the context of our operatorial scheme, one has the existence and uniqueness of the nonlinear abstract wave equation initial value problem on $H^{(1)}$:

$$\frac{\partial^2 U}{\partial^2 t} = -AU + F(U), \tag{3.20}$$
$$U(0) = U_0; \quad U_t(0) = v_0,$$

for $F(x)$ denoting a Lipschitzian function on the real line ($|F(x) - F(y)| \le C|x - y|$). This result is the consequence of the fact that the application

$$T : C([0, T], H^{(1)}) \to C([0, T]; H^{(1)})$$

$$T(\varphi, \pi) = \int_0^t ds\{e^{i(t-s)\bar{L}_A}(0, F(\varphi))\} \tag{3.21}$$

is a contraction on $H^{(1)}$, i.e. $((A\varphi, \varphi)_H \geq C_0(\varphi, \varphi)_H)$:

$$\sup_{0 \leq s \leq T} \|(0, F(\varphi_1(s)) - F(\varphi_2(s))\|_{H^{(1)}}$$

$$\leq \sup_{0 \leq s \leq T} \{(A0, 0)^{(0)} + (F(\varphi_1(s)) - F(\varphi_2(s)), F(\varphi_1(s)) - F(\varphi_2(s)))^{(0)}\}$$

$$\leq C \sup_{0 \leq s \leq T} (\varphi_1(s) - \varphi_2(s), \varphi_1(s) - \varphi_2(s))^{(0)}$$

$$\leq CC_0 \sup_{0 \leq s \leq T} \{\|(\varphi_1, \pi_1) - (\varphi_2, \pi_2)\|_{H^{(1)}}(s)\}, \tag{3.22}$$

which after successive iterations produces bounds leading to the global *weak* solutions of Eq. (3.20) in $C((-\infty, \infty), H^{(1)})$, a nontrivial result easily obtained in this abstract spectral framework (see Sec. 3.7 for further studies on this problem).

3.3. The Abstract Spectral Method — The Dissipative Case

In this section we generalize the Hilbert space method of Sec. 3.2 to the case of a wave equation in the presence of damping terms, a very important problem on statistical continuum mechanics.[4]

We, thus, consider the abstract problem in the notation of Sec. 3.2:

$$\frac{\partial^2 U}{\partial^2 t} + \left(B \frac{\partial U}{\partial t}\right) + (AU)(t) = 0, \tag{3.23}$$

$$U(0) = U_0 \in D(A), \tag{3.24}$$

$$U_t(0) = v_0 \in H^{(0)}, \tag{3.25}$$

where A and B are essential positive self-adjoint operators such that $D(B) \supset D(A)$.

Note that we can write Eqs. (3.23)–(3.25) in the two-component Hilbert space $H^{(1)}$ of Sec. 3.2 as

$$\frac{d}{dt} \begin{pmatrix} U \\ \pi \end{pmatrix} = - \begin{bmatrix} 0 & -1 \\ A & B \end{bmatrix} \begin{pmatrix} U \\ \pi \end{pmatrix}. \tag{3.26}$$

Let us observe that the operator $\hat{A} = \begin{bmatrix} 0 & -1 \\ A & B \end{bmatrix}$ is positive on $H^{(1)}$ as a consequence of the operator B positive definiteness, namely,

$$\langle \hat{A}(U_1, \pi_1), (U_1, \pi_1) \rangle = (B\pi_1, \pi_1)^{(0)} > 0. \tag{3.27}$$

Since $\exists \lambda \in R \geq 1$ such that $\lambda 1 + \hat{A}$ is onto and bijective on $H^{(1)}$ (a topological isomorphism), one can apply, thus, the well-known Hille–Yoshida theorem[1] to the operator $\hat{A}$ in order to consider a contractive C_0 semigroup on $H^{(1)}$ in the interval $[0, \infty)$ possessing $\hat{A}$ as its generator, and naturally producing the unique global weak solution of the abstract wave equation with damping on $C^\infty([0,\infty), H^{(1)})$ if the initial conditions are chosen to be in $H^{(1)}$.

$$\begin{pmatrix} U(x,t) \\ \frac{dU}{dt}(x,t) \end{pmatrix} = \exp\left\{ -t \begin{bmatrix} 0 & -1 \\ A & B \end{bmatrix} \begin{pmatrix} U_0 \\ v_0 \end{pmatrix} \right\} \tag{3.28}$$

as much as in our study presented in Sec. 3.2.

For example, let us consider A as a strongly positive elliptic operator of order $2m$ on a suitable domain $\Omega \subset R^N$ and satisfying the Gärding condition there.[3]

$$A = \sum_{\substack{|a| \leq m \\ |\beta| \leq m}} (-1)^{|\alpha|} D_x^\alpha (a_{\alpha\beta}(x) D_x^\beta, \tag{3.29}$$

$$D(A) = H^{2m}(\Omega) \cap H_0^m(\Omega), \tag{3.30}$$

$$\text{Real}(Af, f)_{L^2(\Omega)} \geq C_0 \|f\|_{H^{2m}(\Omega)}. \tag{3.31}$$

The damping operator B is taken to be explicitly a strictly positive function $\nu(x)$ of $C^\infty(\Omega)$.

As a result of this section, we have weak global solution on $C^\infty([0,\infty), H^{2m}(\Omega) \cap H_0^m(\Omega))$ of the damped wave equation problem in Ω with $f(x,t) \in L^\infty([0,\infty), L^2(\Omega))$.[4]

$$\frac{d^2 U}{d^2 t} = -(AU)(x,t) - \nu(x)\frac{dU}{dt}(x,t) + f(x,t) \tag{3.32}$$

$$U(x,0) = f(x) \in H^{2m}(\Omega) \cap H_0^m(\Omega) \tag{3.33}$$

$$U_t(x,0) = g(x) \in L^2(\Omega). \tag{3.34}$$

Finally, let us show that the operator on $D(A) \oplus H^{(0)} = H^{(1)}$:

$$\hat{A} = \begin{pmatrix} 0 & -I \\ A & B \end{pmatrix} \tag{3.35}$$

is a topological isomorphism from its (dense) domain to the complete Hilbert space $H^{(1)}$ as a consequence of the fact that the self-adjoint operator $A + B$ on $D(A)$ satisfies the Hille–Yoshida criterion: For any real number $\lambda > 0$, the equation $(A + B + \lambda 1)y = f$ has a unique solution on $D(A)$ for each $f \in H^{(0)}$. The same Hille–Yoshida criterion applied to the full two-component operator Eq. (3.35), shows us that for each $(f^1, f^2) \in H^{(1)}$, one can straightforwardly find $(U^1, U^2) \in D(A) \times D(A)$ with $\lambda = 1$, namely, $U^1 = (f^1 + U_2)$ and $(A + B + 1)U^2 = f^2 - Af^1 \in H^{(0)}$ $(Af^1 \in H^{(0)},$ since $f^1 \in D(A)!)$. The uniqueness of (U^1, U^2) is a direct consequence of the strict positivity of the operator $A + B$ on $D(A)$.

3.4. The Wave Equation "Path-Integral" Propagator

In this section of a more physicist-oriented nature, we consider the classical unsolved problem of writing a rigorous Trotter like path-integral formulae for the Cauchy problem on R^3 associated with the elastic vibrations of an infinite length string with a specific mass $(\rho(x) - \rho_0) \in C_c^\infty(R^3)$ and a Young elasticity modulus $(E(x) - E_0) \in C_c^\infty(R^3)$. The general wave evolution equation governing such vibrations is given by[3,4]

$$\frac{\partial}{\partial t}\left(\rho(x)\frac{\partial U(x,t)}{\partial t}\right) = \nabla(E(x)\nabla \cdot U(x,t)), \tag{3.36}$$

$$U(x,0) = f(x) \in H^2(\Omega), \tag{3.37}$$

$$U_t(x,0) = g(x) \in L^2(\Omega). \tag{3.38}$$

Just for simplicity of our mathematical arguments, we consider the case of a string with a constant specific mass $\rho(x) = \rho_0$ in our next discussions. Firstly, let us show that the operator

$$A = \frac{E_0}{\rho_0}\Delta + \nabla\left(\frac{(E(x) - E_0)}{\rho_0}\nabla\right) = A^{(0)} + B \tag{3.39}$$

is a self-adjoint operator on the domain $H^2(\Omega)$. This result is an immediate consequence of the fact that the "perturbation" operator $B = \nabla\left(\frac{E(x) - E_0}{\rho_0}\nabla\right)$ is $A^{(0)}$-bounded with $A^{(0)}$-bound less than 1. As a consequence, one can apply straightforwardly the Kato–Rellich theorem to guarantee the self-adjointness of A on $H^2(\Omega)$. The $A^{(0)}$-boundedness of B is a result of the following estimates for $\varphi \in H^2(\Omega)$ (where $E(x) \ll E_0$ and $|\nabla E(x)| \ll E_0$):

$$\left\| \nabla \left(\frac{E(x) - E_0}{\rho_0} \nabla \right) \varphi \right\|_{L^2(\Omega)}$$

$$\leq \sup_{x \in \Omega} \left\{ \frac{|E(x) - E_0|}{E_0}, \gamma(\Omega) \frac{\nabla E(x)}{E_0} \right\} \left\| \frac{E_0}{\rho_0} \Delta \varphi \right\|_{L^2(\Omega)}, \quad (3.40)$$

where $\gamma(\Omega)$ is the universal constant on the Poincaré relation for function $\varphi(x)$ on $H^2(\Omega)$

$$\|\nabla \varphi\|_{L^2(\Omega)} \leq \gamma(\Omega) \|\Delta \varphi\|_{L^2(\Omega)}. \quad (3.41)$$

By the direct application of Theorem 3.1, we have the rigorous solution for the Cauchy problem Eqs. (3.36)–(3.38) if $f(x) \in H^2(\Omega)$ and $g(x) \in L^2(\Omega)$ for any $-\infty \leq t \leq +\infty$, namely,

$$\begin{pmatrix} U(x,t) \\ \frac{\partial U}{\partial t}(x,t) \end{pmatrix} = \exp \left\{ it \begin{bmatrix} 0 & 1 \\ \nabla \left(\frac{E(x)}{\rho_0} \nabla \right) & 0 \end{bmatrix} \right\} \begin{pmatrix} f(x) \\ g(x) \end{pmatrix}. \quad (3.42)$$

At this point, we apply the Trotter–Kato theorem to rewrite Eq. (3.42) in the short-time propagation form

$$\begin{pmatrix} U(x,t) \\ \frac{\partial U(x,t)}{\partial t} \end{pmatrix} = \underset{(H^2(\Omega) \oplus L^2(\Omega))}{\text{Strong limit}} \underset{n \to \infty}{} \left\{ \exp i \left(\frac{t}{n} \right) \begin{bmatrix} 0 & 1 \\ \nabla \left(\frac{E(x)}{\rho_0} \nabla \right) & 0 \end{bmatrix} \right\}^n \begin{pmatrix} f(x) \\ g(x) \end{pmatrix}. \quad (3.43)$$

However, for short-time propagation, we have the usual asymptotic result

$$\lim_{\varepsilon \to 0^+} \left\{ \exp \left(i\varepsilon \begin{bmatrix} 0 & 1 \\ \frac{\partial}{\partial x} \left(\frac{E(x)}{\rho_0} \frac{\partial}{\partial x} \right) & 0 \end{bmatrix} \right) \begin{pmatrix} f \\ g \end{pmatrix} \right\} (x)$$

$$= \lim_{\varepsilon \to 0^+} \int_{-\infty}^{+\infty} dp\, e^{ipx} \left\{ e^{\left(i\varepsilon \begin{bmatrix} 0 & 1 \\ -\left(\frac{E(x)}{\rho_0} \right) p^2 + \frac{E'(x)}{\rho_0}(ip) & 0 \end{bmatrix} \right)} \begin{pmatrix} \tilde{f} \\ \tilde{g} \end{pmatrix}(p) \right\}$$

$$(3.44)$$

with $\tilde{f}(p)$ and $\tilde{g}(p)$ denoting the Fourier transforms of the functions $f(x)$ and $g(x)$, respectively, and $E'(x)p$ means the scalar product $\vec{\nabla}E(x)\vec{p}$ for short.

It is worth calling to the reader's attention that

$$\exp\left(i\varepsilon\begin{bmatrix} 0 & 1 \\ -\left(\dfrac{E(x)}{\rho_0}\right)p^2 + \dfrac{E'(x)}{\rho_0}ip & 0 \end{bmatrix}\right)\begin{pmatrix} \tilde{f}(p) \\ \tilde{g}(p) \end{pmatrix}$$

$$= \cos\left[\varepsilon\left(-\frac{E(x)}{\rho_0}p^2 + \frac{E'(x)}{\rho_0}ip\right)^{\frac{1}{2}}\right]\begin{pmatrix} \tilde{f}(p) \\ \tilde{g}(p) \end{pmatrix}$$

$$\frac{-i\sin\left[\varepsilon\left(-\frac{E(x)}{\rho_0}p^2 + \frac{E'(x)}{\rho_0}ip\right)^{\frac{1}{2}}\right]}{\sqrt{-\frac{E(x)}{\rho_0}p^2 + \frac{E'(x)}{\rho_0}ip}}\begin{pmatrix} i\varepsilon\tilde{g}(p) \\ \varepsilon\left[-\frac{E(x)}{\rho_0}p^2 + \frac{E'(x)}{\rho_0}ip\right]^{(1/2)}\tilde{f}(p) \end{pmatrix}.$$

$$(3.45)$$

As a consequence of the above formulae we have our discretized path-integral representation for the initial-value Green function of Eqs. (3.36)–(3.38), i.e. by writing the evolution operator as an integral operator-Feynman Propagator

$$\left\{ e^{it\begin{bmatrix} 0 & 1 \\ \frac{\partial}{\partial x}\left(\frac{E(x)}{\rho_0}\frac{\partial}{\partial x}\right) & 0 \end{bmatrix}} \right\}\begin{pmatrix} f \\ g \end{pmatrix}$$

$$= \int_{-\infty}^{+\infty} dx'\begin{bmatrix} G_{11}(x,x',t) & G_{12}(x,x',t) \\ G_{21}(x,x',t) & G_{22}(x,x',t) \end{bmatrix}\begin{pmatrix} f(x') \\ g(x') \end{pmatrix} \qquad (3.46)$$

we have (for $A = 1,2; B = 1,2$) the "path-integral" product representation

$$G_{AB}(x,x',t) = \text{strong} - \lim_{n\to\infty}\left\{\left(\frac{1}{2\pi}\right)^n\int_{-\infty}^{+\infty} dp_1\cdots dp_n\int_{-\infty}^{+\infty} dx_{n-1}\cdots\right.$$

$$\left.\times\int_{-\infty}^{+\infty} dx_1 \times \left(\prod_{j=1}^{n}\left[G_{A_jA_{j+1}}\left(x_j,p_j;\frac{t}{n}\right)\right]\right)\right\}$$

$$\times \exp(ip_j(x_j - x_{j-1})), \qquad (3.47)$$

where $G_{A_jB_j}(x_j,p_j,\frac{t}{n})$ is the 2×2 matrix on the integrand of Eq. (3.31) for $x = x_j$ and $p = p_j$. Note that $x_n = x$ and $x_0 = x'$.

3.5. On The Existence of Wave-Scattering Operators

In this section we give a proof for the existence of somewhat reduced wave-scattering operators associated to the following wave dynamics pair $(\tilde{H}, \tilde{H}_0)$ on the $H^{(1)} = H^2(\Omega) \oplus L^2(\Omega)$ (with $\Omega = R^N$, $N \geq 3)$[7]:

$$\tilde{H}_0 = \begin{bmatrix} 0 & 1 \\ \frac{E_0}{\rho_0} \Delta & 0 \end{bmatrix}, \tag{3.48}$$

$$\tilde{H} = \begin{bmatrix} 0 & 1 \\ \nabla \cdot \frac{E(x)}{\rho_0} \nabla & 0 \end{bmatrix}. \tag{3.49}$$

We need, thus, to show the strong convergence on $H^{(1)}$ of the following family of one-parameter evolution wave operators (s = strong) for T arbitrary (but finite):

$$W^{\pm}(\tilde{H}, \tilde{H}_0) = S - \lim_{t \to T}(\exp(it\tilde{H})\exp(-it\tilde{H}_0)). \tag{3.50}$$

Let us remark that the existence of the strong limit of Eq. (3.50) is a direct consequence of the validity of the famous Cook–Kuroda–Jauch integral criterion[6] applied to the pair $(\tilde{H}, \tilde{H}_0)$ in a dense subset of the domain of $\tilde{H}$. This integral criterion means that there exists a $\delta > 0$, such that the following integral is finite for any (f, g) belonging to a dense set of $H^{(1)}$, i.e.

$$\int_{\delta}^{T} dt \left\| (\tilde{H} - \tilde{H}_0)e^{i\tilde{H}_0 t}\begin{pmatrix} f \\ g \end{pmatrix} \right\|_{H^{(1)}} < \infty. \tag{3.51}$$

Let us denote $\tilde{E}(x) = (E(x) - E_0) \in C_c^{\infty}(R^N)$, the scattering wave potential in what follows.

We can, thus, rewrite Eq. (3.51) in the following explicit form:

$$\int_{\delta}^{T} dt((-\nabla(\tilde{E}(x)\nabla)g(x,t), g(x,t)))_{L^2(R^N)} < \infty \tag{3.52}$$

or

$$\int_{\delta}^{T} dt((+\nabla(\tilde{E}(x)\nabla)f(x,t), +\nabla(\tilde{E}(x)\nabla)f(x,t)))_{L^2(R^N)} < \infty, \tag{3.53}$$

where the pair of functions $f(x,t)$ and $g(x,t)$ satisfy the free wave equation

$$\frac{d^2 f(x,t)}{d^2 t} = \left(\frac{E_0}{\rho} \Delta \right) f(x,t), \tag{3.54}$$

$$f(x,0) = f^\wedge(x), \tag{3.55}$$

$$f_t(x,0) = g^\wedge(x), \tag{3.56}$$

and

$$g(x,t) = \frac{d}{dt} f(x,t). \tag{3.57}$$

Let us choose the initial dates $f^\wedge(x)$ and $g^\wedge(x)$ as the Fourier transforms of $L^2(R^N)$ functions polynomial-exponentially bounded ($\ell \in \mathbb{Z}^+$ and $\Lambda \in R^+$):

$$f^\wedge(x) = \int_{R^N} d^N k e^{ikx} \tilde{f}_\Lambda(k); \quad |\tilde{f}_\Lambda(k)| \le M e^{-\Lambda |k|} |k|^\ell, \tag{3.58}$$

$$g^\wedge(x) = \int_{R^N} d^n k e^{ikx} \tilde{g}_\Lambda(k); \quad |\tilde{g}_\Lambda(k)| \le M e^{-\Lambda |k|} |k|^\ell. \tag{3.59}$$

Note that we have an explicit representation in momentum space for the free waves $f(x,t)$ and $g(x,t)$, namely,

$$\tilde{f}(k,t) = \tilde{f}_\Lambda(|k|) \cos(|k|t) + \frac{\tilde{g}_\Lambda(|k|)}{|k|} \sin(|k|t), \tag{3.60}$$

$$\tilde{g}(k,t) = \frac{\partial}{\partial t} (\tilde{f}(k,t)). \tag{3.61}$$

At this point we observe the estimates for any $\delta > 0$:

$$\int_\delta^T dt |(\tilde{E}\Delta f + \nabla \tilde{E} \cdot \nabla f, \tilde{E}\Delta f + \nabla \tilde{E} \cdot \nabla f)_{L^2(R^N)}(t)|$$

$$\le \int_\delta^T dt \Big\{ \sup_{x \in R^N} |\tilde{E}(x)|^2 (k^4 \tilde{f}(k,t), \tilde{f}(k,t))_{L^2(R^N)}$$

$$+ \sup_{x \in R^N} (\max(\nabla_i \tilde{E}(x), \tilde{E}(x))) (k^2 k_i \tilde{f}(k,t), \tilde{f}(k,t))_{L^2(R^N)}$$

$$+ \sup_{x \in R^N} \max(\nabla_i \tilde{E}(x), \nabla_j \tilde{E}(x)) (k_i k_j \tilde{f}(k,t), \tilde{f}(k,t))_{L^2(R^N)} \Big\}. \tag{3.62}$$

By using the explicit representations Eqs. (3.60) and (3.61) and the simple estimate below for $\alpha > 0$ with $t \in [\delta, T]$ and $\delta > 0$ (with $T < \infty$)

$$I(t) = \int_0^\infty d|k| |k|^{\alpha+\ell} \cos^2(|k|t) e^{-\Lambda|k|}$$

$$\leq \frac{1}{t^{\alpha+1+\ell}} \times \left(\int_0^\infty dU U^{\alpha+\ell} e^{-\frac{\Lambda U}{T}} \right) \leq \frac{\bar{M}}{t^{1+\alpha}} \in L^1([\delta, \infty)), \qquad (3.63)$$

one can see that by just choosing $\delta > 0$, the existence of the reduced $(T < \infty)$ wave operators of our pair Eqs. (3.48)–(3.49) is a direct consequence of the application of the Cook–Kuroda–Jauch criterion,[6] since the set of functions polynomial-exponentially bounded is dense on $C_0(R^N)$ (continuous functions vanishing at the infinite), and so, dense in $L^2(R^N)$.[7] The complete wave operator (with $T \to +\infty$) can be obtained by means of Zorn theorem applied to the chain of reduced wave operators $\{W_\pm(T)\}$ ordered by inclusion, i.e. $T_1 < T_2 \Rightarrow W_\pm(T_1) \subset W_\pm(T_2)$.

3.6. Exponential Stability in Two-Dimensional Magneto-Elasticity: A Proof on a Dissipative Medium

One interesting question in the magneto-elasticity property of an isotropic (uncompressible) medium vibration interacting with an external magnetic is whether the energy of the total system should decay with an exponential uniform bound as time reaches infinity.[8]

We aim in this section to give a rigorous proof of this exponential bound for a model of imaginary medium electric conductivity as an interesting example of abstract techniques developed in the previous sections. (See Appendix A for a complementary analysis on this problem.)

The governing differential equations for the electric–magnetic medium displacement vector $U = (U^1, U^2, 0) \equiv \vec{U}(x, t)$ depending on the time variable $t \in [0, \infty)$ with $x \in \Omega$ and the intrinsic two-dimensional magnetic field $\vec{h}(x, t) = (h^1, h^2, 0)$ are given by

$$\frac{\partial^2 \vec{U}(x, t)}{\partial^2 t} = \Delta \vec{U}(x, t) + ([\vec{\nabla} \times \vec{h}](x, t) \times \vec{B}), \qquad (3.64)$$

$$\beta \frac{\partial \vec{h}(x, t)}{\partial t} = \Delta \vec{h}(x, t) + \beta \left(\vec{\nabla} \times \left[\frac{\partial U}{\partial t}(x, t) \times \vec{B} \right] \right). \qquad (3.65)$$

Here, $\vec{B} = (B, 0, 0)$ is a constant vector external magnetic field, and β is the (constant) medium electric conductivity.

Additionally, one has the initial conditions

$$\vec{U}(0, x) = \vec{U}_0(x); \quad \vec{U}_t(0, x) = \vec{U}_1(x); \quad \vec{h}(0, x) = h_0(x) \tag{3.66}$$

and the (physical) Dirichlet-type boundary conditions with $\vec{n}(x)$ being the normal of the boundary of the medium Ω^9:

$$\vec{U}(\vec{x}, t)|_\Omega = 0; \quad (\vec{n} \times (\vec{\nabla} \times \vec{h}))(\vec{x}, t)|_\Omega = 0. \tag{3.67}$$

Let us follow the previous studies of Sec. 3.1 to consider the contraction semigroup defined by the following essentially self-adjoint operators in the H-energy space:

$$\hat{H}_\Omega = \{(\vec{U}, \vec{\pi}, \vec{h}) \in (L^2(\Omega))^3 \mid (\nabla \vec{U}, \nabla \vec{U})_{L^2(\Omega)}$$

$$+ (\vec{\pi}, \vec{\pi})_{L^2(\Omega)} + (\vec{h}, \vec{h})_{L^2(\Omega)} < \infty\},$$

namely,

$$\mathcal{L}_0 = \begin{bmatrix} 0 & i & 0 \\ i\Delta & 0 & 0 \\ 0 & 0 & -\Delta \end{bmatrix} \tag{3.68}$$

and

$$V = \begin{bmatrix} 0 & 0 & 0 \\ 0 & 0 & [\vec{\nabla} \times (\ )] \times \vec{B} \\ 0 & -(\vec{\nabla} \times [(\ ) \times \vec{B}]) & 0 \end{bmatrix}. \tag{3.69}$$

We, thus, have the contractive C_0-semigroup acting on $\hat{H}_\Omega$ (see Appendix A for technical proof details):

$$T_t(\vec{U}, \vec{\pi}, \vec{h}) = \exp(t(+i\mathcal{L}_0 - V))(\vec{U}, \vec{\pi}, \vec{h})(0). \tag{3.70}$$

It is worth to call the reader's attention that we have proved the essential self-adjointness of the operator V by using explicitly the boundary conditions on the relationships below:

$$([\vec{\nabla} \times (\vec{h})] \times \vec{B}) = (B, B(\partial_1 h_2 - \partial_2 h_1)), \tag{3.71}$$

$$([\vec{\nabla} \times (\vec{h})] \times B) = (-B\partial_2 \pi^2, B\partial_1 \pi^2, 0). \tag{3.72}$$

By standard theorems on contraction semigroup theory,[8] one has that the left-hand side of Eq. (3.70) satisfies the magneto-elasticity equations in the strong sense with $\beta = i = \sqrt{-1}$, a pure imaginary electric conductivity medium, physically meaning that the medium has "electromagnetic dissipation" and in the context that the initial values belong to the subspace $(H^2(\Omega) \cap H_0^1(\Omega))^2 \oplus (L^2(\Omega))^2 \oplus (L^2(\Omega))^2 \subset \hat{H}_\Omega$.

Let us note that Eq. (3.70) still produces a weak integral solution on the full Hilbert space $\hat{H}_\Omega = (H_0^1(\Omega))^2 \oplus (L^2(\Omega))^2 \oplus (L^2(\Omega))^2$, the result suitable when one has initial random conditions sampled on $L^2(\Omega)$ by the Minlos theorem (see Appendix B).

Thus, for any $(\vec{U}_0, \vec{\pi}_0, \vec{h}_0) \in \hat{H}_\Omega$ we have the following energy estimate:

$$\|(\vec{U}, \partial_t \vec{U}, \vec{h})\|_{\tilde{H}_\Omega}$$

$$\equiv \int_\Omega dx (|\nabla \vec{U}|^2 + |\partial_t \vec{U}|^2 + |\vec{h}|^2)(x, t)$$

$$\leq \|(\exp(t(i\mathcal{L}_0 - V)))(\vec{U}(0), \partial_t \vec{U}(0), \vec{h}(0))\|_{\tilde{H}_\Omega}$$

$$\leq \left\| S - \lim_{n \to \infty} \left\{ \left[\exp\left(\frac{it}{n} \mathcal{L}_0 \right) \exp\left(-\frac{t}{n} V \right) \right]^n (\vec{U}_0, \vec{U}_1, \vec{h}_0) \right\} \right\|_{\tilde{H}_\Omega}$$

$$\leq \exp(-t\omega(V)) \|\vec{U}_0, \partial_t \vec{U}(0), \vec{h}(0)\|_{\tilde{H}_\Omega}, \tag{3.73}$$

where $\omega(V)$ is the infimum of the spectrum of the self-adjoint operator V on the space $\hat{H}$.

In order to determine the spectrum of the self-adjoint operator V, which is a discrete set since Ω is a compact region of R^2, we consider the associated V-eigenfunctions problem

$$[\nabla \times (\vec{h}_n)] \times \vec{B} = \lambda_n \vec{\pi}_n, \tag{3.74}$$

$$-\nabla \times [(\vec{\pi}_n) \times \vec{B}] = \lambda_n \vec{h}_n, \tag{3.75}$$

which, by its turn, leads to the usual spectral problem for the Laplacean with the usual Dirichlet boundary conditions on Ω:

$$\Delta \pi_2^n = -\left(\frac{\lambda_n^2}{B} \right) \pi_2^n, \tag{3.76}$$

$$\pi_2^n(x)|_{\partial\Omega} = 0. \tag{3.77}$$

As a result of Eqs. (3.75) and (3.76), one finally gets the exponential bound for the total magneto-elastic energy equation (3.73):

$$\|(\vec{U}, \partial_t \vec{U}, \vec{h})\|_{\tilde{H}_\Omega}(t)$$

$$\leq \exp\left(-t\sqrt{B\lambda_0(\Omega)}\right)\|(\vec{U}_0, \partial_t \vec{U}(0), \vec{h}(0)\|_{\tilde{H}_\Omega}, \qquad (3.78)$$

where $\lambda_0(\Omega) = \inf\{\mathrm{spec}(-\Delta)\}$ em $H^2(\Omega) \cap H^1_0(\Omega)$.

3.7. An Abstract Semilinear Klein Gordon Wave Equation — Existence and Uniqueness

Let A be a positive self-adjoint operator as in Sec. 3.2 and (B_1, B_2) a pair of bounded operators with domain containing $D(A)$ (the A-operator domain). We can, thus, associate to these operatorial objects the following semilinear Klein–Gordon equation with initial dates (zero time field Hilbert-valued configurations):

$$\frac{\partial^2 U}{\partial^2 t} = -AU - B_1 \frac{\partial U}{\partial t} + B_2 U + F(U, t)$$

$$U(0) = U_0 \in D(A) \qquad (3.79)$$

$$U_t(0) = V_0 \in H^0.$$

Here, $F(U, t)$ is a Lipschitzian function on the Hilbert space $H^{(0)}$ with a Lipschitz-bounded being a function $L(t)$ on $\mathcal{L}^q(R^+) = \{$set of mensurable functions $f(t)$ on R^+ such that $\|f\|_{\mathcal{L}^q} = \left(\int_0^\infty ds|f(s)|^q\right)^{1/q} < \infty$ and $q \geq 1\}$, namely,

$$|F(U_1, t) - F(U_2, t)|_{H^{(0)}} \leq L(t)|U_1 - U_2|_{H^{(0)}}, \qquad (3.80a)$$

$$F(0, t) \equiv 0. \qquad (3.80b)$$

It is an important problem in the mathematical physics associated with Eq. (3.79) to find a functional space, where one can show the existence and uniqueness of the Cauchy problem for the global time interval $[0, \infty)$.[11]

Let us address this problem through the abstract techniques developed in the previous section of this work and the Banach fixed-point theorem.

First, let us write Eq. (3.79) in the integral operatorial form

$$\begin{pmatrix} U \\ \pi \end{pmatrix}(t) = e^{t(iL_A + \hat{V})} \begin{pmatrix} U_0 \\ V_0 \end{pmatrix} + \int_0^t ds e^{(t-s)(iL_A + \hat{V})} \begin{bmatrix} 0 \\ \frac{1}{i} F(U, s) \end{bmatrix}, \qquad (3.81)$$

where $\hat{V}$ is the bounded operator with operatorial norm μ acting on the two-component space $H^{(1)}$ by means of the rule:

$$\hat{V}(U,\pi) = (V_1 U, V_2 \pi), \tag{3.82}$$

where the (bounded) component operators are written explicitly in terms of the "damping" operator B_1 and "mass" operator B_2, as given below:

$$V_1 = B_1 - V_2, \tag{3.83}$$

$$2V_2 = -B_1 \pm (B_1^2 - 4B_2)^{1/2}. \tag{3.84}$$

Let us introduce the following exponential weighted space of $H^{(1)}$-valued continuous functions:

$$C_k([0,\infty), H^{(1)}) = \Big\{ g\colon R^+ \to H^{(1)}, \text{with } K > 0$$

$$\text{and } ||g||_{H^{(1)}}^{(k)} = \sup_{0 \le t < \infty} (e^{-kt}||g(t)||_{H^{(1)}}) < \infty \Big\}. \tag{3.85}$$

We are going to show next that there exists a value of the exponential weight parameter k defining the above-considered space such that the application $\hat{T}$ written below is a contraction between the space of Hilbert-valued continuous functions (Eq. (3.85)), i.e.

$$\hat{T}\colon C_k([0,\infty), H^{(1)}) \to C_k([0,\infty), H^{(1)})$$

$$\Phi = (\varphi_1, \varphi_2) \to \int_0^t ds\, e^{(t-s)(iL_A + \hat{V})} \begin{bmatrix} 0 \\ \frac{1}{i}F(\varphi_1(s), s) \end{bmatrix}. \tag{3.86}$$

In order to show this result, let us point out the usual exponential bound for the free Klein–Gordon abstract propagator (operator norm) (see Eq. (A.23)):

$$||\exp[t(iL_A + \hat{V})]||_{\text{op}} \le ||\exp t\hat{V}||_{\text{op}}$$

$$\le \exp(t||\hat{V}||_{\text{op}}) \equiv \exp(t\mu). \tag{3.87}$$

As a consequence of this bound Eq. (3.87), we have the contraction estimate for Φ, Φ' elements of $C_k([0,\infty), H^{(1)})$:

$$||\hat{T}\Phi - \hat{T}\Phi'||_{C_k([0,\infty), H^{(1)})} \le \sup_{0 \le t < \infty} \left\{ \int_0^t ds\, e^{-k(t-s)} e^{\mu(t-s)} L(s) \right\}$$

$$\times ||\Phi - \Phi'||_{C_k([0,\infty), H^{(1)})}. \tag{3.88}$$

At this point, it is a straightforward step to obtain the general expression for the contraction coefficient on Eq. (3.88) (where $\frac{1}{p} + \frac{1}{q} = 1$):

$$\sup_{0 \leq t < \infty} \left\{ \int_0^t ds\, e^{-K(t-s)} e^{\mu(t-s)} L(s) \right\}$$

$$\leq \sup_{0 \leq t < \infty} \left\{ \left(\int_0^t ds\, e^{-p(k-\mu)(t-s)} \right)^{1/p} \right\} \|L\|_{\mathcal{L}^q(R^+)}$$

$$= \left(\frac{1}{p(k-\mu)} \right)^{(1/p)} \cdot \|L\|_{\mathcal{L}^q(R^+)} \equiv C(k,\mu). \tag{3.89}$$

So, T has a fixed point on $C_k([0,\infty), H^{(1)})$ if one makes the choice of the value k such that $(C(k,\mu)) < 1$, namely,

$$K > \left(\frac{1}{p} (\|L\|_{\mathcal{L}^q(R^+)})^p + \mu \right). \tag{3.90}$$

A very important result obtained as a consequence of the above "fixed-point theorem" analysis is that the sequence of iteratives

$$U^{(n)}(t) = (\hat{T})^{(n)} \begin{pmatrix} U_0 \\ V_0 \end{pmatrix} \tag{3.91}$$

converges in $C_k([0,\infty), H^{(1)})$ to the solution of Eq. (3.81) with the error estimate

$$\sup_{0 \leq t < \infty} (e^{-kt} \|U^{(n)} - U\|_{H^{(1)}}(t))$$

$$\leq \frac{(C(k,\mu))^{n-1}}{1 - C(k,\mu)} \left\{ \sup_{0 \leq t < \infty} \left\| \exp t(iL_A + \hat{V}) \begin{pmatrix} U_0 \\ V_0 \end{pmatrix} \right\|_{H^{(1)}} \right\}. \tag{3.92}$$

References

1. M. Reed and B. Simon, *Methods of Modern Mathematical Physics*, Vol. 2, (Academic Press, 1975); Khandekar and Lawande, *Path Integral Methods and their Applications* (World Scientific, Singapore, 1993).
2. L. C. L. Botelho and R. Vilhena, *Phys. Rev.* **49E**(2), R1003–R1004 (1994); L. C. L. Botelho, *Mod. Phys. Lett.* **13B**, 363–370 (1999); L. C. L. Botelho, *Mod. Phys. Lett.* **14B**, 73–78 (2000); L. C. L. Botelho, *J. Phys. A: Math. Gen.* **34**, L131–L137 (2001); L. C. L. Botelho, *Mod. Phys. Lett.* **16B**, 793–806 (2002).
3. S. G. Mikhlin, *Mathematical Physics, An Advanced Course*, Series in Applied Mathematics and Mechanics (North-Holland, 1970).

4. A. P. S. Selvadurai, *Partial Differential Equations in Mechanics 2* (Springer-Verlag, 2000).
5. R. Dautray and J.-L. Lions, Analyse Mathématique et calcul numérique, Vol. 7 (Massun, Paris, 1988).
6. S. T. Kuroda, *Il Nuovo Cimento* **12**, 431–454 (1959).
7. V. Enss, *Commun. Math. Phys.* **61**, 285–291 (1978).
8. G. P. Menzala and E. Zuazua, *Asymptotic Anal.* **18**, 349–367 (1998).
9. J. E. M. Riviera and R. Racue, *IMA J. Appl. Math.* **66**, 269–283 (2001).
10. L. Schwartz, *Random Measures on Arbitrary Topological Spaces and Cylindrical Measures* (Tata Institute, Oxford University Press, 1973).
11. T. Cazennave and A. Haraux, *An Introduction to Semi-Linear Evolution Equations*, Vol. 13 (Clarendon Press, Oxford, 1998).

Appendix A. Exponential Stability in Two-Dimensional Magneto-Elastic: Another Proof

In this complementary technical appendix to Sec. 3.6 on the exponential decay of the magneto-elastic energy associated with the (imaginary electric medium conductivity) magneto-elastic wave

$$\begin{pmatrix} U \\ \pi \\ h \end{pmatrix}(t) = \exp\{it(+\mathcal{L}_0 + iV)\} \begin{pmatrix} U \\ \pi \\ h \end{pmatrix}(0) \tag{A.1}$$

with the essential self-adjointness operators on the Hilbert energy space $\tilde{H}^1(\Omega)$

$$\mathcal{L}_0 = \begin{bmatrix} 0 & +i & 0 \\ +i\Delta & 0 & 0 \\ 0 & 0 & \Delta \end{bmatrix} \tag{A.2}$$

and

$$iV = \begin{bmatrix} 0 & 0 & 0 \\ 0 & 0 & i[\vec{\nabla} \times (\ ) \times \vec{B}] \\ 0 & i\vec{\nabla} \times [(\ ) \times \vec{B}] & 0 \end{bmatrix}, \tag{A.3}$$

we have used the fact that the operator $-V + i\mathcal{L}_0$ is the generator of a contraction semigroup on this energy Hilbert space $\tilde{H}^1(\Omega) = (H^1(\Omega))^3 \oplus (L^2(\Omega))^3 \oplus (L^2(\Omega))^3$ in order to write Eq. (A.1) in a mathematically rigorous way.

Let us give a proof of this mathematical result by means of a direct application of the Hille–Yoshida theorem[9] to the operator $(-V + i\mathcal{L}_0)$.

First, the domain of $(-V + i\mathcal{L}_0)$ is everywhere dense on $\tilde{H}^1(\Omega)$ as a consequence of the self-adjointness of the operators V and $\mathcal{L}_0$ on $\tilde{H}^1(\Omega)$.

Second, the existence of a solution for the elliptic problem

$$(-V + i\mathcal{L}_0)x = \alpha y \tag{A.4}$$

for $x \in D(\mathcal{L}_0) \subset D(V)$ and every $y \in \tilde{H}^2(\Omega)$ with $\alpha > 0$ is a standard result even if Ω has a nontrivial topology (holes inside!), i.e. Ω is a multiple-connected planar region.[9]

The unicity of the solution x is a straightforward consequence of the fact that the spectrum $\{\lambda_n\}$ of the self-adjoint operator V coincides with the positive Laplacean $\omega_n(-\Delta)$, i.e. $\lambda_n^2/B = \omega_n(-\Delta)$. As a consequence, the solution of the equation

$$(-V + i\mathcal{L}_0)x = 0 \tag{A.5}$$

leads to the relation below due to the self-adjointness of the operators V and $\mathcal{L}_0$ on $\tilde{H}^1(\Omega)$:

$$\overline{\langle Vx, x\rangle}_{\tilde{H}^1(\Omega)} = \langle Vx, x\rangle_{\tilde{H}^1(\Omega)} = i\langle \mathcal{L}_0 x, x\rangle_{\tilde{H}^1(\Omega)} = i\overline{\langle \mathcal{L}_0 x, x\rangle}_{\tilde{H}^1(\Omega)}, \tag{A.6}$$

from which we conclude that

$$\langle Vx, x\rangle_{\tilde{H}^1(\Omega)} = 0, \tag{A.7}$$

$$\langle \mathcal{L}_0 x, x\rangle_{\tilde{H}^1(\Omega)} = 0, \tag{A.8}$$

or equivalently

$$x \in \mathrm{Ker}\{V\} \cap \mathrm{Ker}\{\mathcal{L}_0\}, \tag{A.9}$$

which is zero if Ω is a *simply* connected planar domain as supposed in the text.

Finally, for any $x > 0$, we have that

$$\|(\alpha 1 - (-V + i\mathcal{L}_0))^{-1}\|_{\tilde{H}^1(\Omega)} \leq \frac{1}{\alpha}, \tag{A.10}$$

since for every $z \in \mathrm{Dom}(\mathcal{L}_0) \subset \mathrm{Dom}(V)$,

$$\alpha^2 \|z\|_{\tilde{H}^1(\Omega)}^2 \leq \|((\alpha 1 - (-V + i\mathcal{L}_0)))z\|_{\tilde{H}^1(\Omega)}^2$$

$$= \alpha^2 \|z\|_{\tilde{H}^1(\Omega)}^2 + \|\mathcal{L}_0 z\|_{\tilde{H}^1(\Omega)}^2$$

$$+ \|Vz\|_{\tilde{H}^1(\Omega)}^2 + 2\alpha(z, Vz)_{\tilde{H}^1(\Omega)}. \tag{A.11}$$

Note that we have used the fact that V is a positive operator on $\tilde{H}^1(\Omega)$.

As a consequence of the above-exposed results we have that $-V + i\mathcal{L}_0$ is a generator of a contractive semigroup on the space $(1 - P_{\text{Ker}(V) \cap \text{Ker}(\mathcal{L}_0)})$ $\tilde{H}^1(\Omega)$, where $P_{\text{Ker}(V) \cap \text{Ker}(\mathcal{L}_0)}$ is the orthogonal projection on the Kernel subspaces of the self-adjoint operators V and $\mathcal{L}_0$.

As a final remark to be made in this Appendix A, let us call the physicist-oriented reader's attention to the following (somewhat formal) abstract Lemmas, alternatives to the Banach space methods used in Ref. 9.

Lemma A.1. *Suppose that the matrix-valued operator with self-adjoint operators entries on suitable Hilbert spaces H_1, H_2*

$$L_0 = \begin{bmatrix} 0 & A & 0 \\ B & 0 & 0 \\ 0 & 0 & C \end{bmatrix}. \tag{A.12}$$

Then, it will be a self-adjoint on the Hilbert space $\tilde{H} = H_1 \oplus H_1 \oplus H_2$ with an inner product given by (D is a positive self-adjoint operator on H_1)

$$\langle (U, \pi, h); (U', \pi', h') \rangle_{\tilde{H}} = (D^* D U, U')_{H_1} + (\pi, \pi)_{H_1} + (h, h)_{H_2}, \tag{A.13}$$

if we have the constraints below for the operators on the Hilbert component spaces:

$$C = C^* \quad \text{on } H_2$$
$$A^* D^* D = B \quad \text{on } H_1 \tag{A.14}$$
$$\text{Range}(A) \subset \text{Dom}(D).$$

Proof. Let us consider the inner product on $\tilde{H}$

$$\left\langle L_0 \begin{pmatrix} U \\ \pi \\ h \end{pmatrix} ; (U', \pi', h') \right\rangle_{\tilde{H}} = (DA\pi, DU')_{H_1} + (BU, \pi')_{H_1} + (Ch, h')_{H_2} \tag{A.15}$$

and

$$\langle (U, \pi, h); L_0(U', \pi', h') \rangle_{\tilde{H}} = (DU, DA\pi')_{H_1} + (\pi, BU')_{H_1} + (h, Ch')_{H_2}. \tag{A.16}$$

On the basis of Eqs. (A.15) and (A.16) one can see that L_0 is a closed symmetric operator. Besides, if there is a $(U_0, \pi_0, h_0) \in \tilde{H}$ such that for every (U, π, h) we have the orthogonality relation

$$\langle (U_0, \pi_0, h_0); (i + L_0)(U, \pi, h) \rangle_{\tilde{H}} = 0, \tag{A.17}$$

then

$$\langle (U_0, \pi_0, h_0); (A\pi, +iU, BU + i\pi, Ch + ih) \rangle_{\tilde{H}} = 0. \tag{A.18}$$

As a result

$$(D^* DU_0, (A\pi + iU))_{H_1} = 0 \tag{A.19}$$

$$(\pi_0, BU + i\pi)_{H_1} = (h_0, Ch + ih)_{H_2} = 0. \tag{A.20}$$

We thus have by self-adjointness of the operators A, B, and C that $U_0 = \pi_0 = h_0 = 0$. As a consequence, the deficiency indices of L_0 vanish what formally concludes the self-adjointness of L_0 on vectors of $\tilde{H}_1$ such that (the domain of L_0 since L_0 is a closed symmetric operator):

$$\langle L_0(U, \pi, h), (U, \pi, h) \rangle_{\tilde{H}} < \infty. \tag{A.21}$$

Lemma A.2. *Let U be the matrix-valued operator acting on $\tilde{H}$ given by*

$$V = \begin{bmatrix} 0 & 0 & 0 \\ 0 & 0 & V_1 \\ 0 & V_2 & 0 \end{bmatrix}. \tag{A.22}$$

Then V is self-adjoint on $\tilde{H}$ if and only if on H_2 we have the constraint relationship below between the adjoints of the closable symmetric operators V_1 and V_2

$$V_1^* = V_2. \tag{A.23}$$

Proof. One can use similar arguments of the proof of Lemma A.1 to arrive at such a result.

Let us now take into account the case of the spectrum of L_0 contained on the negative real line $(-\infty, 0]$ and that one associated with V is by its turn, contained only on the upper-bounded negative infinite interval $(-\infty, -c]$ (with $c > 0$); then we have by a straightforward application of the

Trotter–Kato formulae for $t \in [0, \infty)$ the following estimate on the norm of semigroup evolution, operator-generated by $L_0 + V_0$:

$$||e^{t(L_0 + V)}|| \leq \lim_n ||e^{\frac{t}{n} L_0} e^{\frac{t}{n} V}||^n$$

$$\leq \lim_n ||e^{\frac{t}{n} V}||^n \leq ||e^{tV}|| \leq e^{-tc}. \qquad (A.24)$$

In the following magneto-elastic wave problem with *real* conductivity $\beta > 0$,

$$\partial_t \begin{pmatrix} U \\ \pi \\ h \end{pmatrix}(t) = \begin{bmatrix} 0 & -1 & 0 \\ -\Delta & 0 & 0 \\ 0 & 0 & \frac{1}{\beta}\Delta \end{bmatrix} \begin{pmatrix} U \\ \pi \\ h \end{pmatrix}(t)$$

$$\times \begin{bmatrix} 0 & 0 & 0 \\ 0 & 0 & -\vec{B} \times [\vec{\nabla} \times (\ )] \\ 0 & -\nabla \times [(\ ) \times \vec{B}] & 0 \end{bmatrix} \begin{pmatrix} U \\ \pi \\ h \end{pmatrix}(t).$$

$$(A.25)$$

One can see that it satisfies the conditions of the above lemmas with the operator $D = \nabla$ and with the Hilbert space

$$\tilde{H} = (L^2(\Omega))^3 \oplus (L^2(\Omega))^3 \oplus (L^2(\Omega))^3$$

and $c = \inf \ \mathrm{spec}(-\Delta)$ and $H^2(\Omega) \cap H_0^1(\Omega)$ (see Sec. 3.5), and thus, leading to the expected total energy exponential decay showed in Sec. 3.5. $\qquad \square$

Appendix B. Probability Theory in Terms of Functional Integrals and the Minlos Theorem

In this complementary appendix, we discuss briefly for the mathematics-oriented reader the mathematical basis and concepts of functional integrals formulation for stochastic processes and the important Minlos theorem on Hilbert space support of probabilistic measures.

The first basic notion of Kolmogorov's probability theory framework is to postulate a convenient topological space (Polish spaces) Ω formed by the phenomenal random events. The chosen topology of Ω should possess a rich set of nontrivial *compact* subsets. After that, we consider the σ-algebra generated by all open sets of this — the so-called topological sample space, denoted by $\mathcal{F}_\Omega$. Thirdly, one introduces a regular measure $d\mu$ on this set algebra $\mathcal{F}_\Omega$, assigning values in $[0, 1]$ to each member of $\mathcal{F}_\Omega$.

The (abstract) triplet $(\Omega, \mathcal{F}_\Omega, d\mu)$ is thus called a probability space in the Kolmogorov–Schwartz scheme.[10]

Let $\{X_\alpha\}_{\alpha \in \Lambda}$ be a set of measurable real functions on Ω, which may be taken as continuous injective functions of compact support on Ω (without loss of generality).

It is a standard result that one can "immerse" (represent) the abstract space Ω on the "concrete" infinite product compact space $R^\infty = \prod_{\alpha \in \Lambda}(\dot{R}_\alpha)$, where $\dot{R}$ is a compact copy of the real line. In order to achieve such a result we consider the following injection of Ω in R^∞ defined by the family $\{X_\alpha\}_{\alpha \in \Lambda}$:

$$I : \Omega \to R^\infty \qquad \qquad \text{(B.1)}$$
$$\omega \to \{X_\alpha(\omega)\}_{\alpha \in \Lambda}.$$

A new σ-algebra of events on Ω can be induced on Ω and affiliated to the family $\{X_\alpha\}_{\alpha \in \Lambda}$.

It is the σ-algebra generated by all the "finite-dimensional cylinder" subsets, explicitly given in its generic formulas by

$$C_{\Lambda_{\text{fin}}} = \left\{ \omega \in \Omega \mid \Lambda_{\text{fin}} \text{ is a finite subset of } \Lambda \right.$$

$$\left. \text{and } [(X_\alpha)_{\alpha \in \Lambda_{\text{fin}}}(\Omega)] \subset \prod_{\alpha \in \Lambda_{\text{fin}}} [a_\alpha, b_\alpha] \right\}.$$

The measure restriction of the initial measure μ on this cylinder set of Ω will be still denoted by μ on what follows. Now it is a basic theorem of Probability theory that μ induces a measure $\nu^{(\infty)}$ on $\prod_{\alpha \in \Lambda}(\dot{R})_\alpha$, which can be identified with the space of all functions of the index set Λ to the real compactified line $\dot{R}$; $F(\Lambda, R)$ with the topology of Pontual convergence.

At this point, it is straightforward to see that the average of any Borelian (mensurable) function $G(\omega)$ (a random variable) is given by the following functional integral on the functional space $F(\Lambda, R)$:

$$\int_\Omega d\mu(\omega) G(\omega) = \int_{F(\Lambda,R)} d\nu^{(\infty)}(f) \cdot G(I^{-1}(f)). \qquad \text{(B.2)}$$

It is worth to recall that the real support of the measure $d\nu^{(\infty)}(f)$ in most practical cases is not the whole space $F(\Lambda, R)$ but only a functional subspace of $F(\Lambda, R)$. For instance, in most practical cases, Λ is the index set of an algebraic vectorial base of a vector space E and $\prod_{\alpha \in \Lambda}(R)_\alpha$ (without the compactification) is the space of all sequences with only a finite number of nonzero entries as it is necessary to consider in the case of the famous extension theorem of Kolmogorov. It turns out that one can consider the support of the probability measure as the set of all linear functionals on E, the so-called algebraic dual of E. This result is a direct consequence of considering the set of random variables given by the family $\{e_\alpha^*\}_{\alpha\in\Lambda}$. For applications one is naturally lead to consider the probabilistic object called characteristic functional associated with the measure $d\nu^{(\infty)}(f)$ when $F(\Lambda, R)$ is identified with the vector space of all algebraic linear functionals E^{alg} of a given vector space E as pointed out above, namely,

$$Z[j] = \int_{E^{\mathrm{alg}}} d\nu^{(\infty)}(f)\exp(if(j)), \qquad (B.3)$$

where j are elements of E written as $j = \sum_{\alpha\in\Lambda_{\mathrm{fin}}} x_\alpha e_\alpha$ with $\{e_\alpha\}_{\alpha\in\Lambda}$ denoting a given Hammel (vectorial) basis of E.

One can show that if there is a subspace H of E with a norm $\|\ \|_H$ coming from an inner product $(\,,)_H$ such that

$$\int_{E\cap H} d\nu^{(\infty)}(f)\cdot\|f\|_H^2 < \infty \qquad (B.4)$$

one can show the support of the measure of exactly the Hilbert space $(H,(\,,)_H)$ — the famous Minlos theorem.

Another important theorem (the famous Wienner theorem) is that when we have the following condition for the family of random variables with the index set Λ being a real interval of the form $\Lambda = [a, b]$:

$$\int_E d\nu^{(\infty)}(f)|X_{t+h}(f) - X_t(f)|^p \le c|h|^{1+\varepsilon}, \qquad (B.5)$$

for p, ε, c positive constants. We have that it is possible to choose as the support of the functional measure $d\nu^{(\infty)}(\cdot)$ the space of the real continuous function on Λ the well-known famous Brownian path space

$$\int_\Omega d\mu(\omega)G(\omega) = \int_{C([a,b],R)} d\nu^{(\infty)}(f(x))\tilde{G}(f(x)). \qquad (B.6)$$

For rigorous proofs and precise formulations of the above-sketched deep mathematical results, the interested reader should consult the basic mathematical book by Schwartz.[10]

Let us finally give a concrete example of such results considering a generating functional $Z[J(x)]$ of the exponential quadratic functional form on $L^2(R^N)$

$$Z[J(x)] = \exp\left\{-\frac{1}{2}\int_{R^N} d^N x \int_{R^N} d^N y\, j(x) K(x,y) j(y)\right\} \qquad \text{(B.7)}$$

associated with a Gaussian random field with two-point correlation function given by the kernel of Eq. (B.7),

$$E\{v(x)v(y)\} = K(x,y). \qquad \text{(B.8)}$$

In the case of the above-written kernel being a class-trace operator on $L^2(R^N)$ ($\int_{R^N} d^N x K(x,x) < \infty$), one can represent Eq. (B.7) by means of a Gaussian measure with the support of the Hilbert space $L^2(R^N)$. This result was called to the reader's attention for the bulk of this paper.

Chapter 4

Nonlinear Diffusion and Wave-Damped Propagation: Weak Solutions and Statistical Turbulence Behavior[*]

4.1. Introduction

One of the most important problems in the mathematical physics of the nonlinear diffusion and wave-damped propagation is to establish the existence and uniqueness of weak solutions in some convenient Banach spaces for the associated nonlinear evolution equation (see Refs. 1, 3–5). Another important class of initial-value problems for nonlinear diffusion-damping came from statistical turbulence as modeled by nonlinear diffusion or damped hyperbolic partial differential equations with random initial conditions associated to the Gaussian processes sampled in certain Hilbert spaces[2,5] and simulating the turbulence physical phenomena.[6]

The purpose of this chapter is to contribute to such mathematical physics studies by using functional spaces compacity arguments in order to produce proofs for the existence and uniqueness of weak solutions for a class of special nonlinear diffusion equations on a "smooth" C^∞ domain with compact closure $\Omega \subset R^3$ with Dirichlet boundary conditions and initial conditions belonging to the space $L^2(\Omega)$. This study is presented in Sec. 4.2.

In Sec. 4.3, we present a similar analysis for the wave equation with a nonlinear damping, analogous in its form to the nonlinear term studied in Sec. 4.2 for the diffusion equation.

In Sec. 4.4, we present in some details the solution to the associated problem for random initial conditions in terms of our proposed cylindrical functional measures representations previously proposed in the literature[6] and defined by a cylindrical measure on the Banach space $L^\infty((0,T), L^2(\Omega))$, a new result on the subject.

Finally, in Sec. 4.5, we present a complementary semigroup analysis on the very important problem of anomalous diffusion for random conditions in the path-integral framework.

---

[*]Author's original results.

4.2. The Theorem for Parabolic Nonlinear Diffusion

Let us consider the following nonlinear diffusion equation in some strip $\Omega \times [0, T] \subset R^4$:

$$\frac{\partial U(x,t)}{\partial t} = (-AU)(x,t) + \Delta(F(U(x,t))) + f(x,t) \qquad (4.1)$$

with the initial and Dirichlet boundary conditions

$$U(x,0) = g(x) \in L^2(\Omega) \subset L^1(\Omega), \qquad (4.2)$$

$$U(x,t)\big|_{\partial\Omega} \equiv 0, \qquad (4.3)$$

where A denotes a second-order self-adjoint uniform elliptic positive differential operator, $F(x)$ is a real function continuously differentiable on the extended real line $(-\infty, +\infty)$ with its derivative $F'(x)$ strictly positive on $(-\infty, +\infty)$. The external source $f(x,t)$ is supposed to be on the space $L^\infty([0,T] \times L^2(\Omega)) = L^\infty([0,T], L^2(\Omega))$.

We now state the existence and uniqueness theorem of ours.

Theorem 4.1. *In the initial-value nonlinear diffusion equations (4.1)–(4.3), for any $g(x) \in L^2(\Omega)$, and $A = -\Delta$ (minus Laplacean) there exists a unique solution $\bar{U}(x,t)$ on $L^\infty([0,T] \times L^2(\Omega))$ satisfying this problem in a certain weak sense with a test functional space given by $C_0^\infty([0,T], H^2(\Omega) \cap H_0^1(\Omega))$.*

Proof. The existence proof will be given for a general A as stated below Eq. (4.3). Let $\{\varphi_i(x)\}$ be the spectral eigenfunctions associated with the operator A. Note that each $\varphi_i(x) \in H^2(\Omega) \cap H_0^1(\Omega)$ and this set is complete on $L^2(\Omega)$ as a result of the Gelfand generalized spectral theorem applied for A^1 since $H^2(\Omega) \cap H_0^1(\Omega)$ is compactly immersed in $L^2(\Omega)$.

Consider the following system of nonlinear ordinary differential equations associated with Eqs. (4.1)–(4.3) — the well-known Galerkin system:[1]

$$\left(\frac{\partial U^{(n)}(x,t)}{\partial t}, \varphi_j(x) \right)_{L^2(\Omega)} + (AU^{(n)}(x,t), \varphi_j(x))_{L^2(\Omega)}$$

$$= (\nabla \cdot [(F'(U^{(n)}(x,t)))\nabla U^{(n)}(x,t)], \varphi_j(x))_{L^2(\Omega)}$$

$$+ (f^{(n)}(x,t), \varphi_j(x))_{L^2(\Omega)} \qquad (4.4)$$

subject to the initial condition

$$U^{(n)}(x,0) = \sum_{i=1}^{n} (g, \varphi_i)_{L^2(\Omega)} \cdot \varphi_i(x), \qquad (4.5)$$

where the finite-dimensional Galerkin approximants are given exactly in terms of the spectral basis $\{\varphi_i(x)\}$ as

$$U^{(n)}(x,t) = \sum_{i=1}^{n} U_i^{(n)}(t)\varphi_i(x), \qquad (4.6)$$

$$f^{(n)}(x,t) = \sum_{i=1}^{n} (f(x,t), \varphi_i(x))_{L^2(\Omega)} \varphi_i(x), \qquad (4.7)$$

and $(,)_{L^2(\Omega)}$ is the usual inner product on $L^2(\Omega)$. Note that

$$U^{(n)}(x,t)\big|_{\partial\Omega} = \sum_{i=1}^{n} U_i^{(n)}(t)(\varphi_i(x))\big|_{\partial\Omega} = 0. \qquad (4.8)$$

Let us introduce the short notations

$$U^{(n)}(x,t) \equiv U^{(n)}, \qquad (4.9)$$

$$(g, \varphi_i)_{L^2(\Omega)} \equiv g_i^{(n)}, \qquad (4.10)$$

$$(f(x,t), \varphi_i(x))_{L^2(\Omega)} = f_i^{(n)}. \qquad (4.11)$$

By multiplying the Galerkin system equation (4.4) by $U^{(n)}$ as usual[3] we get the *a priori* identity

$$\frac{1}{2}\frac{d}{dt}\|U^{(n)}\|_{L^2(\Omega)}^2 + (AU^{(n)}, U^{(n)})_{L^2(\Omega)}$$

$$+ \int_{\Omega} dx F'(U^{(n)})(\nabla U^{(n)}\overline{\nabla U}^{(n)}) = (f, U^{(n)})_{L^2(\Omega)}. \qquad (4.12)$$

By a direct application of the Gärding–Poincaré inequality with the quadratic form associated with the operator A, one has that there is a positive constant $\gamma(\Omega)$ such that[3]

$$(AU^{(n)}, U^{(n)})_{L^2(\Omega)} \geq \gamma(\Omega)\|U^{(n)}\|_{L^2(\Omega)}^2. \qquad (4.13)$$

This yields the following estimate for any integer positive p:

$$\frac{1}{2}\frac{d}{dt}(\|U^{(n)}\|_{L^2(\Omega)}^2) + \gamma(\Omega)\|U^{(n)}\|_{L^2(\Omega)}^2 + \|(F'(U^{(n)}))^{(1/2)}\nabla U^{(n)}\|_{L^2(\Omega)}^2$$

$$\leq \frac{1}{2}\left\{ p\|f(x,t)\|_{L^2(\Omega)}^2 + \frac{1}{p}\|U^{(n)}\|_{L^2(\Omega)}^2 \right\}. \tag{4.14}$$

At this point we observe that the square root of the function $F'(x)$ makes sense since it is always positive: $F'(x) > 0$.

By choosing Eq. (4.14) p big enough such that $a_p \equiv \gamma(\Omega) - \frac{1}{2p} > 0$, we have that

$$\frac{1}{2}\frac{d}{dt}(\|U^{(n)}\|_{L^2(\Omega)}^2) + a_p\|U^{(n)}\|_{L^2(\Omega)}^2 \leq \frac{1}{2}p\|f\|_{L^2(\Omega)}^2(t), \tag{4.15}$$

or by the Gronwall lemma, that there is a uniform constant M [even for $T = +\infty$ if $f \in L^2([0,\infty], L^2(\Omega))$] such that

$$\sup_{0 \leq t \leq T}(\|U^{(n)}(t)\|_{L^2(\Omega)}^2)$$

$$\leq \sup_{0 \leq t \leq T}\left\{ e^{-a_p T}\left[\int_0^T ds\|f\|_{L^2(\Omega)}^2(s)e^{+a_p s} + \|g\|_{L^2(\Omega)}^2 \right] \right\} = M. \tag{4.16}$$

Note that the Galerkin system of (nonlinear) ordinary differential equations for $\{U_i^{(n)}(t)\}$ has a unique global solution on the interval $[0,T]$ as a consequence of Eq. (4.16) — which means that $\sum_{i=1}^n (U_i^{(n)}(t))^2 \leq M$ — and the Peano–Caratheodory theorem since $f(t, \cdot) \in L^\infty([0,T])$.

As a consequence of the Banach–Alaoglu theorem applied to the bounded set $\{U^{(n)}\}$ on $L^\infty([0,T], L^2(\Omega))$, there is a subsequence weak-star convergent to an element (function) $\bar{U}(t,x) \in L^\infty([0,T], L^2(\Omega))$, and this subsequence will still be denoted by $\{U^{(n)}(x,t)\}$ in the analysis that follows.

An important remark is at this point. Since $F(x)$ is a Lipschitzian function on the closed interval where it is defined we have that the set of functions $\{F(U^{(n)})\}$ is a bounded set on $L^\infty([0,T], L^2(\Omega))$ if the set $\{U^{(n)}\}$ has this property.

As a consequence of this remark and of *a priori* inequalities given in detail in Appendix A, we can apply the famous Aubin–Lion theorem[1] to insure the strong convergence of the sequence $\{U^{(n)}(x,t)\}$ to the function $\bar{U}(x,t)$. As a straightforward consequence, our special nonlinearity on Eq. (4.1), namely $F(x)$ is a Lipschitzian function; we obtain, thus, the strong convergence of the sequence $\{F(U^{(n)}(x,t)\}$ to the $L^2(\Omega)$-function $F(\bar{U}(x,t))$.

As a consequence of the above-made comments and by noting that $L^2(\Omega)$ is continuously immersed in $H^{-2}(\Omega)$ and the weak-star convergence definition, we have the weak equation below in the test space function of $v(x,t) \in C_0^\infty([0,T], H^2(\Omega) \cap H_0^1(\Omega))$:

$$\lim_{n\to\infty} \int_0^T dt \left[\left(-U^{(n)}, \frac{dv}{dt} \right)_{L^2(\Omega)} + (AU^{(n)}, v)_{L^2(\Omega)} - (F(U^{(n)}), \Delta v)_{L^2(\Omega)} \right]$$

$$= \lim_{n\to\infty} \int_0^T dt(f^{(n)}, v)_{L^2(\Omega)}, \tag{4.17}$$

or by passing the weak-star limit

$$\int_0^T dt \left[\left(-\bar{U}(x,t), \frac{dv(x,t)}{dt} \right)_{L^2(\Omega)} + (\bar{U}(x,t), (Av)(x,t))_{L^2(\Omega)} \right.$$

$$\left. - (F(\bar{U}(x,t)), \Delta v(x,t))_{L^2(\Omega)} \right] = \int_0^T dt(f(x,t), v(x,t))_{L^2(\Omega)}. \tag{4.18}$$

This concludes the existence proof of Theorem 4.1, since $v(0,x) = v(T,x) = 0$. $\qquad\square$

Let us apply this to a concrete problem of random exponential nonlinear diffusion on a cube $[0,L]^3 \subset R^3$, mainly for the explanation of the above-written abstract results:

$$\frac{\partial U(x,t)}{\partial t} = \nabla \left\{ \left(\frac{k_0}{2} + e^{-U(x,t)} \right) \nabla U(x,t) \right\} + \frac{k_0}{2} \Delta U(x,t) \tag{4.19}$$

with

$$U(x,0) = g(x) \in L^2(\Omega), \tag{4.20}$$

$$U(x,t)\big|_{x\in[0,L]^3} \equiv 0, \tag{4.21}$$

where $g(x)$ are the samples of a stochastic Gaussian process on Ω with correlation function defining an operator of Trace class on $L^2(\Omega)$, namely,

$$E\{g(x)g(y)\} = K(x,y) \in \oint_1 (L^2(\Omega)) \tag{4.22}$$

or in terms of the spectral set associated with the Laplacian, i.e.

$$g(x) = \lim_{n\to\infty} \sum_{i=0}^n g_i^{(n)} \varphi_i(x) \tag{4.23}$$

with

$$E\{g_i g_k\} = K_{ik} = \int_{\Omega \times \Omega} dx dy K(x, y) \varphi_i(x) \varphi_k(y). \qquad (4.24)$$

As a consequence of Theorem 4.1, an explicit solution of Eq. (4.19) must be taken in $L^\infty([0, \infty), L^2(\Omega))$ in the weak sense of Eq. (4.18) and given thus by the *Trigonometrical series sequence of functions*:

$$U(x, t, [g])$$
$$= \lim_{\substack{n_1 \to \infty \\ n_2 \to \infty \\ n_3 \to \infty}} \left\{ \sum_{i,j,k=1}^{n_1, n_2, n_3} d^{i,k,j}(t, [g_\ell]) \sin\left(\frac{i\pi}{L} x\right) \sin\left(\frac{j\pi}{L} y\right) \sin\left(\frac{k\pi}{L} z\right) \right\},$$
$$(4.25)$$

where the set of *absolutely continuous function* $\{d^{i,j,k}(t, [g_0])\}$ on $[0, T]$ satisfies the Galerkin set of ordinary differential equations with "random" Gaussian initial conditions

$$d^{i,j,k}(0, [g_\ell]) = \left(g(x), \sin\left(\frac{i\pi}{L} x\right) \sin\left(\frac{j\pi}{L} y\right) \sin\left(\frac{k\pi}{L} z\right) \right)_{L^2(\Omega)} = g_\ell.$$
$$(4.26)$$

Let us now comment on the uniqueness of solution problem for the nonlinear initial-value diffusion equations (4.1)–(4.3).

In the case under study, the uniqueness comes from the following technical lemma.

Lemma 4.1. *If $\bar{U}_{(1)}$ and $\bar{U}_{(2)}$ in $L^\infty([0, T] \times L^2(\Omega))$ are two functions satisfying the weak relationship, for any $v \in C_0^\infty([0, T], H^2(\Omega) \cap H_0^1(\Omega))$*

$$\int_0^T dt \left\{ \left(\bar{U}_{(1)} - \bar{U}_{(2)}, -\frac{\partial v}{\partial t} \right)_{L^2(\Omega)} + (\bar{U}_{(1)} - \bar{U}_{(2)}, Av)_{L^2(\Omega)} \right.$$
$$\left. - (F(\bar{U}_{(1)}) - F(\bar{U}_{(2)}); \Delta v)_{L^2(\Omega)} \right\} \equiv 0, \qquad (4.27a)$$

then $\bar{U}_{(1)} = \bar{U}_{(2)}$ a.e. in $L^\infty([0, T] \times L^2(\Omega))$.[5]

Let us notice that Eq. (4.27a) is a weak statement on the assumption of vanishing initial values and it is a consequence of

$$\lim_{t \to 0^+} \text{supp ess} \left\{ \int_\Omega d^3 x |\bar{U}_{(1)}(x, t) - \bar{U}_{(2)}(x, t)|^2 \right\} = 0. \qquad (4.27b)$$

The proof of Eq. (4.27) is a direct consequence of the fact that $F(x)$ is a Lipschitz function satisfying an inequality of the form

$$|F(\bar{U}_{(1)}) - F(\bar{U}_{(2)})|(x,t) \leq \left(\sup_{-\infty < x < +\infty} F'(x) \right) |\bar{U}_{(1)} - \bar{U}_{(2)}|(x,t) \quad (4.27c)$$

(for a technical proof see Ref. 5). However in the case of $A = -\Delta$ (minus Laplacian), as stated in our theorem, the identity Eq. (4.27a) means that for the functions of the form $v(x,t) = e^{-\lambda t} v_\lambda(x) \in C_0^\infty([0,T], H^2(\Omega) \cap H_0^1(\Omega))$ with $\Delta v_\lambda(x) = -\lambda v_\lambda(x)$, the following identity must hold true:

$$\int_0^T dt (F(\bar{U}_{(1)}) - F(\bar{U}_{(2)}), v_\lambda)_{L^2(\Omega)} \equiv 0, \quad (4.28)$$

which means that $F(\bar{U}_{(1)}) = F(\bar{U}_{(2)})$ a.e., and thus, $\bar{U}_{(1)} = \bar{U}_{(2)}$ a.e., since $F(x)$ satisfies the lower-bound estimate

$$|F(x) - F(y)| \geq \left(\inf_{-\infty < x < \infty} |F'(x)| \right) |x - y|.$$

This result produces a rigorous uniqueness result for $A = -\Delta$.

Another important example of nonlinear diffusion equation somewhat related to Eq. (4.1) is the density equation for a gas in a porous medium with physical saturation. In this case, the law of conservation of mass

$$\nabla \cdot (\rho \vec{V}) + \frac{\partial \rho}{\partial t} = 0, \quad (4.29)$$

where $\vec{V}$ is the velocity of gas, ρ is the density of the gas, and P is the pressure, together with Darcy's law and the isothermic equation of state, namely,

$$P = c\rho^\gamma \quad (4.30)$$
$$\vec{V} = -k\nabla P$$

leads to the following nonlinear diffusion equation:

$$\begin{cases} \dfrac{\partial \rho(x,t)}{\partial t} = -\gamma ck \nabla(\rho^\gamma \nabla \rho)(x,t) \\ \rho(x,0) = f(x) \in L^2(\Omega). \end{cases} \quad (4.31)$$

Since it is showed that for finite volume open, convex with C^∞-boundary domain Ω as in our study, the function $\rho(x,t)$ which satisfies Eq. (4.31) should be a bounded function for any $t \geq t_0 > 0$ (with t_0 fixed),[5] it is clear that one can replace the above-written mathematical gas porous medium equation, by a more physical equation of the form, taking into account the physical phenomena of saturation,

$$\begin{cases} \dfrac{\partial \rho(x,t)}{\partial t} = -\gamma c \nabla (F_{(\varepsilon)}(\rho) \nabla \rho)(x,t) \\ \rho(x,0) = f(x) \in L^2(\Omega) \end{cases} \tag{4.32}$$

with $F_{(\varepsilon)}(x)$ denoting a differentiable regularizing function of the function x^γ on the interval $[0, M]$, where M is the global upper-bound of the function $\rho(x,t)$.

One can thus apply our Theorem 4.1 to obtain an explicit set of functions $\{\rho_{(\varepsilon)}^{(n)}(x,t)\}$ converging in the weak-star topology of $L^\infty([t_0, \infty), L^2(\Omega))$ to the solution $\rho_{(\varepsilon)}(x,t)$ of the problem equation (4.32). It is a reasonable conjecture that in the nonsaturation case, one should take the $\varepsilon \to 0$ limit on Eq. (4.32), namely, $\rho(x,t) = \lim\limits_{\varepsilon \to 0} \rho_{(\varepsilon)}(x,t)$, since $F_{(\varepsilon)}(x)$ converge to x^γ in the C^∞-topology of $C([0, M])$. As a consequence, it is expected that a solution for the physical equation (4.31) should be produced. A full technical description of this limiting process will appear in an extended mathematics-oriented paper.

4.3. The Hyperbolic Nonlinear Damping

We aim in this third section state a theorem analogous to Theorem 4.1 of Sec. 4.2 on diffusion, but now for the important case of existence of nonlinear damping on the hyperbolic initial-value problem on $\Omega \times [0, T]$ with imposed Dirichlet boundary conditions, including the new global case of $T = +\infty$ (see Sec. 4.2) and a damping positive constant ν,

$$\frac{\partial^2 U(x,t)}{\partial t^2} + (AU)(x,t) = -\nu \frac{\partial U(x,t)}{\partial t} + \Delta \left(F \left(\frac{\partial U}{\partial t}(x,t) \right) \right) + f(x,t). \tag{4.33}$$

The $L^2(\Omega)$-initial conditions are given by

$$U(x,0) = g(x) \in L^2(\Omega), \tag{4.34a}$$

$$U_t(x,0) = h(x) \in L^2(\Omega), \tag{4.34b}$$

$$U(x,t)|_{\partial \Omega} = 0, \tag{4.34c}$$

and now the nonhomogenous term $f(x,t)$ is considered to be a function belonging to the functional space

$$L^2([0,T] \times \Omega) \cap L^\infty([0,T], L^2(\Omega)). \tag{4.35}$$

We thus have the following theorem of existence (without uniqueness).

Theorem 4.2. *There exists a solution $\bar{U}(x,t)$ on $L^\infty([0,T] \times L^2)\Omega))$ for Eqs. (4.33)–(4.35) in the weak sense with a test functional space as given by $C_0^\infty([0,T], H^2(\Omega) \cap H_0^1(\Omega))$.*

In order to arrive at such a theorem, let us consider the analogous of the estimated Eq. (4.12) for Eq. (4.33) with $\dot{U}^{(n)} \equiv \frac{\partial}{\partial t}(U^{(n)}(x,t))$, namely,

$$\begin{aligned}
\frac{1}{2}\frac{d}{dt}\|\dot{U}^{(n)}\|_{L^2(\Omega)}^2 &+ \nu\|\dot{U}^{(n)}\|_{L^2(\Omega)}^2 \\
&+ (A\dot{U}^{(n)}, U^{(n)})_{L^2(\Omega)} + \|(F'(\dot{U}^{(n)}))^{(1/2)}\nabla\dot{U}^{(n)}\|_{L^2(\Omega)}^2 \\
&= (f, \dot{U}^{(n)})_{L^2(\Omega)},
\end{aligned} \tag{4.36}$$

or equivalently

$$\begin{aligned}
\frac{1}{2}\frac{d}{dt}&\left\{\|\dot{U}^{(n)}\|_{L^2(\Omega)}^2 + (AU^{(n)}, U^{(n)})_{L^2(\Omega)}\right\} \\
&+ \nu\left\{\|\dot{U}^{(n)}\|_{L^2(\Omega)}^2 + (AU^{(n)}, U^{(n)})_{L^2(\Omega)}\right\} \\
&+ \|(F'(\dot{U}^{(n)}))^{(1/2)}\nabla\dot{U}^{(n)}\|_{L^2(\Omega)}^2 \leq \nu(AU^{(n)}, U^{(n)}) \\
&+ \frac{1}{2}\left(p\|f\|_{L^2(\Omega)}^2 + \frac{1}{p}\|\dot{U}^{(n)}\|_{L^2(\Omega)}^2\right).
\end{aligned} \tag{4.37}$$

If one chooses here the integer p such that $\frac{1}{2p} = \nu$, we have the simple bound

$$\frac{1}{2}\frac{d}{dt}\left\{\|\dot{U}^{(n)}\|_{L^2(\Omega)}^2 + (AU^{(n)}, U^{(n)})_{L^2(\Omega)}\right\} \leq \frac{1}{4\nu}\|f\|_{L^2(\Omega)}^2. \tag{4.38}$$

As a consequence of Eq. (4.38), there is a constant M such that the uniform bounds hold true (*even if for the case $T = +\infty$ for the case of $f \in L^2([0,\infty), L^2(\Omega))$*):

$$\sup\,\text{ess}_{0 \leq t \leq T}\|\dot{U}^{(n)}\|_{L^2(\Omega)}^2 \leq M, \tag{4.39}$$

$$\sup\,\text{ess}_{0 \leq t \leq T}\|U^{(n)}\|_{L^2(\Omega)}^2 \leq M. \tag{4.40}$$

As a consequence of the bounds Eqs. (4.39) and (4.40), there are two functions $\bar{U}(x,t)$ and $\bar{P}(x,t)$ such that we have the weak-star convergence on $L^\infty([0,T] \times L^2(\Omega))$:

$$\text{weak-star} \quad \lim_{n\to\infty} (U^{(n)}(x,t)) = \bar{U}(x,t), \tag{4.41}$$

$$\text{weak-star} \quad \lim_{n\to\infty} \left(\frac{\partial}{\partial t} U^{(n)}(x,t) \right) = \bar{P}(x,t). \tag{4.42}$$

We thus have that the relationship below hold true for any test function $v(x,t) \in C_c^\infty([0,T] \times H^2(\Omega) \cap H_0^1(\Omega))$ obviously satisfying the relations $v(x,0) = v(x,T) = \Delta v(x,0) = \Delta v(x,T) = v_t(x,0) = v_t(x,T) = v_{tt}(x,0) = v_{tt}(x,T) \equiv 0$ as a consequence of applying the Aubin–Lion theorem in a similar way as it was used on Eq. (4.17):

$$\int_0^T dt \left[\left(\bar{U}, \frac{d^2 v}{d^2 t} \right)_{L^2(\Omega)} + (\bar{U}, Av)_{L^2(\Omega)} \right.$$

$$\left. + \nu \left(\bar{U}, -\frac{dv}{dt} \right)_{L^2(\Omega)} - (F(\bar{P}), \Delta v)_{L^2(\Omega)} \right]$$

$$= \int_0^T dt (f, v)_{L^2(\Omega)}, \tag{4.43}$$

with the initial conditions

$$\bar{U}(x,0) = g(x) \in L^2(\Omega), \tag{4.44}$$

$$\bar{P}(x,0) = h(x) \in L^2(\Omega). \tag{4.45}$$

Let us now show that

$$\bar{U}(x,t) = \int_0^t ds \bar{P}(x,s). \tag{4.46}$$

First, let us remark that integrating on the interval $0 \le t \le T$, in the relationship Eq. (4.36), one obtains the following estimate:

$$\frac{1}{2} \left(\|\dot{U}^n(t)\|_{L^2(\Omega)}^2 - \|\dot{U}^n(0)\|_{L^2(\Omega)}^2 \right)$$

$$+ \left(\nu \int_0^t ds \|\dot{U}_n\|_{L^2(\Omega)}^2(s) \right) + \frac{1}{2} (AU^{(n)}(t), U^{(n)}(t))_{L^2(\Omega)}$$

$$- \frac{1}{2} (AU^{(n)}(0), U^{(n)}(0)) + \int_0^t ds \|(F'(\dot{U}^{(n)}))^{(1/2)} \nabla \dot{U}^{(n)}\|_{L^2(\Omega)}^2$$

$$\le \frac{p'}{2} \int_0^t ds \|f\|_{L^2(\Omega)}^2 + \frac{1}{2p'} \int_0^t \|\dot{U}_n(s)\|^2 ds. \tag{4.47}$$

Since the operator A satisfies the Gärding–Poincaré inequality on $L^2(\Omega)$,

$$(AU^{(n)}, U^{(n)})_{L^2(\Omega)}(t) \geq \gamma(\Omega)\|U^{(n)}\|^2_{L^2(\Omega)}(t), \tag{4.48}$$

one can see straightforwardly from Eq. (4.47) by choosing $2p' > \frac{1}{\nu}$ and the previous bounds Eq. (4.39) that there is a positive constant B such that

$$\int_0^T ds \|\dot{U}^{(n)}(s)\|^2_{L^2(\Omega)} \leq B = MT, \tag{4.49}$$

which by its turn yields that $\frac{dU_n(t,x)}{dt}$ is weakly convergent on $\bar{P}(x,t)$ in $L^2([0,T], L^2(\Omega))$.

As a consequence of general theorems of function convergence on the space of integrable functions (Aubin–Lion theorem[1] again), one has that $\bar{P}(x,t)$ is the time-derivative of the function $\bar{U}(x,t)$ almost everywhere on Ω, since it is expected that $U^{(n)}(x,t)$ should be a strongly convergent sequence to $\bar{U}(x,t)$ on the separable and reflexive Banach space $L^\infty([0,T] \times L^2(\Omega))$.[1] (See Appendix B for mathematical details.)

4.4. A Path-Integral Solution for the Parabolic Nonlinear Diffusion

Let us start the physicist-oriented section of our note by writing the abstract scalar parabolic equations (4.1)–(4.3) in the integral (weak) form

$$U(t) = e^{-At}U(0) + \int_0^t ds \, e^{-(t-s)A}\Delta F(U(s)), \tag{4.50}$$

where $F(x)$ is a general nonlinear scalar functional such that $F \colon L^2(\Omega) \to H^2(\Omega)$.

Let us suppose that the initial conditions for Eq. (4.1) are samples of Gaussian stochastic processes belonging to the $L^2(\Omega)$ space. For instance, this mathematical fact is always true when the processes correlation function defines an integral operator of the trace class on $L^2(\Omega)$,

$$E\{U(0,x)U(0,y)\} = K(x,y), \tag{4.51}$$

where

$$\int_\Omega dx K(x,x) < \infty. \tag{4.52}$$

At a *formal path-integral method*, one aims to compute the initial condition average of arbitrary product of the function $U(x,t) \equiv U_x(t)$ for arbitrary space–time points. This task is achieved by considering the associated characteristic functional for Eq. (4.51), namely,

$$Z[j(t)] = \frac{1}{Z(0)} E \left\{ \exp \left(i \int_0^T dt \langle j(t), U(t) \rangle_{L^2(\Omega)} \right) \right\}. \qquad (4.53)$$

In order to write a path-integral representation for Eq. (4.53), we follow our previous studies on the subject[6] by realizing the random initial conditions $U(0) = U_0$ average as a Gaussian functional integral on $L^2(\Omega)$, namely,

$$Z[f(t)] = \frac{1}{Z(0)} \left\{ \int_{L^2(\Omega)} d\mu[U_0] \times \int_{L^2([0,T], L^2(\Omega))} \left(\prod_{x \in \Omega, t \in [0,T]} dU(x,t) \right) \right.$$

$$\times \delta^{(F)} \left[U(t) - \left(e^{-At} U_0 + \int_0^t ds \, e^{-(t-s)A} \Delta F(U(s)) \right) \right]$$

$$\left. \times \exp \left(i \int_0^\infty dt \langle j(t), U(t) \rangle_{L^2(\Omega)} \right) \right\}, \qquad (4.54)$$

where formally

$$d\mu[U_0] = \left(\prod_{x \in \Omega} (dU_0(x)) \right) \times \exp \left\{ -\frac{1}{2} \int_\Omega dx dy U_0(x) K^{-1}(x,y) U_0(y) \right\}. \qquad (4.55)$$

By using the well-known Fourier functional integral representation for the delta-functional inside the identity equation (4.54)[6] (see this old idea in A. S. Monin and A. M. Yaglom, *Statistical Fluid Mechanics*, Vol. 2 (MIT Press, Cambridge, 1971),[6] one can rewrite Eq. (4.54) in the following form:

$$Z[j(t)] = \frac{1}{Z(0)} \left\{ \int_{L^\infty([0,T], L^2(\Omega))} \prod (dU(x,t)) \times \int_{L^2([0,T], L^2(\Omega))} d\lambda(t) \right.$$

$$\times \exp \left\{ i \int_0^t \left\langle \lambda(t), U(t) + \int_0^t ds \, e^{-(t-s)A} \Delta F(U)(s) \right\rangle_{L^2(\Omega)} \right\}$$

$$\left. \times \exp \left\{ -\frac{1}{2} \int_0^t ds \left\langle \lambda(s)((e^{-As} K^{-1} e^{+As}) \lambda)(s) \right\rangle_{L^2(\Omega)} \right\}. \qquad (4.56)$$

After evaluating the Gaussian cylindrical measure associated with the Lagrange multiplier fields $\lambda(x,t)$, one obtains our formal path-integral representation, however, with a weight well-defined mathematically as showed in Sec. 4.2, as the main conclusion of Secs. 4.2–4.4

$$Z[j(t)] = \frac{1}{Z(0)} \left\{ \int_{L^\infty((0,T),L^2(\Omega))} \prod dU(x,t) \right.$$

$$\times \exp\left\{ -\frac{1}{2} \int_0^T ds \int_0^T ds' \right.$$

$$\times \left\langle \left(\frac{\partial U}{\partial s} + AU - \Delta F(U) \right)(s); \right.$$

$$\times \left. \left(\frac{\partial U}{\partial s'} + AU - \Delta F(U) \right)(s') \right\rangle_{L^2(\Omega)} \right\}$$

$$\times \left. \exp\left(i \int_0^T ds \langle j(s), U(s) \rangle_{L^2(\Omega)} \right). \right. \tag{4.57}$$

It is worth rewriting the result equation (4.57) in the usual physicist notation of Feynman path-integrals[6]

$$Z[f(x,t)] = \frac{1}{Z(0)} \left\{ \int_{L^\infty((0,T),L^2(\Omega))} D^F[U(x,s)] \right.$$

$$\times \exp\left\{ -\frac{1}{2} \int_0^T ds \int_0^T ds' \int_\Omega dx \left(\frac{\partial U}{\partial t} + AU - \Delta F(U) \right)(x,s) \right.$$

$$\times \left. \left(\frac{\partial U}{\partial t} + AU - \Delta F(U) \right)(x,s') \right\}$$

$$\times \left. \exp\left\{ i \int_0^T ds \int_\Omega dx\, j(s,x) U(s,x) \right\}. \right. \tag{4.58}$$

4.5. Random Anomalous Diffusion, A Semigroup Approach

Recently, it has been an important issue on mathematical physics of diffusion on random porous medium, the study of the anomalous diffusion equation on the full space R^D but under the presence of a positive random potential as modeled by the parabolic quasilinear equation ($0 < t < \infty$) as written below[7]:

$$\frac{\partial U(x,t)}{\partial t} = -D_0(-\Delta)^\alpha U(x,t) + V(x)U(x,t) + F(U(x,t)), \quad (4.59a)$$

$$U(x,0) = f(x), \quad (4.59b)$$

where $U(x,t)$ is the diffusion field, D_0 is the medium diffusibility constant, $V(x)$ denotes the stochastic samples (positive functions) associated with a general random field process with realizations on the space of square-integrable functions (by the Minlo's theorem — see Appendix D), $F(x)$ denotes the problem's nonlinearity represented by a Lipschitzian function as in the previous sections and, finally, $(-\Delta)^\alpha$ represents the effect of the anomalous diffusion of the field $U(x,t)$ on the ambient R^D where the diffusion takes place and it is represented here by a fractional power of the Laplacean operator. The constant α here is called the anomalous diffusion exponent.

In order to give a precise mathematical meaning we will present a different mathematical scheme for the previous sections for Eq. (4.59). Let us, thus, proceed for a moment by rewriting it formally in the weak-integral form for those initial-values $f(x) \in L^2(R^D)$ as in Eq. (4.54), namely,

$$U(t,[V]) = \exp[-t(D_0(-\Delta)^\alpha + V)]f$$

$$+ g \int_0^t ds\{\exp[-(t-s)(D_0(-\Delta)^\alpha + V)]\}F(U(s)), \quad (4.60)$$

where we have used a notation emphasizing the stochastic variable nature of the diffusion field $U(x,t,[V])$ as a functional of the samples $V(x) \in L^2(R^D)$ associated with our random diffusion potential.

Physical quantities are functionals of the diffusion field $U(t,[V])$, and after its determination one should average over these random potential samples. The whole averaging information is contained in the space–time characteristic functional as pointed out in Sec. 4.4:

$$Z[j(x,t)] = \int_{L^2(R^D)} d\mu[V] \times \exp i\left\{\int_{R^D} dx \int_0^\infty dt\, j(x,t) \cdot U(x,t,[V])\right\}. \quad (4.61)$$

Here, $d\mu[V]$ means the random potential cylindrical measure associated with the $V(x)$-characteristic functional

$$Z[h(x)] = \int_{L^2(R^D)} d\mu[V] \exp i(\langle h, v \rangle_{L^2(R^D)}). \quad (4.62)$$

Note that we do not suppose the Gaussianity of the field statistics in Eq. (4.62).

In order to write a functional integral representation for the characteristic functional equation (4.61) as much as similar representation obtained in Sec. 4.4, let us rewrite Eq. (4.61) into the weak-integral form as done in Eq. (4.56):

$$Z[j(x,t)] = \int_{L^2(R^D)} d\mu[V] \int_{L^2(R^D)} D^F[\lambda] \int_{L^2(R^D)} D^F[U]$$

$$\times \exp i\{(\lambda, U + I(F(U)) + e^{-tA}f)_{L^2(R^D)}\} \qquad (4.63)$$

where $I(F(U))$ and e^{-tA} denote the objects on the right-hand side of Eq. (4.60) and the (formal at this point of our study) contractive generator semigroup below:

$$A = D_0(-\Delta)^\alpha + V. \qquad (4.64)$$

At this point one could proceed in a physicist way by rewriting Eq. (4.63) as a path-integral associated with the dynamics of three fields (the well-known Martin–Siggin–Rose component path-integral — Ref. 6):

$$Z[j(x,t)]$$

$$= \int_{L^2(R^D)} d\mu[V] \int_{L^2(R^D)} D^F[\lambda] \int_{L^2(R^D)} D^F[U]$$

$$\times \exp i \left([\lambda, U], \begin{bmatrix} 0 & \frac{1}{2}\left(\frac{\partial}{\partial t} - D_0(-\Delta)^\alpha\right) \\ \frac{1}{2}\left(-\frac{\partial}{\partial t} - D_0(-\Delta)^\alpha\right) & 0 \end{bmatrix} \begin{bmatrix} \lambda \\ U \end{bmatrix} \right)_{L^2(R^D)}$$

$$\times \exp i\, g(\lambda, F(U))_{L^2(R^D)}$$

$$\times \exp i \left([\lambda, U], \begin{bmatrix} 0 & V \\ V & 0 \end{bmatrix} \begin{bmatrix} \lambda \\ U \end{bmatrix} \right)_{L^2(R^D)}. \qquad (4.65)$$

It is worth remarking that in the usual case of the cylindrical measures $d\mu[V]$ is a purely Gaussian measure, formally written in the physicist notation as

$$d\mu[V] = D^F[V] \exp\left\{-\frac{1}{2}\int_{L^2(R^D)} dx \int_{L^2(R^D)} dy\, V(x) K^{-1}(x,y) V(y)\right\}$$

$$(4.66)$$

with $K(x, y)$ defining an integral operator of the trace class on $L^2(R^D)$, one can easily make a further simplification by integrating out exactly the $V(x)$-Gaussian functional integral and producing the effective quartic interaction term as written below:

$$\int_{L^2(R^D)} d\mu[V] \exp i \left([\lambda, U], \begin{bmatrix} 0 & V \\ V & 0 \end{bmatrix} \begin{bmatrix} \lambda \\ U \end{bmatrix} \right)_{L^2(R^D)}$$

$$= \exp \left\{ -\frac{1}{2} \int_{R^D} dx \int_0^\infty dt \int_{R^D} dy \int_0^\infty ds \right.$$

$$\left. \times (\lambda(x,t)U(x,t))K(x,y)(\lambda(y,s)U(y,s)) \right\}. \qquad (4.67)$$

Let us call the reader's attention that at this point one can straightforwardly implement the usual Feynman–Wild–Martin–Siggia–Rosen diagramatics with the free propagator given explicitly in the momentum space by[6]

$$[\partial_t - D_0(-\Delta)^\alpha]^{-1} \leftrightarrow \frac{1}{iw - D_0(|\vec{k}|^{2\alpha})}. \qquad (4.68)$$

It is important to remark that all the above analyses are still somewhat formal at this point of our study since it is based strongly on the hypothesis that the operator A as given by Eq. (4.64) is a C_0-generator of a contractive semigroup on $L^2(R^D)$, which straightforwardly leads to the existence and uniqueness of global solution for Eq. (4.59). Let us give a rigorous proof of ours of such a self-adjointness result, which by its turn will provide a strong connection between the parameter α of the Laplacean power, related to the anomalous diffusion coefficient, and the underline dimension D of the space where the anomalous diffusion is taking place.

Let us first recall the famous Kato–Rellich theorem on self-adjointness perturbative of self-adjoint operators.

Theorem of Kato–Rellich. *Suppose that A is a self-adjoint operator on $L^2(R^D)$, B is a symmetric operator with $Dom(B) \supset Dom(A)$, such that $(0 \leq a < 1)$,*

$$\|B\varphi\|_{L^2(R^D)} \leq a\|A\varphi\|_{L^2(R^D)} + b\|\varphi\|_{L^2(R^D)}, \qquad (4.69)$$

then, $A + B$ is self-adjoint on $Dom(A)$ and essentially self-adjoint on any core of A.

In order to apply this powerful theorem to our general case under study, let us recall that the fractional power of the Laplacean operator, $(-\Delta)^\alpha$, has as an operator domain the (Sobolev) space of all square-integrable functions $\varphi(x)$ such that its Fourier transform $\tilde{\varphi}(k)$ satisfies the condition $|k|^\alpha \tilde{\varphi}(k) \in L^2(R^D)$ and as a core domain the Schwartz space $S(R^D)$.[8]

As a consequence, we have the straightforward estimate for $\varphi \in S(R^D)$ as

$$\|\tilde{\varphi}\|_{L^2(R^D)} \leq \left\|\frac{1}{1+k^{2\alpha}}\right\|_{L^2(\dot{R}^D)} \|(k^{2\alpha}+1)\tilde{\varphi}\|_{L^2(R^D)}$$

$$\leq C(\alpha, D)(\|k^{2\alpha}\hat{\varphi}\|_{L^2(R^D)} + \|\hat{\varphi}\|_{L^2(R^D)}), \qquad (4.70)$$

where the finite constant

$$C(\alpha, D) = \int_{R^D} \frac{dk}{(1+k^{2\alpha})^2} < \infty, \qquad (4.71)$$

if $\alpha > \frac{D}{4}$, a condition relating the "anomalous" diffusion coefficient α and the intrinsic space–time dimensionality D as said before in the introduction. Note that for the physical case of $D = 3$, we have that $\alpha = \frac{3}{4} + \varepsilon$ with $\varepsilon > 0$.

Let us rescale the function $\tilde{\varphi}(k)$ $(r > 0)$ (for $\beta > 0$ arbitrary)

$$\tilde{\varphi}_r(k) = r^\beta \tilde{\varphi}(rk). \qquad (4.72)$$

We thus have

$$\|\tilde{\varphi}_r\|_{L^1(R^D)} = \int_{R^D} dk\, r^\beta \tilde{\varphi}(rk)$$

$$= r^{\beta-D}\|\tilde{\varphi}\|_{L^1(R^D)} \qquad (4.73)$$

and

$$\|\tilde{\varphi}_r\|_{L^2(R^D)} = r^{(2\beta-D)/2}\|\tilde{\varphi}\|_{L^2(R^D)} \qquad (4.74)$$

together with

$$\|K^{2\alpha}\tilde{\varphi}_r\|_{L^2(R^D)} = r^{(2\beta-D-4\alpha)/2}\|K^{2\alpha}\tilde{\varphi}\|_{L^2(R^D)}. \qquad (4.75)$$

Let us substitute Eqs. (4.73)–(4.75) into Eq. (4.70). We arrive at the estimate

$$\|\tilde{\varphi}_r\|_{L^1(R^D)} \leq C(\alpha, D)\big[r^{(2\beta-D-4\alpha)/2}\|K^{2\alpha}\tilde{\varphi}\|_{L^2(R^D)}$$

$$+ r^{(2\beta-D)/2}\|\tilde{\varphi}\|_{L^2(R^D)}\big] \qquad (4.76)$$

or equivalently

$$\|\tilde{\varphi}\|_{L^1(R^D)} \le C(\alpha, D)\Big(r^{\frac{D}{2}-2\alpha}\|K^{2\alpha}\tilde{\varphi}\|_{L^2(R^D)}$$
$$+ r^{\frac{D}{2}}\|\tilde{\varphi}\|_{L^2(R^D)}\Big). \tag{4.77}$$

Let us now estimate the function $V\varphi$ on the space $L^2(R^D)$:

$$\|V\varphi\|_{L^2(R^D)} \le \|V\|_{L^2(R^D)}\|\varphi\|_{L^\infty(R^D)}$$
$$\le \|V\|_{L^2(R^D)}\|\tilde{\varphi}\|_{L^1(R^D)}$$
$$\le (\|V\|_{L^2(R^D)}C(\alpha, D) \cdot r^{\frac{D}{2}-2\alpha})\|K^{2\alpha}\tilde{\varphi}\|_{L^2(R^D)}$$
$$+ (\|V\|_{L^2(R^D)}C(\alpha, D) \cdot r^{\frac{D}{2}})\|\tilde{\varphi}\|_{L^2(R^D)}. \tag{4.78}$$

Now one can just choose r such that

$$\|V\|_{L^2(R^D)} \times C(\alpha, D) \times r^{(D-4\alpha)/2} < 1, \tag{4.79}$$

in order to obtain the validity of the bound $a < 1$ on Eq. (4.69) and, thus, concluding that $(-\Delta)^\alpha + V$ is an essentially self-adjoint operator on $S(R^D)$. So, its closure on $L^2(R^D)$ produces the operator used in the above-exposed semigroup C_0-construction for $V(x)$ being a positive function almost everywhere. This result of ours is a substantial generalization of Theorem X.15 presented in Ref. 8. Note that the $V(x)$ sample positivity is always the physical case of random porosity in a Porous medium. In such a case, the random potential is of the exponential form $V(x) = V_0(\exp\{ga(x)\})$, where g is a positive small parameter and V_0 is a background porosity term. The effective cylindrical measure on the random porosity parameters is written as (see Eq. (4.66))

$$d\mu[V] \underset{g\ll 1}{\cong} D^F[a(x)]\exp\left\{-\frac{1}{2}\int_{R^D} dx \int_{R^D} dy\, e^{ga(x)}K^{-1}(x,y)e^{ga(y)}\right\}. \tag{4.80}$$

The above-exposed C_0-semigroup study complements the study on purely nonlinear diffusion made in the previous sections purely by compacity arguments.

Finally, let us briefly sketch on the important case of wave propagation $U(x,t)$ in a random medium described by a small stochastic damping positive $L^2(R^D)$ function $\nu(x)$, such that $\frac{\nu^2(x)}{4}$ is a Gaussian process sampled in the $L^2(R^D)$ space. The hyperbolic linear equation governing such physical wave propagation is given by (see Sec. 4.3)

$$\frac{\partial^2(U(x,t))}{\partial^2 t} + \nu(x)\frac{\partial U(x,t)}{\partial t} = -(-\Delta)^\alpha U(x,t), \tag{4.81}$$

$$U(x,0) = f(x) \in L^2(R^D), \tag{4.82}$$

$$U_t(x,0) = g(x) \in L^2(R^D), \tag{4.83}$$

where

$$E\left(\frac{\nu^2(x)}{4}\frac{\nu^2(y)}{4}\right) = K(x,y) \in \oint_1(L^2(R^D)), \tag{4.84}$$

One can see that after the variable change

$$U(x,t,[\nu]) = e^{-\frac{\nu(x)}{2}t} \cdot \Phi(x,t,[\nu]), \tag{4.85}$$

the damped-stochastic wave equation takes the suitable form of a wave in the presence of a random potential

$$\frac{\partial^2\Phi(x,t)}{\partial^2 t} = -(-\Delta)^\alpha\Phi(x,t) + \left(\frac{\nu^2(x)}{4}\right)\Phi(x,t),$$

$$\Phi(x,0) = f(x), \tag{4.86}$$

$$\Phi_t(x,0) = \frac{\nu(x)}{2}f(x) + g(x) = \bar{g}(x).$$

Equation (4.86) has an operatorial scalar weak-integral solution for $t > 0$ of the following form:

$$\Phi(t) = \cos(t\sqrt{A})f + \frac{\sin(t\sqrt{A})}{\sqrt{A}}\bar{g}, \tag{4.87}$$

where the nonpositive self-adjoint operator A is given explicitly by

$$A = (-\Delta)^\alpha - \frac{\nu^2(x)}{4}. \tag{4.88}$$

As a conclusion, one can see that anomalous diffusion can be handled mathematically in the framework of our previous technique of path-integrals with constraints.

References

1. J. Wloka, *Partial Differential Equations* (Cambridge University Press, 1987).
2. Y. Yamasaki, *Measures on Infinite Dimensional Spaces*, Series in Pure Mathematics, Vol. 5 (World Scientific, 1985); B. Simon, *Functional Integration and Quantum Physics* (Academic Press, 1979); J. Glimn and A. Jaffe, *Quantum*

Physics — A Functional Integral Point of View (Springer Verlag, 1981); Xia Dao Xing, *Measure and Integration Theory on Infinite Dimensional Space* (Academic Press, 1972); L. Schwartz, *Random Measures on Arbitrary Topological Space and Cylindrical Measures* (Tata Institute, Oxford University Press, 1973); Ya G. Sinai, *The Theory of Phase Transitions: Rigorous Results* (London, Pergamon Press, 1981).

3. O. A. Ladyzenskaja, V. A. Solonninkov and N. N. Ural'ceva, *Linear and Quasilinear Equations of Parabolic Type* (Amer. Math. Soc. Transl., A.M.S., Providence, RI, 1968).

4. S. G. Mikhlin, *Mathematical Physics, An Advanced Course*, Series in Applied Mathematics and Mechanics (North-Holland, 1970).

5. L. C. L. Botelho, *J. Phys. A: Math. Gen.*, **34**, L131–L137 (2001); *Mod. Phys. Lett.* **16B**(21), 793–8-6, (2002); A. Bensoussan and R. Teman, *J. Funct. Anal.* **13**, 195–222 (1973); A. Friedman, *Variational Principles and Free-Boundary Problems*, Pure & Applied Mathematics (John-Wiley & Sons, NY, 1982).

6. L. C. L. Botelho, *Il Nuovo Cimento*, **117B**(1), 15 (2002); *J. Math. Phys.* **42**, 1682 (2001); A. S. Monin and A. M. Yaglon, *Statistical Fluid Mechanics*, Vol. 2 (MIT Press, Cambridge, 1971); G. Rosen, *J. Math. Phys.* **12**, 812 (1971); L. C. L. Botelho, *Mod. Phys. Lett.* **13B**, 317 (1999); A. A. Migdal, *Int. J. Mod. Phys. A* **9**, 1197 (1994); V. Gurarie and A. Migdal, *Phys. Rev. E* **54**, 4908 (1996); U. Frisch, *Turbulence* (Cambridge Press, Cambridge, 1996); W. D. Mc-Comb, *The Physics of Fluid Turbulence* (Oxford University, Oxford, 1990). L. C. L. Botelho, *Mod. Phys. Lett. B* **13**, 363 (1999).

7. L. C. L. Botelho, *Nuov. Cimento* **118B**, 383 (2004); S. I. Denisov and W. Horsthemke, *Phys. Lett. A* **282**(6), 367 (2001); L. C. L. Botelho, *Int. J. Mod. Phys. B* **13**, 1663 (1999); L. C. L. Botelho, *Mod. Phys. Lett. B* **12**, 301 (1998); J. C. Cresson and M. L. Lyra, *J. Phys. Condens. Matter* **8**(7), L83 (1996); L. C. L. Botelho, *Mod. Phys. Lett. B* **12**, 569 (1998); L. C. L. Botelho, *Mod. Phys. Lett. B* **12**, 1191 (1998); L. C. L. Botelho, *Mod. Phys. Lett. B* **13**, 203 (1999); L. C. L. Botelho, *Int. J. Mod. Phys. B* **12**, 2857 (1998); L. C. L. Botelho, *Phys. Rev.* **58E**, 1141 (1998).

8. M. Reed and B. Simon, *Methods of Modern Mathematical Physics*, Vol 2 (Academic Press, New York, 1980).

Appendix A

In this mathematical appendix, we intend to give a rigorous mathematical proof of the result used on Sec. 4.2 about parabolic nonlinear diffusion in relation to the weak continuity of the nonlinearity $\Delta F(U)$ used in the passage of Eq. (4.17) for Eq. (4.18).

First, we consider the physical hypothesis on the nonlinearity of the parabolic equation (4.1) that the Laplacean operator has a cutoff in its spectral range, namely,

$$\Delta F(U(x,t)) \to \Delta^{(\Lambda)} F(U(x,t)), \tag{A.1}$$

where the regularized Laplacean $\Delta^{(\Lambda)}$ means the bounded operator of norm Λ, i.e. in terms of the spectral theorem of the Laplacean

$$\Delta = \int_{-\infty}^{+\infty} \lambda \, dE_s(\lambda), \tag{A.2}$$

it is given by

$$\Delta^{(\Lambda)} = \int_{|\lambda| < \Lambda} \lambda \, dE_s(\lambda). \tag{A.3}$$

Let us impose either the well-known path-integral Sturm–Liouville time-boundary conditions imposed on the elements of the path-integral domain (Eq. (4.58)), when one is defining the path-integral by means of fluctuations around classical configurations

$$\bar{U}(x,T) = \bar{U}(x,0) = 0. \tag{A.4}$$

It is straightforward to see the validity of the following chain of inequalities with $a > 1$:

$$\int_0^T dt \left\| \frac{dU_n(t)}{dt} \right\|_{L^2(\Omega)}^2 \le \text{Real}((AU_n(t), U_n(t)) - (AU_n(0), U_n(0)))$$

$$+ \int_0^T dt \left\| \Delta^{(\Lambda)} F(U_n(t)) \cdot \frac{dU_n}{dt} \right\|_{L^2(\Omega)}$$

$$\le 0 + 0 + a \left(\int_0^T \|\Delta F(U_n(t))\|_{L^2(\Omega)}^2 \, dt \right)$$

$$+ \frac{1}{a} \left(\int_0^T dt \left\| \frac{dU_n}{dt} \right\|_{L^2(\Omega)}^2 \right). \tag{A.5}$$

In other words,

$$\left(1 - \frac{1}{a} \right) \left(\int_0^T dt \left\| \frac{dU_n}{dt} \right\|_{L^2(\Omega)}^2 \right)$$

$$\le ca\Lambda^2 \cdot \int_0^T \|U_n(t)\|_{L^2(\Omega)}^2$$

$$\le Ca\Lambda^2 T M, \tag{A.6}$$

where $c = \sup_{a \le x \le b} |F'(x)|$, and M is the bound given by Eq. (4.16).

Now, by a direct application of the well-known Aubin–Lion theorem,[1] one has that the sequence $\{U_n\}$ is a compact set on $L^2(\Omega)$. So, it converges strongly to $\bar{U}$ on $L^2(\Omega)$ as a consequence of the Lipschitzian property of $F(x)$.

We have either the strong convergence of $F(U_n)$ on $F(\bar{U})$, since for each t we have that the inequality below holds true:

$$\int_\Omega dx |F(U_n) - F(\bar{U})|^2_{L^2(\Omega)}(t)$$

$$\leq \left(\sup_{-\infty < x < +\infty} (F'(x)) \right)^2 \left(\int_\Omega dx |U_n - \bar{U}|^2_{L^2(\Omega)}(t) \right) \to 0. \tag{A.7}$$

It is, thus, an immediate result of the validity of the weak convergence on $L^2(\Omega)$ used on Eq. (4.18):

$$\lim_{n\to\infty} \int_0^T dt (F(U_n), \Delta v)_{L^2(\Omega)} = \int_0^T dt (F(\bar{U}), \Delta v)_{L^2(\Omega)}. \tag{A.8}$$

Appendix B

In this appendix, we show that the function $U(x,t) \in L^\infty([0,T], L^2(\Omega))$, the weak solution of Eq. (4.1) in the test space $D((0,T), L^2(\Omega))$ (the Schwart space of $L^2(\Omega)$-valued test functions), satisfies the initial condition (Eq. (4.2)). In order to show such a result, let us consider a test function possessing the following form for each $\varepsilon > 0$: $v_i^{(\varepsilon)}(x,t) = \varphi_i(x) \left(\chi_{[0,\varepsilon]}(t)/\varepsilon \right)$, where $\varphi_i(x)$ is a member of the spectral set associated with the self-adjoint operator $A \colon H^2(\Omega) \cap H^1_0(\Omega) \to L^2(\Omega)$. The notation $\chi_{[a,b]}(t)$ denotes the characteristic function of the interval $[a,b]$.

Since $U_m(x,t)$ converges to $U(x,t)$ in the weak-star topology of $L^\infty((0,T), L^2(\Omega))$ and $D((0,T), L^2(\Omega)) \subset L^1((0,T), L^2(\Omega))$, we have the relation below for each $v_i^{(\varepsilon)}(x,t)$ that holds true

$$\int_0^T dt \langle U_m(x,t), v_i^{(\varepsilon)}(x,t) \rangle_{L^2(\Omega)} \to \int_0^T dt \langle U(x,t), v_i^{(\varepsilon)}(x,t) \rangle_{L^2(\Omega)}, \tag{B.1}$$

or equivalently

$$\frac{1}{\varepsilon} \int_0^\varepsilon dt \langle U_m(x,t), \varphi_i(x) \rangle_{L^2(\Omega)} \to \frac{1}{\varepsilon} \int_0^\varepsilon dt \langle U(x,t), \varphi_i(x) \rangle_{L^2(\Omega)}. \tag{B.2}$$

By means of the mean-value theorem applied to both sides of Eq. (A.2), and taking the limit $\varepsilon \to 0$, we have for each $i \in \mathbb{Z}$ that

$$\langle U_m(x,0), \varphi_i(x) \rangle_{L^2(\Omega)} \to \langle U(x,0), \varphi_i(x) \rangle_{L^2(\Omega)} \qquad (\text{B.3})$$

as a direct consequence of our choice of the Galerkin elements Eqs. (4.6) whose functional form does not change on the function of the order $m \in \mathbb{Z}$.

As a consequence of Eqs. (4.5) and (4.6), it yields the result

$$\langle g(x), \varphi_i \rangle_{L^2(\Omega)} = \langle U(x,0), \varphi_i(x) \rangle_{L^2(\Omega)}, \qquad (\text{B.4})$$

or by means of Parserval–Fourier theorem,

$$\lim_{t \to 0} \| U(x,t) - g(x) \|_{L^2(\Omega)} = 0 \qquad (\text{B.5})$$

result, which by its turn, means that the function $U(x,t)$ obtained by means of our compacity technique satisfies the initial-condition Eq. (4.2) in the $L^2(\Omega)$-mean sense.

Appendix C

Let us show that the functions $\bar{U}(x,t) \in L^\infty([0,T], L^2(\Omega))$ given by Eq. (4.41) and $\bar{P}(x,t) \in L^\infty([0,T], L^2(\Omega))$ — Eq. (4.42) are *coincident as elements* of the above-written functional space of $L^2(\Omega)$ — valued essential bounded functions on $(0,T)$.

To verify such a result, let us call the reader's attention that since $U_m(x,t)$ is weakly convergent to $U(x,t)$ in $L^\infty([0,T], L^2(\Omega))$, we have that the set $\{U_m(x,t)\}$ is convergent to the function $U(x,t)$ as a Schwartz distribution on $L^2(\Omega)$ since $D([0,T], L^2(\Omega)) \subset L^1([0,T], L^2(\Omega))$.

This means that

$$U_m(x,t) \to U(x,t) \quad \text{in } D'([0,T], L^2(\Omega)). \qquad (\text{C.1})$$

Analogous result holds true for the time-derivative of the above-written equation as a result of $U(x,t)$ being a function

$$\frac{\partial}{\partial t} U_m(x,t) \to \frac{\partial}{\partial t} U(x,t) \quad \text{in } D'([0,T], L^2(\Omega)). \qquad (\text{C.2})$$

On the other side, the set $\left\{ \frac{\partial U_m(x,t)}{\partial t} \right\}$ converges weakly in $L^1([0,T],$ $L^2(\Omega))$ to $\bar{P}(x,t) \in L^\infty([0,T], L^2(\Omega))$, which, by its turn, means that

$$\frac{\partial U_m(x,t)}{\partial t} \to \bar{P}(x,t) \quad \text{in } D'([0,T], L^2(\Omega)). \qquad (\text{C.3})$$

By the uniqueness of the limit on the distributional space $D'([0,T],L^2(\Omega))$, we have the coincidence of $\frac{\partial U(x,t)}{\partial t}$ and $\bar{P}(x,t)$ as elements of $D'([0,T],L^2(\Omega))$. However, $\bar{P}(x,t)$ is a function; so by general theorems on Schwartz distribution theory $\left\{\frac{\partial U(x,t)}{\partial t}\right\}$ must be a function either since $L^2(\Omega)$ is a separable Hilbert space. As a consequence we have that $\frac{\partial U(x,t)}{\partial t} = \bar{P}(x,t)$ as elements of $L^\infty([0,T],L^2(\Omega))$, which is the result searched:

$$\frac{\partial U(x,t)}{\partial t} = \bar{P}(x,t) \quad \text{a.e. in } ([0,T] \times \Omega. \tag{C.4}$$

Appendix D. Probability Theory in Terms of Functional Integrals and the Minlos Theorem — An Overview

In this complementary appendix, we discuss briefly for the mathematics oriented reader the mathematical basis and concepts of functional integrals formulation for stochastic process and the important Minlos theorem on Hilbert space support of probabilistic measures.

The first basic notion of Kolmogorov's probability theory framework is to postulate a convenient topological space (Polish spaces) Ω formed by the phenomena random events. The chosen topology of Ω should possess a rich set of nontrivial *compact* subsets. After that, we consider the σ-algebra generated by all open sets of this — the so-called topological sample space, denoted by $\mathcal{F}_\Omega$. Thirdly, one introduces a regular measure $d\mu$ on this set algebra $\mathcal{F}_\Omega$, assigning values in $[0,1]$ to each member of $\mathcal{F}_\Omega$.

The (abstract) triplet $(\Omega, \mathcal{F}_\Omega, d\mu)$ is thus called a probability space in the Kolmogorov–Schwartz scheme.[10]

Let $\{X_\alpha\}_{\alpha\in\Lambda}$ be a set of measurable real functions on Ω, which may be taken as continuous injective functions of compact support on Ω (without loss of generality).

It is a standard result that one can "immerse" (represent) the abstract space Ω on the "concrete" infinite product compact space $R^\infty = \prod_{\alpha\in\Lambda}(\dot{R}_\alpha)$, where $\dot{R}$ is a compactified copy of the real line. In order to achieve such result we consider the following injection of Ω in R^∞ defined by the family $\{X_\alpha\}_{\alpha\in\Lambda}$:

$$\begin{aligned} I: \Omega &\to R^\infty \\ \omega &\to \{X_\alpha(\omega)\}_{\alpha\in\Lambda}. \end{aligned} \tag{D.1}$$

A new σ-algebra of events on Ω can be induced on Ω and affiliated to the family $\{X_\alpha\}_{\alpha \in \Lambda}$.

It is the σ-algebra generated by all the "finite-dimensional cylinders" subsets, which are explicitly given in its generic formulas by

$$C_{\Lambda_{\text{fin}}} = \Big\{ \omega \in \Omega \mid \Lambda_{\text{fin}} \text{ is a finite subset of } \Lambda$$

$$\text{and } [(X_\alpha)_{\alpha \in \Lambda_{\text{fin}}}(\Omega)] \subset \prod_{\alpha \in \Lambda_{\text{fin}}} [a_\alpha, b_\alpha] \Big\}.$$

The measure restriction of the initial measure μ on this cylinder sets of Ω will be still denoted by μ on what follows. Now it is a basic theorem of Probability theory that μ induces a measure $\nu^{(\infty)}$ on $\prod_{\alpha \in \Lambda}(\dot{R})_\alpha$, which can be identified with the space of all functions of the index set Λ to the real compactified line $\dot{R}$; $F(\Lambda, R)$ with the topology of pontual convergence.

At this point, it is straightforward to see that the average of any Borelian (mensurable) function $G(\omega)$ (a random variable) is given by the following functional integral on the functional space $F(\Lambda, R)$:

$$\int_\Omega d\mu(\omega) G(\omega) = \int_{F(\Lambda, R)} d\nu^{(\infty)}(f) \cdot G(I^{-1}(f)). \qquad (D.2)$$

It is worth recalling that the real support of the measure $d\nu^{(\infty)}(f)$ in most practical cases is not the whole space $F(\Lambda, R)$ but only a functional subspace of $F(\Lambda, R)$. For instance, in the most practical cases Λ is the index set of an algebraic vectorial base of a vector space E, and $\prod_{\alpha \in \Lambda}(R)_\alpha$ (without the compactification) is the space of all sequences with only a finite number of nonzero entries as it is necessary to consider in the case of the famous extension theorem of Kolmogorov. It turns out that one can consider the support of the probability measure as the set of all linear functionals on E, the so-called algebraic dual of E. This result is a direct consequence of considering the set of random variables given by the family $\{e_\alpha^*\}_{\alpha \in \Lambda}$. For applications, one is naturally lead to consider the probabilistic object called characteristic functional associated to the measure $d\nu^{(\infty)}(f)$ when $F(\Lambda, R)$ is identified with the vector space of all algebraic linear functionals E^{alg} of a given vector space E as pointed out above, namely,

$$Z[j] = \int_{E^{\text{alg}}} d\nu^{(\infty)}(f) \exp(if(j)), \qquad (D.3)$$

where j are elements of E written as $j = \sum_{\alpha \in \Lambda_{\text{fin}}} x_\alpha e_\alpha$ with $\{e_\alpha\}_{\alpha \in \Lambda}$ denoting a given Hammel (vectorial) basis of E.

One can show that if there is a subspace H of E with a norm $\| \ \|_H$ coming from an inner product $(,)_H$ such that

$$\int_{E \cap H} d\nu^{(\infty)}(f) \cdot \|f\|_H^2 < \infty \tag{D.4}$$

one can show that the support of the measure of exactly the Hilbert space $(H, (,)_H)$ is the famous Minlos theorem.

Another important theorem (the famous Wiener theorem) is when we have the following condition for the family of random variables with the index set Λ being a real interval of the form $\Lambda = [a, b]$

$$\int_E d\nu^{(\infty)}(f) |X_{t+h}(f) - X_t(f)|^p \le c|h|^{1+\varepsilon} \tag{D.5}$$

for p, ε, c positive constants. We have that it is possible to choose as the support of the functional measure $d\nu^{(\infty)}(\cdot)$ the space of the real continuous function on Λ the well-known famous Brownian path space

$$\int_\Omega d\mu(\omega)G(\omega) = \int_{C([a,b],R)} d\nu^{(\infty)}(f(x))\tilde{G}(f(x)). \tag{D.6}$$

For rigorous proofs and precise formulations of the above-sketched deep mathematical results, the interested reader should consult the basic mathematical book on *Random Measures on Arbitrary Topological Spaces and Cylindrical Measures* by L. Schwartz (Tata Institute, Oxford University Press, 1973).[10]

Let us now finally give a concrete example of such results considering a generating functional $Z[J(x)]$ of the exponential quadratic functional form on $L^2(R^N)$:

$$Z[J(x)] = \exp\left\{-\frac{1}{2}\int_{R^N} d^N x \int_{R^N} d^N y j(x) K(x,y) j(y)\right\} \tag{D.7}$$

and associate to a Gaussian random field with two-point correlation function given by the kernel of Eq. (D.7):

$$E\{v(x)v(y)\} = K(x,y). \tag{D.8}$$

In the case of the above-written kernel being a class-trace operator on $L^2(R^N)$ ($\int_{R^N} d^N x K(x,x) < \infty$), one can represent Eq. (B.7) by means of a Gaussian measure with the support of the Hilbert space $L^2(R^N)$. This result was called on for the reader's attention for the bulk of this paper.

Chapter 5

Domains of Bosonic Functional Integrals and Some Applications to the Mathematical Physics of Path-Integrals and String Theory*

5.1. Introduction

Since the result of R. P. Feynman on representing the initial value solution of Schrodinger equation by means of an analytically time continued integration on an infinite-dimensional space of functions, the subject of Euclidean functional integrals representations for quantum systems has become the mathematical-operational framework to analyze quantum phenomena and stochastic systems as shown in the previous decades of research on theoretical physics.[1-3]

One of the most important open problem in the mathematical theory of Euclidean functional integrals is related to the implementation of sound mathematical approximations to these infinite-dimensional integrals by means of finite-dimensional approximations outside of the always used [computer oriented] space-time lattice approximations (see Refs. 2 and 3 — Chap. 9). As a first step to tackle the above-cited problem, there is a need to mathematically characterize the functional domain where these functional integrals are defined.

The purpose of this chapter is to present, in Sec. 5.2, the formulation of Euclidean quantum field theories as functional Fourier transforms by means of the Bochner–Martin–Kolmogorov theorem for topological vector spaces (Refs. 4 and 5 — Theorem 4.35) and suitable to define and analyze rigorously the functional integrals by means of the well-known Minlos theorem (Ref. 5, Theorem 4.31 and Ref. 6, Part 2) and presented in full details in Sec. 5.3.

In Sec. 5.4, we present new results on the difficult problem of defining rigorously infinite-dimensional quantum field path-integrals in general space times $\Omega \subset R^{\nu}(\nu = 2, 4, \dots)$ by means of the analytical regularization scheme.

---

*Author's original results.

5.2. The Euclidean Schwinger Generating Functional as a Functional Fourier Transform

The basic object in a scalar Euclidean quantum field theory in R^D is the Schwinger generating functional (see Refs. 1 and 3).

$$Z[j(x)] = \left\langle \Omega_{VAC} \left| \exp\left(i \int d^D x j(\vec{x}, it) \phi^{(m)}(\vec{x}, it) \right) \right| \Omega_{VAC} \right\rangle, \qquad (5.1)$$

where $\phi^{(m)}(\vec{x}, it)$ is the supposed self-adjoint Minkowski quantum field analytically continued to imaginary time and $j(x) = j(\vec{x}, it)$ is a set of functions belonging to a given topological vector space of functions denoted by E which topology is not specified yet and will be called the Schwinger classical field source space. It is important to remark that $\{\phi^m(\vec{x}, it)\}$ is a commuting algebra of self-adjoints operators as Symanzik has pointed out.[7]

In order to write Eq. (5.1) as an integral over the space E^{alg} of all linear functionals on the Schwinger source space E (the so-called algebraic dual of E), we take the following procedure, different from the usual abstract approach (as given — for instance — in the proof of IV-11-2), by making the hypothesis that the restriction of the Schwinger generating functional equation (5.1) to any finite-dimensional R^N of E is the Fourier transform of a positive continuous function, namely

$$Z\left(\sum_{\alpha=1}^{N} C_\alpha \vec{j}_\alpha(x) \right) = \int_{R^N} \exp\left(i \sum_{\alpha=1}^{N} C_\alpha P_\alpha \right) \tilde{g}(P_1, \ldots, P_N) dP_1, \ldots, dP_N.$$
$$(5.2)$$

Here, $\{\vec{j}_\alpha(x)\}_{\alpha=1,\ldots,N}$ is a fixed vectorial basis of the given finite-dimensional subspace (isomorphic to R^N) of E.

As a consequence of the above-made hypothesis (based physically on the renormalizability and unitary of the associated quantum field theory), one can apply the Bochner–Martin–Kolmogorov theorem (Ref. 5, Theorem 4.35) to write Eq. (5.1) as a functional Fourier transform on the space E^{alg} (see Appendix A)

$$Z[j(x)] = \int_{E^{alg}} \exp(ih(j(x))d\mu(h), \qquad (5.3)$$

where $d\mu(h)$ is the Kolmogorov cylindrical measure on $E^{alg} = \Pi_{\lambda \in A}(R^\lambda)$ with A denoting the index set of the fixed Hamel vectorial basis used in

Eq. (5.2) and $h(j(x))$ is the action of the given linear (algebric) functional (belonging to E^{alg}) on the element $j(x) \in E$.

At this point, we relate the mathematically non-rigorous physicist point of view to the Kolmogorov measure $d\mu(h)$ (Eq. (5.3)) over the algebraic linear functions on the Schwinger source space. It is formally given by the famous Feynman formulae when one identifies the action of h on E by means of an "integral" average

$$h(j) = \int_{R^D} dx^D j(x) h(x). \qquad (5.4)$$

Formally, we have the equation

$$d\mu(h) = \left(\prod_{x \in R^D} dh(x) \right)' \exp\{-S(h(x))\}, \qquad (5.5)$$

where S is the classical action of the classical field theory under quantization, but with the necessary coupling constant renormalizations needed to make the associated quantum field theory well-defined.

Let us outline these proposed steps on a $\lambda\phi^4$-field theory on R^4.

At first we will introduce the massive free field theory generating functional directly into the infinite volume space R^4.

$$Z[j(x)] = \exp\left\{ -\frac{1}{2} \int d^4x d^4x' j(x)((-\Delta)^\alpha + m^2)^{-1}(x,x')j(x') \right\}, \qquad (5.6)$$

where the free field propagator is given by

$$((-\Delta)^\alpha + m^2)^{-1}(x,x') = \int d^4k \frac{e^{ik(x-x')}}{k^{2\alpha} + m^2}, \qquad (5.7)$$

with α a regularizing parameter with $\alpha > 1$.

As the source space, we will consider the vector space of all real sequences on $\prod_{\lambda \in (-\infty,\infty)}(R)^\lambda$, but with only a finite number of non-zero components. Let us define the following family of finite-dimensional positive linear functionals $\{L_{\Lambda_f}\}$ on the functional space $C(\prod_{\lambda \in (-\infty,\infty)} R^\lambda; R)$

$$L_{\Lambda_f}\left(g(P_{\lambda_{s_1}}, \ldots, P_{\lambda_{s_N}})\right) = \int_{\left(\prod_{\lambda \in \Lambda_f} R^\lambda\right)} g(P_{\lambda_{s_1}}, \ldots, P_{\lambda_{s_N}})$$

$$\times \exp\left\{-\frac{1}{2} \sum_{\lambda \in \Lambda_f} (\lambda^{2\alpha} + m^2)(P_\lambda)^2\right\}$$

$$\times \left(\prod_{\lambda \in \Lambda_f} d(P_\lambda \sqrt{\pi(\lambda^{2\alpha} + m^2)})\right). \tag{5.8}$$

Here, $\Lambda_f = \{\lambda_{s_1}, \ldots, \lambda_{s_N}\}$ is an ordered sequence of real number of the real line which is the index set of the Hamel basis of the algebraic dual of the proposed source space.

Note that we have the generalized eigenproblem expansion

$$((-\Delta)^\alpha + m^2)e^{i\lambda x} = (\lambda^{2\alpha} + m^2)e^{i\lambda x}. \tag{5.9}$$

By the Stone–Weierstrass theorem or the Kolmogoroff theorem applied to the family of finite-dimensional measure in Eq. (5.8), there is a unique extension measure $d\mu(\{P_\lambda\}_{\lambda \in (-\infty,\infty)})$ to the space $\Pi_{\lambda \in (-\infty,\infty)} R^\lambda = E^{\text{alg}}$ and representing the infinite-volume generating functional on our chosen source space (the usual Riesz–Markov theorem applied to the linear functional $L = \lim_{\{\Lambda_f\}} \sup L_{\Lambda_f}$ on $C(\Pi_{\lambda \in (-\infty,\infty)} R^\lambda, R)$ leads to this extension measure).[10]

$$Z[j(x)] = Z[\{j_\lambda\}_{\lambda \in \Lambda_f}]$$

$$= \int_{\left(\prod_{\lambda \in (-\infty,\infty)} R^\lambda\right)} d\mu^{(0)\cdot(\alpha)}(\{P_\lambda\}_{\lambda \in (-\infty,\infty)}) \exp\left(i \sum_{\lambda \in (-\infty,\infty)} j_\lambda P_\lambda\right)$$

$$= \exp\left\{-\frac{1}{2} \sum_{\lambda \in \Lambda_f} \frac{(j_\lambda)^2}{\lambda^{2\alpha} + m^2}\right\}. \tag{5.10}$$

At this point it is very important to remark that the generating functional equation (5.10) has continuous natural extension to any test space $(S(R^N), D(R^N)$, etc.) which contains the continuous functions of compact support as a dense subspace.

At this point we consider the following quantum field interaction functional which is a measurable functional in relation to the above-constructed Kolmogoroff measure $d\mu^{(0)\cdot(\alpha)}(\{P_\lambda\}_{\lambda \in (-\infty,\infty)})$ for α non-integer in the original field variable $\phi(x)$

$$V^{(\alpha)}(\phi) = \lambda_R \phi^4 + \frac{1}{2}(Z_\phi^{(\alpha)}(\lambda_R, M) - 1)\phi((-\Delta)^\alpha)\phi - \frac{1}{2}[(m^2 Z_\phi^{(\alpha)}(\lambda_R, m) - 1$$
$$- (\delta m^2)^{(\alpha)}(\lambda_R)]\phi^2 - [Z_\phi^{(\alpha)}(\lambda_R, m)(\delta^{(\alpha)}\lambda)(\lambda_R, m)]\phi^4. \qquad (5.11)$$

Here, the renormalization constants are given in the usual analytical finite-part regularization form for a $\lambda\phi^4$-field theory. It is still an open problem in the mathematical-physics of quantum fields to prove the integrability in some distributional space of the cut-off removing $\alpha \to 1$ limit of the interaction lagrangean $\exp(-V^{(\alpha)}(\phi))$ (see Sec. 5.4 for the analysis of this cut-off removing on space functions).

5.3. The Support of Functional Measures — The Minlos Theorem

Let us now analyze the measure support of quantum field theories generating functional equation (5.3).

For higher dimensional space-time, the only available result in this direction is the case that we have a Hilbert structure on E.[4-6]

At this point, we introduce some definitions. Let $\varphi : \mathbb{Z}^+ \to R$ be an increasing fixed function (including the case $\varphi(\infty) = \infty$). Let E be denoted by H and H^Z be the subspace of $H^{\text{alg}} = (\Pi_{\lambda \in A = [0,1]} R^\lambda)$ (with A being the index set of a Hamel basis of H), formed by all sequences $\{x_\lambda\}_{\lambda \in A} \in H^{\text{alg}}$ with coordinates different from zero at most a countable number

$$H^Z = \{(x_\lambda)_{\lambda \in A} | x_\lambda \neq 0 \text{ for } \lambda \in \{\lambda_\mu\}_{\mu \in \mathbb{Z}}\}. \qquad (5.12)$$

Consider the following weighted subset of H^{alg}:

$$H_{(e)}^Z = \{\{x_\lambda\}_{\lambda \in A} \in H^Z\}$$

and

$$\lim_{N \to \infty} \left\{ \frac{1}{\varphi(N)} \sum_{n=1}^{N} (x_{\lambda\sigma(\mu)})^2 \right\} < \infty$$

for any $\sigma : N \to N$, a permutation of the natural numbers.

We now state our generalization of the Minlos theorem.

Theorem 5.1. *Let T be an operator with domain $D(T) \subset H$, and $T; D(T) \to H$ such that for any finite-dimensional space $H^N \subset H$, the sum is bounded by the function $\varphi(N)$*

$$\left(\sum_{(i,j)=1}^{N} \langle Te_i, Te_j \rangle^{(0)} \right) \le \varphi(N). \tag{5.13}$$

Here $\langle , \rangle^{(0)}$ is the inner product of H and $\{e_p\}_{1 \le p \le N}$ is a vectorial basis of the subspace H^N with dimension N.

Suppose that $Z[j(x)]$ is a continuous function on $D(T) = \overline{(D(T), \langle , \rangle^{(1)})}$, where $\langle , \rangle^{(1)}$ is a new inner product defined by the operator $T(\langle j, \bar{j} \rangle^{(1)} = \langle Tj, T\bar{j} \rangle^{(0)})$. We have, thus, that the support of the cylindrical measure equation (5.3) is the measurable set H_e^Z.

Proof. Following closely Refs. 1, Theorems 2.2 and 4, let us consider the following representation for the characteristic function of the measurable set $H_e^Z \subset H^{\text{alg}}$

$$\begin{cases} \cdot X_{H_e^Z}(\{x_\lambda\}_{\lambda \in A}) \\ = \lim_{\alpha \to 0} \lim_{N \to \infty} \exp \left\{ -\frac{1\alpha}{2\varphi(N)} \sum_{\ell=1}^{N} x_{\lambda_l}^2 \right\} = 1 \quad \text{if} \quad \lim_{N \to \infty} \frac{1}{\varphi(N)} \sum_{\ell=1}^{N} x_{\lambda_l}^2 < \infty \\ 0 \qquad\qquad\qquad\qquad\qquad\qquad\qquad\qquad\qquad\qquad \text{otherwise} \end{cases} \tag{5.14}$$

Now, its measure satisfies the following inequality:

$$\int_{H^{\text{alg}}} d\mu(h) = \mu(H^{\text{alg}}) = 1 > \mu(H_e^Z). \tag{5.15}$$

But

$$\mu(H_e^Z) = \lim_{\alpha \to 0} \lim_{N \to \infty} \int_{H^{\text{alg}}} d\mu(h) \exp - \left\{ \frac{\alpha}{2\varphi(N)} \sum_{\ell=1}^{N} x_{\lambda_\ell}^2 \right\}$$

$$= \lim_{\alpha \to 0} \lim_{N \to \infty} \left\{ \frac{1}{(\frac{2\pi\alpha}{\varphi(N)})^{N/2}} \right\} \int_{R^N} dj_1, \ldots, dj_N$$

$$\times \exp \left(-\frac{1}{2} \left(\frac{\varphi(N)}{\alpha} \right) \sum_{\ell=1}^{N} j_\ell^2 \right) \tilde{Z}(j_1, \ldots, J_N), \tag{5.16}$$

where

$$\tilde{Z}(j_1, \ldots, j_N) = \int_{\substack{\pi R^\lambda \\ \lambda \in A}} d\mu(\{x_\lambda\}) \exp\left(i \sum_{\ell=1}^{N} x_{\lambda_\ell} j_\ell \right)$$

$$= \int_{H^{\text{alg}}} d\mu(h) \exp\left(i \sum_{\ell=1}^{N} x_{\lambda_\ell} j_\ell \right). \tag{5.17}$$

Now due to the continuity and positivity of $Z[j]$ in $D(T)$; we have that for any $\varepsilon > 0 \to \exists \delta$ such that the inequality below is true since we have that: $Z(j_1, \ldots, j_N) \geq 1 - \varepsilon - \frac{2}{\delta^2}(j,j)^{(1)}$

$$\frac{1}{\left(\frac{2\pi\alpha}{\varphi(N)} \right)^{N/2}} \int_{R^N} dj_1 \cdots dj_N \exp\left(-\frac{1}{2}\frac{\varphi(N)}{\alpha} \sum_{\ell=1}^{N} j_\ell^2 \right) \tilde{Z}(j_1, \ldots, j_N)$$

$$\geq 1 - \varepsilon - \frac{2}{\delta^2} \left\{ \sum_{(m,n)=1}^{N} \frac{1}{\left(\frac{2\pi\alpha}{\varphi(N)} \right)^{N/2}} \int_{R^N} dj_1 \cdots dj_N \right.$$

$$\left. \times \exp\left(-\frac{1}{2} \left(\frac{\varphi(N)}{\alpha} \right) \sum_{\ell=1}^{N} j_\ell^2 \right) j_m j_n \langle e_m, e_n \rangle^{(1)} \right\}$$

$$= 1 - \varepsilon - \frac{2}{\delta^2} \left\{ \left(\frac{\alpha}{\varphi(N)} \right) \sum_{(m,n)=1}^{N} \delta_{mn} \langle Te_n, Te_m \rangle^{(0)} \right\}$$

$$\geq 1 - \varepsilon - \frac{2}{\delta^2} \left(\frac{\alpha}{\varphi(N)} \right) \varphi(N) \geq 1 - \varepsilon - \frac{2}{\delta^2}\alpha. \tag{5.18}$$

By substituting Eq. (5.18) into Eq. (5.15), we get the result

$$1 \geq \mu(H_e^Z) \geq 1 - \varepsilon - \frac{2}{\delta^2} \left(\lim_{\alpha \to 0} \alpha \right) = 1 - \varepsilon. \tag{5.19}$$

Since ε was arbitrary we have the validity of our theorem.

As a consequence of this theorem in the case of $\varphi(N)$ being bounded (so TT^* is an operator of Trace Class), we have that $H_e^Z = H$ which is the usual topological dual of H.

At this point, a simple proof may be given to the usual Minlos theorem on Schwartz spaces.[5,6]

Let us consider $S(R^D)$ represented as the countable normed spaces of sequences[8]

$$S(R^D) = \bigcap_{m=0}^{\infty} \ell_m^2, \qquad (5.20)$$

where

$$\ell_m^2 = \left\{ (x_n)_{n \in \mathbb{Z}}, x_n \in R \left| \sum_{n=0}^{N} (x_n)^2 n^m < \infty \right. \right\} \qquad (5.21)$$

The topological dual is given by the nuclear structure sum[8]

$$S'(R^D) = \bigcup_{n=0}^{\infty} \ell_{-n}^2 = \bigcup_{n=0}^{\infty} (\ell_n^2)^* \qquad (5.22)$$

We, thus, consider $E = S(R^D)$ in Eq. (9.3) and $Z[j(x)] = Z[\{j_n\}_{n \in \mathbb{Z}}]$ is continuous on $\bigcap_n^{\infty} \ell_n$. Since $Z[\{j_n\}_{n \in \mathbb{Z}}] \in C(\bigcap_{n=0}^{\infty} \ell_n^2, R)$ we have that for any fixed integer p, $Z[\{j_n\}_{n \in \mathbb{Z}}]$ is continuous on the Hilbert space ℓ_p^2 which, by its turn, may be considered as the domain of the following operator:

$$T_p : \ell_p^2 cl_0 \to l_0$$
$$\{j_n\} \to \{n^{p/2} j_n\}. \qquad (5.23)$$

It is straightforward to have the estimate

$$\left| \sum_{(m,n)=1}^{N} \langle T_p e_m, T_p e_n \rangle^{(0)} \right| \le N^{(B_p)} \qquad (5.24)$$

for some positive integer B and $\{e_i\}$ being the canonical orthonormal basis of l_0^2. By an application of our theorem for each fixed p; we get that the support of measure is given by the union of weighted spaces

$$\operatorname{supp} d\mu(h) = \bigcup_{p=0}^{\infty} (\ell_p^2)^* = \bigcup_{p=0}^{\infty} \ell_{-p}^2 = S'(R^D). \qquad (5.25)$$

$\square$

At this point we can suggest, without a proof of straightforward (non-topological) generalization of the Minlos theorem.

Theorem 5.2. *Let $\{T_\beta\}_{\beta \in C}$ be a family of operators satisfying the hypothesis of Theorem 5.3. Let us consider the Locally Convex space $\bigcup_{\beta \in C} \overline{Dom(T_\beta)}$ (supposed non-empty) with the family of norms $\| \psi \|_\beta = \langle T_\beta \psi, T_\beta \psi \rangle^{1/2}$.*

If the Functional Fourier transform is continuous on this locally convex space, the support of the Kolmogoroff measure equation (5.3) is given by the following subset of $[\bigcup_{\beta \in C} \overline{Dom(T_\beta)}]^{\mathrm{alg}}$, namely

$$\operatorname{supp} d\mu(h) = \bigcup_{\beta \in C} H^2_{\varphi_\beta}, \tag{5.26}$$

where φ_β are the functions given by Theorem 5.1. This general theorem will not be applied in what follows.

Let us now proceed to apply the above displayed results by considering the Schwinger generating functionals for two-dimensional Euclidean quantum eletrodynamics in Bosonized parametrization[9]

$$Z[j(x)] = \exp\left\{ -\frac{1}{2} \int_{R^2} d^2x \int_{R^2} d^2y\, j(x) \left((-\Delta)^2 + \frac{e^2}{\pi}(-\Delta) \right)^{-1} (x,y) j(y) \right\}, \tag{5.27}$$

where in Eq. (5.27), the electromagnetic field has the decomposition in Landau Gauge

$$A_\mu(x) = (\varepsilon_{\mu\nu}\partial_\nu \phi)(x) \tag{5.28}$$

and $j(x)$ is, thus, the Schwinger source for the $\phi(x)$ field taken as a basic dynamical variable.[9]

Since Eq. (5.27) is continuous in $L^2(R^2)$ with the inner product defined by the trace class operator $((-\Delta)^2 + \frac{e^2}{\pi}(-\Delta))^{-1}$, we conclude on the basis of Theorem 5.1 that the associated Kolmogoroff measure in Eq. (5.3) has its support in $L^2(R^2)$ with the usual inner product. As a consequence, the quantum observable algebra will be given by the functional space $L^1(L^2(R^2), d\mu(h))$ and usual orthonormal finite-dimensional approximations in Hilbert spaces may be used safely, i.e. if one considers the basis expansion $h(x) = \sum_{n=1}^{\infty} h_n e_n(x)$ with $e_n(x)$ denoting the eigenfunctions of the operator in Eq. (10.27) we get the result

$$\bigcup_{n=1}^{\infty} L^1(R^N, d\mu(h_1, \ldots, h_N)) = L^1(L^2(R^2), d\mu(h)). \tag{5.29}$$

It is worth mentioning that if one uses the Gauge vectorial field parametrization for the $(Q.E.D)_2$ — Schwinger functional

$$Z[j_1(x), j_2(x)]$$

$$= \exp\left\{ -\frac{1}{2} \int_{R^2} d^2x \int_{R^2} d^2y j_i(x) \left(-\Delta + \frac{e^2}{\pi} \right)^{-1} (x, y) \delta_{il} j_l(x) \right\} \quad (5.30)$$

the associated measure support will now be the Schwartz space $S'(R^2)$ since the operator $(-\Delta + \frac{e^2}{\pi})^{-1}$ is an application of $S(R^2)$ to $S'(R^2)$. As a consequence, it will be very cumbersome to use Hilbert finite-dimensional approximations[8] as in Eq. (5.29).

An alternative to approximate tempered distributions is the use of its Hermite expansion in $S'(R)$ distributional space associated with the eigenfunctions of the Harmonic-oscillator $V(x) \in L^{\infty}(R) \bigcup L^2(R)$ potential pertubation (see Ref. 3 for details with $V(x) \equiv 0$).

$$\left(-\frac{d^2}{dx^2} + x^2 + V(x) \right) H_n(x) = \lambda_n H_n(x). \quad (5.31)$$

Another important class of Bosonic functionals integrals are those associated with an Elliptic positive self-adjoint operator A^{-1} on $L^2(\Omega)$ with suitable boundary conditions. Here Ω denotes a D-dimensional compact manifold of R^D with volume element $d\nu(x)$.

$$Z[j(x)] = \exp\left\{ -\frac{1}{2} \int_{\Omega} d\nu(x) \int_{\Omega} d\nu(y) j(x) A^{+1}(x, y) j(y) \right\}. \quad (5.32)$$

If A is an operator of trace class on $(L^2(\Omega), d\nu)$ we have, thus, the validity of the usual eigenvalue functional representation

$$Z[\{j_n\}_{\mu \in \mathbb{Z}}]$$

$$= \int \left(\prod_{\ell=1}^{\infty} d(c_\ell \sqrt{\lambda_\ell}) \right) \exp\left(-\frac{1}{2} \sum_{\ell=1}^{\infty} \lambda_\ell c_\ell^2 \right) X_{\ell^2}(\{c_n\}_{n \in \mathbb{Z}}) \exp\left(i \sum_{\ell=1}^{\infty} c_\ell j_\ell \right)$$

with the spectral set

$$A^{-1} \sigma_\ell = \lambda_\ell \sigma_\ell,$$
$$j_\ell = \langle j, \sigma_\ell \rangle \quad (5.33)$$

and the characteristic function set

$$X_{\ell^2}(\{c_n\}_{n\in\mathbb{Z}}) = \begin{cases} 1 & \text{if } \sum_{n=0}^{\infty} c_n^2 < \infty, \\ 0 & \text{otherwise.} \end{cases} \tag{5.34}$$

It is instructive to point out the usual Hermite functional basis (see 5.4, Ref. 5) as a complete set in $L^2(E^{\mathrm{alg}}, d\mu(h))$, only if the Gaussian Kolmogoroff measure $d\mu(h)$ is of the class above studied.

A criticism to the usual framework to construct Euclidean field theories is that it is very cumbersome to analyze the infinite volume limit from the Schwinger generating functional defined originally on compact space times. In two dimensions the use of the result that the massive scalar field theory generating functional

$$\exp\left\{-\frac{1}{2}\int_{R^2} d^2x \int_{R^2} d^2y\, j(x)(-\Delta + m^2)^{-1}(x,y)j(y)\right\} \tag{5.35}$$

with $j(x) \in S(R^2)$ is given by the limit of finite volume Dirichlet field theories

$$\lim_{\substack{L\to\infty \\ T\to\infty}} \exp\left\{-\frac{1}{2}\int_{-L}^{L} dx^0 \int_{-T}^{T} dx^1 \int_{-L}^{L} dy^0 \right.$$
$$\left. \times \int_{-T}^{T} dy^1 j(x^0, x^1)(-\Delta_D + m^2)^{-1}(x^1, y^1, x^0, y^0)j(y^0, y^1)\right\} \tag{5.36}$$

may be considered, in our opinion, as the similar claim made that is possible from a mathematical point of view to deduce the Fourier transforms from Fourier series, a very, difficult mathematical task (see Appendix B).

Let us comment on the functional integral associated with Feynman propagation of fields configurations used in geometrodynamical theories in the scalar case

$$G[\beta^{\mathrm{in}}(x); \beta^{\mathrm{out}}(x), T](j)$$

$$= \int_{\substack{\phi(x,0)=\beta^{\mathrm{in}}(x) \\ \phi(x,T)=\beta^{\mathrm{out}}(x)}} \exp\left\{-\frac{1}{2}\int_0^T dt \int_{-\infty}^{+\infty} d^\nu x \left(\phi\left(-\frac{d^2}{dt^2} + A\right)\phi\right)(x,t)\right\}$$

$$\times \exp\left(i\int_0^T dt \int_{-\infty}^{+\infty} d^\nu x\, j(x)t)\phi(x,t)\right). \tag{5.37}$$

If we define the formal functional integral by means of the eigenfunctions of the self-adjoint Elliptic operator A, namely:

$$\phi(x,t) = \sum_{\{k\}} \phi_k(t)\psi_k(x), \tag{5.38}$$

where

$$A\psi_k(x) = (\lambda_k)^2 \psi_k(x) \tag{5.39}$$

it is straightforward to see that Eq. (5.36) is formally exactly evaluated in terms of an infinite product of usual Feynman Wiener path measures

$$G[\beta^{\text{in}}(x); \beta^{\text{out}}(x), T](j)$$

$$= \prod_{\{k\}} \int_{\substack{c_k(0)=\phi_k(0) \\ c_k(T)=\phi_k(T)}} D^F[c_k(t)]$$

$$\times \exp\left\{ -\frac{1}{2} \int_0^T \left(c_k \left(-\frac{d^2}{dt^2} + \lambda_k^2 \right) c_k \right)(t)dt \exp\left(i \int_0^T dt j_k(t) c_k(t) \right) \right\}$$

$$= \prod_{\{k\}} \left\{ \sqrt{+\frac{\lambda_k}{\sin(\lambda_k T)}} \exp\left\{ -\frac{\lambda_k}{2\sin(\lambda_k T)} \left[(\phi_k^2(T) \right. \right. \right.$$

$$+ \phi_k^2(0)) \cos(\lambda_k T) - 2\phi_k(0)\phi_k(T) \Big\} \Big\}$$

$$- \frac{2\phi_k(T)}{\lambda_k} \int_0^T dt j_k(t) \sin(\lambda_k t) - \frac{2\phi_k(0)}{\lambda_k} \int_0^T dt j_k(t) \sin(\lambda_k(T-t))$$

$$\times \left\{ -\frac{2}{(\lambda_k)^2} \int_0^T dt \int_0^t ds j_k(t) j_k(s) \sin(\lambda_k(T-t)) \sin(\lambda_k s) \right] \Big\} . $$

$$\tag{5.40}$$

Unfortunately, our theorems do not apply in a straightforward way to infinite (continuum) measure product of Wiener measures in Eq. (5.40) to produce a sensible measure theory on the functional space of the infinite product of Wiener trajectories $\{c_k(t)\}$ (note that for each fixed x, a sample field configuration $\phi(t,0)$ in Eq. (5.36) is a Hölder continuous function, result opposite to the usual functional integral representation for the Schwinger generating functional equations (5.1)–(5.5)), where it does not make mathematical sense to consider a fixed point distribution $\phi(t,0)$ — see Sec. 5.4, Eq. (5.74).

Let us call to attention that there is still a formal definition of the above Feynman path propagator for fields equation (5.37) which at large time $T \to +\infty$ gives formally the quantum field functional integral equation (5.5) associated with the Schwinger generating functional.

We thus consider the functional domain for Eq. (5.37) as composed of field configurations which has a classical piece added with another fluctuating component to be functionally integrated out, namely

$$\sigma(x,t) = \sigma_{\mathrm{CL}}(x,t) + \sigma_q(x,t). \tag{5.41}$$

Here, the classical field configuration problem (added with all zero modes of the free theory) defined by the kinetic term $\pounds$

$$\left(-\frac{d^2}{dt^2} + \pounds\right)\sigma^{\mathrm{CL}}(x,t) = j(x,t), \tag{5.42}$$

with

$$\sigma^{\mathrm{CL}}(x,-T) = \beta_1(x); \quad \sigma^{\mathrm{CL}}(x,T) = \beta_2(x), \tag{5.43}$$

namely,

$$\sigma_{\mathrm{CL}}(x,t) = \left(-\frac{d^2}{dt^2} + \pounds\right)^{-1} j(x,t) + (\text{all projection on zero modes of } \pounds). \tag{5.44}$$

As a consequence of the decomposition equation (5.43), the formal geometrical propagator with an external source is given by

$$G[\beta_1(x), \beta_2(x), T, [j]] = \int_{\substack{\sigma(x,-T)=\beta_1(x)\\ \sigma(x,+T)=\beta_2(x)}} D[\sigma(x,t)]$$

$$\times \exp\left(-\frac{1}{2}\int_{-T}^{T} dt d^\nu x \sigma(x,t)\left(-\frac{d^2}{dt^2} + \pounds\right)\sigma(x,t)\right)$$

$$\times \exp\left(i\int_{-T}^{T} dt \int d^\nu x j(x,t)\sigma(x,t)\right) \tag{5.45}$$

may be defined by the following mathematically well-defined Gaussian functional measure

$$\exp\left\{-\frac{1}{2}\int_{-T}^{T} dt \int d^\nu x j(x,t)\sigma^{\mathrm{CL}}(x,t)\right\} \int_{\substack{\sigma_q(x,-T)=0\\ \sigma_q(x,+T)=0}} d\sigma_q(x,t)$$

$$\times \exp\left\{-\frac{1}{2}\int_{-T}^{T} dt \int d^\nu x \sigma_q(x,t)\left(-\frac{d^2}{dt^2} + \pounds\right)\sigma_q(x,t)\right\}. \tag{5.46}$$

The above claim is a consequence of the result below

$$
\int_{\substack{\sigma_q(x,-T)=0 \\ \sigma_q(x,T)=0}} D[\sigma_q(x,t)] \exp\left\{ -\frac{1}{2} \int_{-T}^{T} dt \int d^D x \sigma_q(x,t) \left(-\frac{d^2}{dt^2} + \pounds \right) \sigma_q(x,t) \right\}
$$

$$
= det_{\text{Dir}}^{-\frac{1}{2}} \left[-\frac{d^2}{dt^2} + \pounds \right], \tag{5.47}
$$

where the subscript Dirichlet on the functional determinant means that one must impose formally the Dirichlet condition on the domain of the operator $(-\frac{d^2}{dt^2} + \pounds)$ on $D'(R^D \times [-T,T])$ (or $L^2(R^D \times [-T,T]$ if $\pounds^{-1}$ belongs to trace class). Note that the operator $\pounds$ in Eq. (5.46) does not have zero modes by the construction of Eq. (5.41).

At this point, we remark that at the limit $T \to +\infty$, Eq. (5.45) is exactly the quantum field functional equation (5.5) if one takes $\beta_1(x) = \beta_2(x) = 0$ (note that the classical vacuum limit $T \to \infty$ of Wiener measures is mathematically ill-defined (see Theorem 5.1 of Ref. 1).

It is an important point to remark that $\sigma_{\text{CL}}(x,t)$ is a regular $C^\infty([-T,T] \times \Omega)$ solution of the Elliptic problem equation (5.42) and the fluctuating component $\sigma_q(x,t)$ is a Schwartz distribution in view of the Minlos-Dao Xing Theorem 1, since the elliptic operator $-\frac{d^2}{dt^2} + \pounds$ in Eq. (5.47) acts now on $D'([-T,T] \times \Omega)$ with the range $D([-T,T] \times \Omega)$, which by its turn shows the difference between this framework and the previous one related to the infinite product of Wiener measures since these objects are functional measures in different functional spaces.

Finally, we comment that functional Schrodinger equation, may be mathematically defined for the above displayed field propagators equation (5.37) only in the situation of Eq. (5.40). For instance, with $\pounds = -\Delta$ (the Laplacean), we have the validity of the Euclidean field wave equation for the geometrodynamical path-integral equation (5.37)

$$
\frac{\partial}{\partial T} G[\beta_1(x), \beta_2(x), T, [j]]
$$

$$
= \int_\Omega d^\nu x \left[+\frac{\delta^2}{\delta^2 \beta_2(x)} - |\nabla \beta_2(x)|^2 + j(x,T) \right] G[\beta_1(x), \beta_2(x), T, [j]] \tag{5.48}
$$

with the functional initial condition

$$
\lim_{T \to 0^+} G[\beta_1(x), \beta_2(x), T] = \delta^{(F)}(\beta_1(x) - \beta_2(x)). \tag{5.49}
$$

5.4. Some Rigorous Quantum Field Path-Integral in the Analytical Regularization Scheme

In this core section, we address the important problem of producing concrete non-trivial examples of mathematically well-defined (in the ultra-violet region!) path-integrals in the context of the exposed theorems in the previous sections of this paper, especially Sec. 5.2, Eq. (5.11).

Let us thus start our analysis by considering the Gaussian measure associated with the (infrared regularized) α-power ($\alpha > 1$) of the Laplacean acting on $L^2(R^2)$ as an operational quadratic form (the Stone spectral theorem)

$$(-\Delta)_\varepsilon^\alpha = \int_{\varepsilon_{\mathrm{IR}} \leq \lambda} (\lambda)^\alpha dE(\lambda), \tag{5.50a}$$

$$Z_{\alpha, \varepsilon_{\mathrm{IR}}}^{(0)}[j] = \exp\left\{ -\frac{1}{2} \langle j, (-\Delta)_\varepsilon^{-\alpha} j \rangle_{L^2(R^2)} \right\}$$

$$= \int d_{\alpha, \varepsilon}^{(0)} \mu[\varphi] \exp(i\langle j, \varphi \rangle_{L^2(R^2)}). \tag{5.50b}$$

Here, $\varepsilon_{\mathrm{IR}} > 0$ denotes the infrared cut-off.

It is worth to call the readers' attention that due to the infrared regularization introduced in Eq. (5.50a), the domain of the Gaussian measure is given by the space of square integrable functions on R^2 by the Minlos theorem of Sec. 3, since for $\alpha > 1$, the operator $(-\Delta)_{\varepsilon_{\mathrm{IR}}}^{-\alpha}$ defines a class trace operator on $L^2(R^2)$, namely

$$\mathrm{Tr}_{\mathfrak{H}_1}((-\Delta)_{\varepsilon_{\mathrm{IR}}}^{-\alpha}) = \int d^2k \frac{1}{(|K|^{2\alpha} + \varepsilon_{\mathrm{IR}})} < \infty. \tag{5.50c}$$

This is the only point of our analysis where it is needed to consider the infrared cut-off considered on the spectral resolution equation (5.50a). As a consequence of the above remarks, one can analyze the ultraviolet renormalization program in the following interacting model proposed by us and defined by an interaction $g_{\mathrm{bare}} V(\varphi(x))$ with $V(x)$ denoting a function on R such that it posseses an essentially bounded Fourier transform and g_{bare} denoting the positive bare coupling constant.

Let us show that by defining a renormalized coupling constant as (with $g_{\mathrm{ren}} < 1$)

$$g_{\mathrm{bare}} = \frac{g_{\mathrm{ren}}}{(1 - \alpha)^{1/2}} \tag{5.51}$$

one can show that the interaction function

$$\exp\left\{-g_{\text{bare}}(\alpha)\int d^2 x V(\varphi(x))\right\} \tag{5.52}$$

is an integrable function on $L^1(L^2(R^2), d^{(0)}_{\alpha,\varepsilon_{\text{IR}}}\mu[\varphi])$ and leads to a well-defined ultraviolet path-integral in the limit of $\alpha \to 1$.

The proof is based on the following estimates:

Since almost everywhere we have the pointwise limit

$$\exp\left\{-g_{\text{bare}}(\alpha)\int d^2 x V(\varphi(x))\right\}$$

$$\times \lim_{N\to\infty}\left\{\sum_{n=0}^{N}\frac{(-1)^n(g_{\text{bare}}(\alpha))^n}{n!}\int_R dk_1\cdots dk_n \tilde{V}(k_1)\cdots\tilde{V}(k_n)\right.$$

$$\left.\times \int_{R^2} dx_1\cdots dx_n e^{ik_1\varphi(x_1)}\cdots e^{ik_n\varphi(x_n)}\right\} \tag{5.53}$$

we have that the upper-bound estimate below holds true

$$|Z^\alpha_{\varepsilon_{\text{IR}}}[g_{\text{bare}}]| \leq \left|\sum_{n=0}^{\infty}\frac{(-1)^n(g_{\text{bare}}(\alpha))^n}{n!}\int_{R^2} dk_1\cdots dk_n \tilde{V}(k_1)\cdots\tilde{V}(k_n)\right.$$

$$\left.\times \int_{R^2} dx_1\cdots dx_n \int d^{(0)}_{\alpha,\varepsilon_{\text{IR}}}\mu[\varphi](e^{i\sum_{\ell=1}^{N}k_\ell\varphi(x_\ell)})\right|. \tag{5.54a}$$

with

$$Z^\alpha_{\varepsilon_{\text{IR}}}[g_{\text{bare}}] = \int d^{(0)}_{\alpha,\varepsilon_{\text{IR}}}\mu[\varphi]\exp\left\{-g_{\text{bare}}(\alpha)\int d^2 x V(\varphi(x))\right\} \tag{5.54b}$$

we have, thus, the more suitable form after realizing $d^2 k_i$ and $d^{(0)}_{\alpha,\varepsilon_{\text{IR}}}\mu[\varphi]$ integrals, respectively

$$|Z^\alpha_{\varepsilon_{\text{IR}}=0}[g_{\text{bare}}]| \leq \sum_{n=0}^{\infty}\frac{(g_{\text{bare}}(\alpha))^n}{n!}\left(\|\tilde{V}\|_{L^\infty(R)}\right)^n$$

$$\times \left|\int dx_1\cdots dx_n \det^{-\frac{1}{2}}[G^{(N)}_\alpha(x_i,x_j)]_{\substack{1\leq i\leq N \\ 1\leq j\leq N}}\right|. \tag{5.55}$$

Here, $[G_\alpha^{(N)}(x_i, x_j)]_{\substack{1 \leq i \leq N \\ 1 \leq j \leq N}}$ denotes the $N \times N$ symmetric matrix with (i,j) entry given by the Green-function of the α-Laplacean (without the infrared cut-off here! and the needed normalization factors!).

$$G_\alpha(x_i, x_j) = |x_i - x_j|^{2(1-\alpha)} \frac{\Gamma(1-\alpha)}{\Gamma(\alpha)}. \tag{5.56}$$

At this point, we call the readers' attention that we have the formulae on the asymptotic behavior for $\alpha \to 1$.

$$\left\{ \lim_{\substack{\alpha \to 1 \\ \alpha > 1}} \det^{-\frac{1}{2}} [G_\alpha^{(N)}(x_i, x_j)] \right\} \sim (1-\alpha)^{N/2} \times \left(\left| \frac{(N-1)(-1)^N}{\pi^{N/2}} \right| \right)^{-\frac{1}{2}}. \tag{5.57}$$

After substituting Eq. (5.57) into Eq. (5.55) and taking into account the hypothesis of the compact support of the nonlinear $V(x)$ (for instance: supp $V(x) \subset [0,1]$), one obtains the finite bound for any value $g_{\text{rem}} > 0$, and producing a proof for the convergence of the perturbative expansion in terms of the renormalized coupling constant.

$$\lim_{\alpha \to 1} |Z_{\varepsilon_{\text{IR}}=0}^\alpha [g_{\text{bare}}(\alpha)]| \leq \sum_{n=0}^\infty \frac{(\|\hat{V}\|_{L^\infty(R)})^n}{n!} \left(\frac{g_{\text{ren}}}{(1-\alpha)^{\frac{1}{2}}} \right)^n \frac{(1^n)}{\sqrt{n}} (1-\alpha)^{n/2}$$

$$\leq e^{g_{\text{ren}} \|\hat{V}\|_{L^\infty(R)}} < \infty. \tag{5.58}$$

Another important rigorously defined functional integral is to consider the following α-power Klein Gordon operator on Euclidean space time

$$\mathcal{L} = (-\Delta)^\alpha + m^2 \tag{5.59}$$

with m^2 a positive "mass" parameters.

Let us note that $\mathcal{L}^{-1}$ is an operator of class trace on $L^2(R^\nu)$ if and only if the result below holds true

$$\text{Tr}_{L^2(R^\nu)}(\mathcal{L}^{-1}) = \int d^\nu k \frac{1}{k^{2\alpha} + m^2} = \bar{C}(\nu) m^{(\frac{\nu}{\alpha} - 2)} \left\{ \frac{\pi}{2\alpha} \text{cosec} \frac{\nu\pi}{2\alpha} \right\} < \infty, \tag{5.60}$$

namely if

$$\alpha > \frac{\nu}{2}. \tag{5.61}$$

In this case, let us consider the double functional integral with functional domain $L^2(R^\nu)$

$$Z[j, k] = \int d_G^{(0)} \beta[v(x)] \int d_{(-\Delta)^\alpha + v + m^2}^{(0)} \mu[\varphi]$$

$$\times \exp\left\{i \int d^\nu x (j(x)\varphi(x) + k(x)v(x))\right\}, \qquad (5.62)$$

where the Gaussian functional integral on the fields $V(x)$ has a Gaussian generating functional defined by a $\oint_1$-integral operator with a positive defined kernel $g(|x - y|)$, namely

$$Z^{(0)}[k] = \int d_G^{(0)} \beta[v(x)] \exp\left\{i \int d^\nu x k(x)v(x)\right\}$$

$$= \exp\left\{-\frac{1}{2} \int d^\nu x \int d^\nu y (k(x)g(|x - y|)k(x))\right\}. \qquad (5.63)$$

By a simple direct application of the Fubbini–Tonelli theorem on the exchange of the integration order in Eq. (5.62), lead us to the effective $\lambda\varphi^4$-like well-defined functional integral representation

$$Z_{\text{eff}}[j] = \int d_{((-\Delta)^\alpha + m^2)}^{(0)} \mu[\varphi(x)]$$

$$\times \exp\left\{-\frac{1}{2} \int d^\nu x d^\nu y |\varphi(x)|^2 g(|x - y|) |\varphi(y)|^2\right\}$$

$$\times \exp\left\{i \int d^\nu x j(x)\varphi(x)\right\}. \qquad (5.64)$$

Note that if one introduces from the begining a bare mass parameters m_{bare}^2 depending on the parameters α, but such that it always satisfies Eq. (5.60) one should obtain again Eq. (5.64) as a well-defined measure on $L^2(R^\nu)$. Of course, that the usual pure Laplacian limit of $\alpha \to 1$ in Eq. (5.59), will need a renormalization of this mass parameters ($\lim_{\alpha \to 1} m_{\text{bare}}^2(\alpha) = +\infty$!) as much as its done in the previous example.

Let us continue our examples by showing again the usefulness of the precise determination of the functional-distributional structure of the domain of the functional integrals in order to construct rigorously these path-integrals without complicated limit procedures.

Let us consider a general R^ν Gaussian measure defined by the generating functional on $S(R^\nu)$ defined by the α-power of the Laplacean operator $-\Delta$ acting on $S(R^\nu)$ with a of small infrared regularization mass parameter μ^2

$$Z_{(0)}[j] = \exp\left\{-\frac{1}{2}\left\langle j, ((-\Delta)^\alpha + \mu_0^2)^{-1}j\right\rangle_{L^2(R^\nu)}\right\}$$

$$= \int_{E^{\mathrm{alg}}(S(R^\nu))} d_\alpha^{(0)}\mu[\varphi]\exp(i\varphi(j)). \tag{5.65}$$

An explicit expression in momentum space for the Green function of the α-power of $(-\Delta)^\alpha + \mu_0^2$ is given by

$$((-\Delta)^{+\alpha} + \mu_0^2)^{-1}(x - y) = \int \frac{d^\nu k}{(2\pi)^\nu} e^{ik(x-y)}\left(\frac{1}{k^{2\alpha} + \mu_0^2}\right). \tag{5.66}$$

Here, $\bar{C}(\nu)$ is a ν-dependent (finite for ν-values!) normalization factor.

Let us suppose that there is a range of α-power values that can be chosen in such a way that one satisfies the constraint below

$$\int_{E^{\mathrm{alg}}(S(R^\nu))} d_\alpha^{(0)}\mu[\varphi](\|\varphi\|_{L^{2j}(R^\nu)})^{2j} < \infty \tag{5.67}$$

with $j = 1, 2, \ldots, N$ and for a given fixed integer N, the highest power of our polynomial field interaction. Or equivalently, after realizing the φ-Gaussian functional integration, with a space-time cut-off volume Ω on the interaction to be analyzed in Eq. (5.70)

$$\int_\Omega d^\nu x[(-\Delta)^\alpha + \mu_0^2]^{-j}(x, x) = \mathrm{vol}(\Omega)\left(\int \frac{d^\nu k}{k^{2\alpha} + \mu_0^2}\right)^j$$

$$= C_\nu(\mu_0)^{(\frac{\nu}{\alpha}-2)}\left(\frac{\pi}{2\alpha}\mathrm{cosec}\frac{\nu\pi}{2\alpha}\right) < \infty. \tag{5.68}$$

For $\alpha > \frac{\nu-1}{2}$, one can see by the Minlos theorem that the measure support of the Gaussian measure equation (5.65) will be given by the intersection Banach space of measurable Lebesgue functions on R^ν instead of the previous one $E^{\mathrm{alg}}(S(R^\nu))$

$$\mathcal{L}_{2N}(R^\nu) = \bigcap_{j=1}^{N}(L^{2j}(R^\nu)). \tag{5.69}$$

In this case, one obtains that the finite-volume $p(\varphi)_2$ interactions

$$\exp\left\{-\sum_{j=1}^{N}\lambda_{2j}\int_\Omega (\varphi^2(x))^j dx\right\} \leq 1 \tag{5.70}$$

is mathematically well-defined as the usual pointwise product of measurable functions and for positive coupling constant values $\lambda_{2j} \geq 0$. As a consequence, we have a measurable functional on $L^1(\mathcal{L}_{2N}(R^\nu); d^{(0)}_\alpha \mu[\varphi])$ (since it is bounded by the function 1). So, it would make sense to consider mathematically the well-defined path-integral on the full space R^ν with those values of the power α satisfying the contraint equation (5.67).

$$Z[j] = \int_{\mathcal{L}_{2N}(R^\nu)} d^{(0)}_\alpha \mu[\varphi] \exp \left\{ -\sum_{j=1}^N \lambda_{2j} \int_\Omega \varphi^{2j}(x)dx \right\}$$

$$\times \exp \left(i \int_{R^\nu} j(x)\varphi(x) \right). \qquad (5.71)$$

Finally, let us consider an interacting field theory in a compact space-time $\Omega \subset R^\nu$ defined by an integer even power $2n$ of the Laplacean operator with Dirichlet boundary conditions as the free Gaussian kinetic action, namely

$$Z^{(0)}[j] = \exp \left\{ -\frac{1}{2} \langle j, (-\Delta)^{-2n} j \rangle_{L^2(\Omega)} \right\}$$

$$= \int_{W_2^n(\Omega)} d^{(0)}_{(2n)} \mu[\varphi] \exp(i\langle j, \varphi \rangle_{L^2(\Omega)}). \qquad (5.72)$$

Here, $\varphi \in W_2^n(\Omega)$ is the Sobolev space of order n which is the functional domain of the cylindrical Fourier transform measure of the generating functional $Z^{(0)}[j]$, a continuous bilinear positive form on $W_2^{-n}(\Omega)$ (the topological dual of $W_2^n(\Omega)$).

By a straightforward application of the well-known Sobolev immersion theorem, we have that for the case of

$$n - k > \frac{\nu}{2} \qquad (5.73)$$

including k a real number of the functional Sobolev space $W_2^n(\Omega)$ is contained in the continuously fractional differentiable space of functions $C^k(\Omega)$. As a consequence, the domain of the Bosonic functional integral can be further reduced to $C^k(\Omega)$ in the situation of Eq. (5.73)

$$Z^{(0)}[j] = \int_{C^k(\Omega)} d^{(0)}_{(2n)} \mu[\varphi] \exp(i\langle j, \varphi \rangle_{L^2(\Omega)}). \qquad (5.74)$$

That is our new result generalizing the Wiener theorem on Brownian paths in the case of $n = 1$, $k = \frac{1}{2}$, and $\nu = 1$.

Since the Bosonic functional domain in Eq. (5.74) is formed by real functions and not by distributions, we can see straightforwardly that any interaction of the form

$$\exp\left\{ -g \int_\Omega F(\varphi(x))d^\nu x \right\} \tag{5.75}$$

with the non-linearity $F(x)$ denoting a lower bounded real function ($\gamma > 0$)

$$F(x) \geq -\gamma \tag{5.76}$$

is well-defined and is an integrable function on the functional space $(C^k(\Omega), d^{(0)}_{(2n)}\mu[\varphi])$ by the direct application of the Lebesque theorem

$$\left| \exp\left\{ -g \int_\Omega F(\varphi(x))d^\nu x \right\} \right| \leq \exp\{+g\gamma\}. \tag{5.77}$$

At this point we make a subtle mathematical remark that the infinite volume limit of Eqs. (5.74) and (5.75) is very difficult, since one loses the Garding–Poincaré inequality at this limit for those elliptic operators and, thus, the very important Sobolev theorem. The probable correct procedure to consider the thermodynamic limit in our Bosonic path-integrals is to consider solely a volume cut-off on the interaction term Gaussian action as in Eq. (5.71) and there search for vol(Ω) $\to \infty$.

As a last remark related to Eq. (5.73) one can see that a kind of "fishnet" exponential generating functional

$$Z^{(0)}[j] = \exp\left\{ -\frac{1}{2} \langle j, \exp\{-\alpha\Delta\}j \rangle_{L^2(\Omega)} \right\} \tag{5.78}$$

has a Fourier transformed functional integral representation defined on the space of the infinitely differentiable functions $C^\infty(\Omega)$, which physically means that all field configurations making the domain of such path-integral has a strong behavior-like purely nice smooth classical field configurations.

As a general conclusion, we can see that the technical knowledge of the support of measures on infinite-dimensional spaces — specially the powerful Minlos theorem of Sec. 3 is very important for a deep mathematical physical understanding into one of the most important problem is quantum field theory and turbulence which is the problem related to the appearance of ultraviolet (short-distance) divergences on perturbative path-integral calculations.

5.5. Remarks on the Theory of Integration of Functionals on Distributional Spaces and Hilbert–Banach spaces

Let us first consider a given vector space E with a Hilbertian structure $\langle,\rangle$, namely $\mathcal{H} = (E, \langle\ \rangle)$, where $\langle\ \rangle$ means a inner product and $\mathcal{H}$ a complete topological space with the metrical structure induced by the given $\langle,\rangle$. We have, thus, the famous Minlos theorem on the support of the cylindrical measure associated with a given quadratic form defined by a positive definite class trace operator $A \in \mathfrak{f}_1(\mathcal{H}, \mathcal{H})$ (see Appendix for a discussion on Fourier transforms in vector spaces of infinite-dimension)

$$\exp\left\{-\frac{1}{2}\langle b, Ab\rangle\right\} = \exp\left\{-\frac{1}{2}\langle |A|^{\frac{1}{2}}b, |A|^{\frac{1}{2}}b\rangle\right\} = \int_{\mathcal{H}} d_A\mu(v) \cdot \exp(i\langle v, b\rangle)$$

(5.79)

since any given class trace operator can always be considered as the composition of two Hilbert–Schmidt, each one defined by a function on $L^2(M \times M, d\nu \otimes d\nu)$.

Here, the cylindrical measure $d_A\mu(v)$, defined already on the vector space of the linear forms of E, with the topology of pontual convergence — the so-called algebraic dual of E — has its support concentrated on the Hilbert spaces $\mathcal{H}$, through the isomorphism of $\mathcal{H}$ and its dual $\mathcal{H}'$ by means of the Riesz theorem.

This result can be understood more easily, if one represents the given Hilbert space $\mathcal{H}$ as a square-integrable space of measurable functions on a complete measure space $(M, d\nu)L^2(M, d\nu)$. In this case, the class trace positive definite operator is represented by an integral operator with a positive-definite Kenel $K(x, y)$. Note that it is worth to re-write Eq. (5.79) in the Feynman path-integral notation as

$$\exp\left\{-\frac{1}{2}\int_{M \times M} d\nu(x)d\nu(y)f(x)K(x, s)\overline{f(s)}\right\}$$

$$= \frac{1}{Z(0)} \int_{L^2(M, d\nu)} \left(\prod_{x \in M} d\varphi(x)\right)$$

$$\times \exp\left[-\frac{1}{2}\int_{M \times M} d\nu(x)d\nu(y)\overline{\varphi(x)}K^{-1}(x, y)\varphi(y)\right]$$

$$\times \exp\left\{\int_M d\nu(x)f(x)\overline{\varphi(x)}\right\}.$$

(5.80)

Here, the "inverse Kenel" of the operator A is given by the relation

$$\int_M d\nu(y)K(x,y)K^{-1}(y,x') = \text{ identity operator} \qquad (5.81)$$

and the path-integral normalization factor is given by the functional determinant $Z(0) = \det^{-\frac{1}{2}}(K) = \det^{\frac{1}{2}}(K^{-1})$.

A more invariant and rigorous representation for the Gaussian path-integral equations (5.79) and (5.80) can be exposed through an eigenfunction–eigenvalue harmonic expansion associated with our given class trace operator A, namely

$$A\beta_n = \lambda_n \beta_\mu, \qquad (5.82)$$

$$v = \sum_{n=0}^{\infty} v_n\beta_n, \qquad \varphi = \sum_{n=0}^{\infty} \varphi_n\beta_n, \qquad (5.83)$$

$$\exp\left\{-\frac{1}{2}\sum_{n=0}^{\infty}\lambda_n|v_n|^2\right\}$$

$$= \limsup_N\left\{\int_{R^N} d\langle\varphi|\varphi_1\rangle\ldots d\langle\varphi|\varphi_N\rangle\right.$$

$$\left. \times \exp\left(-\frac{1}{2}\sum_{n=0}^{N}\frac{|\langle\varphi|\varphi_n\rangle|^2}{\lambda_n}\right)\left(\prod_{n=0}^{N}\frac{1}{2\pi\lambda_n}\right)^{\frac{1}{2}}\exp\left(i\sum_{n=0}^{N}\varphi_n\bar{v}_n\right)\right\}.$$

$$(5.84)$$

The above cited theorem for the support characterization of Gaussian path-integrals can be generalized to the highly non-trivial case of a non-linear functional $Z(v)$ on E, satisfying the following conditions:

(a) $Z(0) = 1$,

(b) $\displaystyle\sum_{j,k}^{N} Z(v_j - v_k)z_j\bar{z}_k \geq 0,$ for any $\{z_i\}_{1\leq i\leq N} : z_i \in \mathbb{N}$

and $\{v_i\}_{1\leq i\leq N} : v_i \in E.$ $\qquad (5.85)$

(c)   there is an H-subspace of E with an inner product $\langle,\rangle$, such that $Z(v)$ is continuous in relation to a given inner product $\langle,\rangle_A$ coming from a quadratic form defined by a positive definite class trace operator A on $\mathcal{H}$. In others words, we have the sequential continuity criterium (if $\mathcal{H}$ is separable):

$$\lim_{n \to \infty} Z(v_n) = 0, \tag{5.86}$$

if

$$\lim_{n \to \infty} \langle v_n, A v_n \rangle = 0 \quad \left(A \in \oint_1 (\mathcal{H}) \right) \tag{5.87}$$

We have, thus, the following path-integral representation:

$$Z(v) = \int_H d\mu(\varphi) \exp(i \langle v, \varphi \rangle), \tag{5.88}$$

where

$$\int_H d\mu(\varphi) = 1. \tag{5.89}$$

Another less mathematically rigorous result is that the one related to an inversible self-adjoint positive-definite operator A in a given Hilbert space $(H, \langle, \rangle)$ — not necessarily a bounded operator in the class trace operator as considered previously. In order to write (more or less) formal path-integrals representations for the Gaussian functional

$$Z(j) = \exp \left\{ -\frac{1}{2} \langle j, A^{-1} j \rangle \right\} \tag{5.90}$$

with $f \in \text{Dom}(A^{-1}) \subset \mathcal{H}$, we start by considering the usual spectral expression for the following quadratic form:

$$\langle \varphi, A\varphi \rangle = \int_{\sigma(A)} \lambda \langle \varphi, dE(\lambda)\varphi \rangle \tag{5.91}$$

with $\sigma(A)$ denoting the spectrum of A (a subset of R^+!) and $dE(\lambda)$ are the spectral projections associated with the spectral representation of A.

In this case one has the result for the path-integral weight as

$$\exp \left\{ -\frac{1}{2} \int_{\sigma(A)} \lambda \langle \varphi, dE(\lambda)\varphi \rangle \right\} = \exp \left\{ -\frac{1}{2} \langle \varphi, A\varphi \rangle \right\}$$

$$= \limsup \left\{ \prod_{\lambda \in \sigma_{\text{Fin}}(A)} \exp \left(-\frac{1}{2} \langle \varphi, \lambda dE(\lambda)\varphi \rangle \right) \right\}. \tag{5.92}$$

Here $\sigma_{\text{Fin}}(A)$ denotes all subsets with a finite number of elements of $\sigma(A)$.

As a consequence, one should define formally the generating functional as

$$Z(j) = \exp\left\{-\frac{1}{2}\langle j, A^{-1}j\rangle\right\}$$

$$= \limsup\left\{\prod_{\lambda\in\sigma_{\mathrm{Fin}}(A)}\int_{-\infty}^{+\infty} dx_\lambda \cdot e^{-\frac{1}{2}\lambda(x_\lambda)^2}e^{ij_\lambda x_\lambda}\sqrt{\frac{\lambda}{2\pi}}\right\} \qquad (5.93)$$

associated with the self-adjoint operator A acting on a Hilbert space $\mathcal{H}$

$$Z(j) = \int_{\mathcal{H}} d_A\mu(\varphi)e^{i\langle j,\varphi\rangle}. \qquad (5.94)$$

Otherwise, one should introduce formal redefinitions of parameters entering in the definition of our action operator A, in such a way to render finite the functional determinant in Eq. (5.80). Let us exemplify such calculational point with the operator $(-\Delta + m^2)$, acting an $L^2(R^N)$ (with domain being given precisely by the Sobolev space $H^2(R^N)$). We note that

$$\mathrm{Tr}_{L^2(R^N)}(\exp(-t(-\Delta + m^2)))$$

$$= \left(\frac{1}{\sqrt{2\pi}}\right)^2\left[\int_{-\infty}^{+\infty} d^n k e^{-tk^2}e^{-tm^2}\right]$$

$$= C(N)\begin{cases}\dfrac{(N-2)!!}{2(2t)^{\frac{N-1}{2}}}\sqrt{\dfrac{\pi}{t}} & \text{if } N-1 \text{ is even,}\\[3mm] \dfrac{((N-2/2))!}{2}\dfrac{1}{(t)^{\frac{N}{2}}} & \text{if } N-1 \text{ is odd.}\end{cases} \qquad (5.95)$$

with $C(N)$ denoting an N-dependent constant.

It is worth to note that one must introduce in the path-integral equation (5.93), some formal definition for the functional determinant of the self-adjoint operator A, which by its term leads to the formal process of the "infinite renormalization" in quantum field path-integrals

$$\int_{\mathcal{H}} d_A\mu(\varphi) = \lim_{(R\text{-valued net})}\sup\left\{\prod_{\lambda\in\sigma_{\mathrm{Fin}}(A)}\int_{-\infty}^{+\infty}\frac{dx_\lambda}{\sqrt{2\pi}}e^{-\frac{1}{2}\lambda(x_\lambda)^2}\right\}$$

$$= \limsup\left(\prod_{\lambda\in\sigma_{\mathrm{Fin}}(A)}\left\{\frac{1}{\sqrt{\lambda'}}\right\}\right) = \det{}_F^{-\frac{1}{2}}[A]$$

$$= \lim_{\varepsilon \to 0^+} \left\{ \prod_{\lambda \in \sigma_{\mathrm{Fin}}(A)} \exp\left[+\frac{1}{2} \int_\varepsilon^{1/\varepsilon} \frac{dt}{t} e^{-t\lambda} \right] \right\}$$

$$= \lim_{\varepsilon \to 0^+} \left\{ \exp\left(+\frac{1}{2} \int_\varepsilon^{1/\varepsilon} \frac{dt}{t} Tr(e^{-tA}) \right) \right\}. \tag{5.96}$$

In the case of the finitude of the right-hand side of Eq. (5.94) (which means that e^{-tA} is a class trace operator and the finitude up the proper-time parameter t), one can proceed as in the appendix to define mathematically the Gaussian path-integral.

Let us now consider the proper-time (Cauchy principal value sense) integration process as indicated by Eq. (5.95), for the case of N an even space-time dimensionality

$$I(m^2, \varepsilon) = \int_\varepsilon^{1/\varepsilon} \frac{dt}{t} \frac{e^{-tm^2}}{t^{N/2}} = \int_\varepsilon^{1/\varepsilon} dt \frac{e^{-tm^2}}{t^{\frac{N+2}{2}}} \tag{5.97}$$

The whole idea of the renormalization/regularization program means a (non-unique) choice of the mass parameter as a function of the proper-time cut-off ε in such a way that the infinite limit of $\varepsilon \to 0^+$ turns out to be finite, namely

$$\lim_{\varepsilon \to 0^+} I(m^2(\varepsilon), \varepsilon) < \infty. \tag{5.98}$$

A slight generalization of the above exposed Minlos–Bochner theorem in Hilbert spaces is the following theorem:

Theorem 5.3. *Let $(\mathcal{H}, \langle, \rangle_1)$ be a separable Hilbert space with the inner product $\langle, \rangle_1$. Let $(\mathcal{H}_0, \langle, \rangle_0)$ be a subspace of $\mathcal{H}$, so there is a trace class operator $T : \mathcal{H} \to \mathcal{H}$, such that the inner product $\langle, \rangle_0$ is given explicitly by $\langle g, h \rangle_1 = \langle g, Th \rangle_0 = \langle T^{1/2}g, T^{1/2}h \rangle_0$. Let us, thus, consider a positive definite functional $Z(j) \in C((\mathcal{H}, \langle, \rangle_1), R)$ [if $j_n \xrightarrow{\|\|_1} j$, then $Z(j_n) \to Z(j)$ on R^+]. We obtain that the Bochner path-integral representation of $Z(j)$ is given by a measure supported at these linear functionals, such that their restrictions in the subspace H_0 are continuous by the norm induced by the "trace-class" inner product $\langle, \rangle_1$.*

With this result in our hands, it became more or less straightforward to analyze the cylindrical measure supports in distributional spaces.

For instance, the basic Euclidean quantum field distributional spaces of tempered distributions in $R^N : S'(R^N)$, can always be seen as the strong topological dual of the inductive limit of Hilbert spaces below considered

$$s_p = \left\{ (x_n) \in \mathbb{C} \,\middle|\, \lim_{n \to \infty} n^p x_n \right.$$

$$\left. = 0, \text{ with the inner product } \langle (x_n), (y_n) \rangle_{s_p} = \sum_{n=1}^{\infty} n^{2p} x_n \bar{y}_n \right\}. \quad (5.99)$$

We note now that

$$S(R^N) = \bigcap_{p \geq 1}^{\infty} s_p \quad (5.100)$$

and

$$S'(R^N) = \bigcup_{p \geq 1}^{\infty} s_{-p}. \quad (5.101)$$

An important property is that $s_p \supset s_{p+1}$ and they satisfy the hypothesis of Theorem 5.3, since

$$\sum_{n=1}^{\infty} n^{2p} |x_n|^2 = \sum_{n=1}^{\infty} n^{(2p+2)} \frac{1}{n^2} |x_n^2| \quad (5.102)$$

and

$$\sum_{n=1}^{\infty} \frac{1}{n^2} = \frac{\pi^2}{6}. \quad (5.103)$$

As a consequence

$$||(x_n)||_{s_p} \leq \frac{\pi^2}{6} ||(x_n)||_{s_{p+1}} \quad (5.104)$$

If one has an arbitrary continuous positive-definite functional in $S(R^N)$, necessarily its measure support will always be on the topological dual of s_{p+1}, for each p. As a consequence, its support will be on the union set $\bigcup_{p=1}^{\infty} s_{-p}$ (since $s_{-p} \subset s_{-(p+1)}$) and thus it will be the whole distribution space $S'(R^N)$.

Similar results hold true in other distributional spaces.

The application of the above-mentioned result in Gaussian path-integrals is always made with the use of the famous result of the kernel theorem of Schwartz–Gelfand.

Theorem 5.4 (Gelfand). *Any continuous bilinear form $B(j,j)$ defined in the test space of the tempered distribution $S'(R^N)$ has the following explicit representation:*

$$B(j,j) = \int d^N x\, d^N y\, j(x)(D_x^m D_y^n F)(x,y)\bar{j}(y) \qquad (5.105)$$

with $j \in S(R^N)$, $F(x,y)$ is a continuous function of polynomial growth and D_x^m, D_y^n are the distributional derivatives of order m and n, respectively.

In all cases of application of this result to our study presented in the previous chapters were made in the context that $(D_x^m D_y^n F)(x,y)$ is a fundamental solution of a given differential operator representing the kinetic term of a given quantum field Lagrangean.

As a consequence, we have the basic result in the Gaussian path-integral in Euclidean quantum field theory

$$e^{-\frac{1}{2}B(j,j)} = \int_{T \in S'(R^N)} d\mu(T)\exp\{i(T(j))\}, \qquad (5.106)$$

where $T(j)$ denotes the action of the distribution T on the test function $j \in S(R^N)$.

At this point of our exposition let us show how to produce a fundamental solution for a given differential operator $P(D)$ with constant coefficients, namely

$$P(D) = \sum_{|\rho| \leq m} a_\rho D^\rho. \qquad (5.107)$$

A fundamental solution for Eq. (5.107) is given by a (numeric) distribution $E \in S'(R^N)$ such that for any $\varphi \in S(R^N)$, we have:

$$(P(D)E)(\varphi) = \delta(\varphi) = \varphi(0) \qquad (5.108)$$

or equivalently

$$E(^tP(D)\varphi) = \delta(\varphi), \qquad (5.109)$$

where $^tP(D)$ is the transpose operator through the duality of $S(R^N)$ and $S'(R^N)$.

By means of the use of a harmonic–Hermite expansion for the searched fundamental solution

$$E \overset{S'(R^N)}{=} \sum_{p=1}^{\infty}(E,H_p)H_p = \sum_{p=1}^{\infty}E_p H_p \qquad (5.110)$$

with H_p denoting the appropriate Hermite polynomials in R^N, together with the test function harmonic expansion

$$\varphi \overset{S(R^N)}{=} \sum_{p=1}^{\infty}(\varphi,H_p)H_p = \sum_{p=1}^{\infty}\varphi_p H_p \qquad (5.111)$$

and the use of the relationship between Hermite polynomials

$$^t P(D)H_p = \sum_{|q|\leq \ell(p)} M_{pq}H_q \qquad (5.112)$$

with $\ell(p)$ depending on the order of H_p and the order of the differential operator $^t P(D)$. (For instance in $S(R)$: $(d/dx)H_n(x) = 2nH_{n-1}(x)$;

$$\frac{d^2}{d^2 x}H_n(x) = 2n\frac{d}{dx}H_{n-1}(x) = 4n(n-1)H_{n-2}(x), \text{ etc.},$$

one obtains the recurrence equations for the searched coefficients E_p in Eq. (5.110)

$$\sum_n \varphi_n \left[\sum_{|q|\leq \ell(n)} M_{nq}E_q \right] = \sum_n \varphi_n H_n(0) \qquad (5.113)$$

for any $(\varphi_n) \in \ell^2$ or equivalently

$$\sum_{|q|\leq \ell(n)} M_{nq}E_q = H_n(0) \qquad (5.114)$$

the solution of the above written infinite-dimensional system produces a set of coefficients $\{E_p\}_{p=1,...,\infty}$ satisfying a condition that it belongs to some space s_{-r}, where r is the order of the fundamental solution being searched (the rigorous proof of the above assertions is left as an exercise to our mathematically oriented reader!).

Finally, let us sketch the connection between path-integrals and the operator framework in Euclidean quantum field theory, both still mathematically non-rigorous from a strict mathematical point of view. In other former approach, one has a self-adjoint operator $\bar{H}(j)$ indexed by a set of functions (the field classical sources of the quantum field theory under analysis) belonging to the distributional space $S(R^N)$. This self-adjoint operator is formally given by the space-time integrated Lagrangean field theory and the basic object is the generating functional as defined by the vacuum–vacuum transition amplitude

$$Z(j) = \langle e^{iH(j)}\Omega_{\text{VAC}}, \Omega_{\text{VAC}}\rangle \tag{5.115}$$

with Ω_{VAC} denoting the theory vacuum state. It is assumed that $Z(j)$ is a continuous positive definite functional on $S(R^N)$. As a consequence of the above exposed theorems of Minlos and Bochner, there is a cylindrical measure $d\mu(T)$ on $S'(R^N)$ such that the generating functional $Z(j)$ is represented by the quantum field path-integral defined by the above-mentioned measure:

$$Z(j) = \int_{S'(R^N)} d\mu(T)\exp\{i(T(j)\} \tag{5.116}$$

in which the Feynman symbolic notation express itself in the following symbolic-operational Feynman notation:

$$Z[j(x,t)] = \frac{1}{Z(0)}\int_{S'(R^N)} D^F[T(x,t)]$$

$$\times\ e^{-\frac{1}{2}\int_{-\infty}^{+\infty} d^{n-1}x dt \mathcal{L}(T,\partial_t T,\partial_x T)} e^{i\int_{-\infty}^{+\infty} d^{n-1}x dt j(x,t)T(x,t)}, \tag{5.117}$$

with $\mathcal{L}(T,\partial_t T,\partial_x T)$ means generically the Lagrangean density of our field theory under quantization.

Let us exemplify the Feynman symbolic Euclidean path-integral as given by Eq. (5.117) in the Gaussian case (free Euclidean field massless theory in R^N, $N \geq 2$)

$$\exp\left\{-\frac{1}{2}\int_{-\infty}^{+\infty} d^N x\int_{-\infty}^{+\infty} d^N y j(x)\left(\frac{(-1)}{(n-2)\Gamma(n)|x-y|^{n-2}}\right)j(y)\right\}$$

$$= \int_{S'(R^N)} D^F[T(x)]$$

$$\times\exp\left\{-\frac{1}{2}\int_{-\infty}^{+\infty} d^N x\int_{-\infty}^{+\infty} d^N y T(x)((-\Delta)_x\delta^{(N)}(x-y)T(y)\right\}$$

$$\times\det{}^{+\frac{1}{2}}((-\Delta)x\delta^{(N)}(x-y))\exp\left\{i\int_{-\infty}^{+\infty} d^N x j(x)T(x)\right\}. \tag{5.118}$$

Or for the heat differential operator in $S'(R^N \times R^+)$

$$\exp\left\{-\frac{1}{2}\int_{-\infty}^{+\infty}d^Nx\int_0^\infty dt\int_{-\infty}^{+\infty}d^Ny\int_0^\infty dt'\right.$$

$$\left.\times\left(\frac{1}{(\sqrt{2\pi(t-t')}))^N}\exp\left(-\frac{|x-y|^2}{4(t-t')}\right)\right)\theta(t-t')j(y,t')\right\}$$

$$=\int_{S'(R^N\times R^+)}D^F[T(x,t)]$$

$$\times\exp\left\{-\frac{1}{2}\int_{-\infty}^{+\infty}d^Nx\int_0^\infty dt\,T(x,t)\left[\left(\Delta_x-\frac{\partial}{\partial t}\right)T(x,t)\right]\right\}$$

$$\times\det{}^{1/2}\left[\Delta_x-\frac{\partial}{\partial t}\right]\times\exp\left\{i\int_{-\infty}^{+\infty}d^Nx\int_0^\infty dt\,J(x,t)T(x,t)\right\}.$$

$$(5.119a)$$

It is worth to point out that the fourth-order path-integral in $S'(R^4)$:

$$\exp\left\{-\frac{1}{2}\int_{-\infty}^{+\infty}d^4x\int_{-\infty}^{+\infty}d^4(x)j(x)\left(\frac{1}{8\pi}|x-y|^2\ell n(x-y)\right)j(y)\right\}$$

$$=\int_{S'(R^4)}D^F[\varphi(x)]e^{-\frac{1}{2}\int_{-\infty}^{+\infty}d^4x(\varphi(x)(\Delta^2)\varphi(x))}$$

$$=\int_{R'(R^4)}d_{\Delta^2}\mu(\varphi)e^{i\varphi(j)}e^{i\int_{-\infty}^{+\infty}d^4xj(x)\varphi(x)}\qquad(5.119b)$$

holds mathematically true since the locally integrable function $\frac{1}{8\pi}|x|^2\ell n|x|$ is a fundamental solution of the differential operator $\Delta^2 : S'(R^4) \to S(R^4)$ when acting on distributional spaces.

As a last important point of this section, we present an important result on the geometrical characterization of massive free field on a Euclidean space-time.

Firstly, we announce a slightly improved version of the usual Minlos theorem.

Theorem 5.5. *Let E be a nuclear space of test functions and $d\mu$ a given σ-measure on its topologic dual with the strong topology. Let $\langle,\rangle_0$ be an inner product in E, inducing a Hilbertian structure on $\mathcal{H}_0 = \overline{(E,\langle,\rangle_0)}$, after its topological completion.*

We assume the following:

(a) There is a continuous positive definite functional in $\mathcal{H}_0$, $Z(j)$, with an associated cylindrical measure $d\mu$.

(b) There is a Hilbert–Schmidt operator $T : \mathcal{H}_0 \to \mathcal{H}_0$; invertible, such that $E \subset \text{Range}(T)$, $T^{-1}(E)$ is dense in $\mathcal{H}_0$ and $T^{-1} : \mathcal{H}_0 \to \mathcal{H}_0$ is continuous.

We have, thus, that the support of the measure satisfies the relationship

$$\text{supp } d\mu \subseteq (T^{-1})^*(\mathcal{H}_0) \subset E^*. \tag{5.120}$$

At this point we give a non-trivial application of Theorem 5.5.

Let us consider a differential inversible operator $\mathcal{L} : S'(R^N) \to S(R)$, together with a positive inversible self-adjoint elliptic operator $P : D(P) \subset L^2(R^N) \to L^2(R^N)$. Let H_α be the following Hilbert space:

$$H_\alpha = \left\{ \overline{S(R^N)}, \langle P^\alpha \varphi, P^\alpha \varphi \rangle_{L^2(R^N)} = \langle , \rangle_\alpha, \quad \text{for } \alpha \text{ a real number} \right\}. \tag{5.121}$$

We can see that for $\alpha > 0$, the operators below

$$P^{-\alpha} : L^2(R^N) \to \mathcal{H}_{+\alpha} \tag{5.122}$$
$$\varphi \to (P^{-\alpha}\varphi)$$

$$P^\alpha : \mathcal{H}_{+\alpha} \to L^2(R^N) \tag{5.123}$$
$$\varphi \to (P^\alpha \varphi)$$

are isometries among the following subspaces

$$\overline{D(P^{-\alpha}), \langle , \rangle_{L^2})} \quad \text{and} \quad H_{+\alpha}$$

since

$$\langle P^{-\alpha}\varphi, P^{-\alpha}\varphi \rangle_{\mathcal{H}_{+\alpha}} = \langle P^\alpha P^{-\alpha}\varphi, P^\alpha P^{-\alpha}\varphi \rangle_{L^2(R^N)} = \langle \varphi, \varphi \rangle_{L^2(R^N)} \tag{5.124}$$

and

$$\langle P^\alpha f, P^\alpha f \rangle_{L^2(R^N)} = \langle f, f \rangle_{H_{+\alpha}}. \tag{5.125}$$

If one considers T a given Hilbert–Schmidt operator on H_α, the composite operator $T_0 = P^\alpha T P^{-\alpha}$ is an operator with domain being $D(P^{-\alpha})$ and its image being the Range(P^α). T_0 is clearly an invertible operator and $S(R^N) \subset$ Range(T) means that the equation $(TP^{-\alpha})(\varphi) = f$ has always a non-zero solution in $D(P^{-\alpha})$ for any given $f \in S(R^N)$. Note that the condition $T^{-1}(f)$ be a dense subset on Range($P^{-\alpha}$) which means that

$$\langle T^{-1}f, P^{-\alpha}\varphi \rangle_{L^2(R^N)} = 0 \tag{5.126}$$

has a unique solution of the trivial solution $f \equiv 0$.

Let us suppose that $T^{-1} : S(R^N) \to H_\alpha$ be a continuous application and the bilinear term $(\mathcal{L}^{-1}(j))(j)$ be a continuous application in the Hilbert spaces $H_{+\alpha} \supset S(R^N)$, namely: if $j_n \xrightarrow{L^2} j$, then $\mathcal{L}^{-1} : P^{-\alpha}j_n \xrightarrow{L^2} \mathcal{L}^{-1}P^{-\alpha}j$, for $\{j_n\}_{n\in\mathbb{Z}}$ and $j_n \in S(R^N)$.

By a direct application of Theorem 5.5, we have the result

$$Z(j) = \exp\left\{ -\frac{1}{2}[\mathcal{L}^{-1}(j)(j)] \right\} = \int_{(T^{-1})^* H_\alpha} d\mu(T) \exp(iT(j)). \tag{5.127}$$

Here, the topological space support is given by

$$\begin{aligned}
(T^{-1})^* \mathcal{H}_\alpha &= [(P^{-\alpha}T_0 P^\alpha)^{-1}]^* (\overline{(P^\alpha(S(R^N)))}) \\
&= [(P^\alpha)^* (T_0^{-1})^* (P^{-\alpha})^*] P^\alpha (S(R^N)) \\
&= P^\alpha T_0^{-1}(L^2(R^N)).
\end{aligned} \tag{5.128}$$

In the important case of $\mathcal{L} = (-\Delta + m^2) : S'(R^N) \to S(R^N)$ and $T_0 T_0^* = (-\Delta + m^2)^{-2\beta} \in \oint_1(L^2(R^N))$ since $\mathrm{Tr}(T_0 T_0^*) = \frac{1}{2(m^2)^\beta}(\frac{m^2}{1})^{\frac{N}{2}}\frac{\Gamma(\frac{N}{2})\Gamma(2\beta-\frac{N}{2})}{\Gamma(\beta)}\langle\infty$ for $\beta\rangle\frac{N}{4}$ with the choice $P = (-\Delta + m^2)$, we can see that the support of the measure in the path-integral representation of the Euclidean measure field in R^N may be taken as the measurable subset below

$$\mathrm{supp}\{d_-(-\Delta + m^2)u(\varphi)\} = (-\Delta + m^2)^{-\alpha} \cdot (-\Delta + m^2)^{+\beta}(L^2(R^N)) \tag{5.129}$$

since $\mathcal{L}^{-1}P^{-\alpha} = (-\Delta + m^2)^{-1-\alpha}$ is always a bounded operator in $L^2(R^N)$ for $\alpha > -1$.

As a consequence each field configuration can be considered as a kind of "fractional distributional" derivative of a square integrable function written as

$$\varphi(x) = \left[(-\Delta + m^2)^{\frac{N}{4} + \varepsilon - 1} f\right](x) \tag{5.130}$$

with a function $f(x) \in L^2(R^N)$ and any given $\varepsilon > 0$, even if originally all fields configurations entering into the path-integral were elements of the Schwartz tempered distribution spaces $S'(R^N)$ certainly very "rough" mathematical objects to characterize from a rigorous geometrical point of view.

We have, thus, made a further reduction of the functional domain of the free massive Euclidean scalar field of $S'(R^N)$ to the measurable subset as given in Eq. (5.130) denoted by $W(R^N)$

$$\exp\left\{-\frac{1}{2}[(-\Delta + m^2)^{-1}j](j)\right\}$$

$$= \int_{S'(R^N)} d_{(-\Delta+m^2)}\mu(\varphi) e^{i\varphi(j)}$$

$$= \int_{W(R^N) \subset S'(R^N)} d_{(-\Delta+m^2)}\tilde{\mu}(f) e^{i\langle f, (-\Delta+m^2)^{\frac{N}{4}+\varepsilon-1}f\rangle_{L^2(R^N)}}. \tag{5.131}$$

References

1. B. Simon, *Functional Integration and Quantum Physics* (Academic Press, 1979).
2. B. Simon, The $P(\phi)_2$ Euclidean (quantum) field theory, in *Princeton Series in Physics* (1974).
3. J. Glimm and A. Jaffe, *Quantum Physics, — A Functional Integral Point of View* (Springer Verlag, 1988).
4. Y. Yamasaki, *Measure on Infinite Dimensional Spaces*, Vol. 5 (World Scientific, 1985).
5. D.-X. Xia, *Measure and Integration Theory on Infinite Dimensional Space* (Academic Press, 1972).
6. L. Schwartz, *Random Measure on Arbitrary Topological Space and Cylindrical Measures* (Tata Institute, Oxford University Press, 1973).
7. K. Symanzik, *J. Math. Phys.* **7**, 510 (1966).
8. B. Simon, *J. Math. Phys.* **12**, 140 (1971).
9. L. C. L. Botelho, *Phys. Rev.* **33**D, 1195 (1986).

10. E. Nelson, Regular probability measures on function space, *Ann. Math.* **69**(2), 630–643 (1959); V. Rivasseau, *From Perturbative to Constructive Renormalization* (Princeton University, 1991).

11. L. C. L. Botelho, *Il Nuovo Cimento* **117**B, 37 (2002); *Il Nuovo Cimento* **117**B, 331 (2002); *J. Phys. A: Math. Gen.* **34**, 131–137 (2001); *Mod. Phys. Lett.* 173, Vol. 13/14, 733 (2003); *Il Nuovo Cimento* **117**B, 15 (2002); Ya. G. Sinai, *Topics in Ergodic Theory* (Princeton University Press, 1994).

12. R. Teman, *Infinite-Dimensional Systems in Mechanics and Physics*, Vol. 68 (Springer-Verlag, 1988); J. M. Mourão, T. Thiemann and J. M. Velhinho, *J. Math. Phys.* **40**(5), 2337 (1995).

13. A. M. Polyakov, *Phys. Lett.* **103**B, 207 (1981); L. C. L. Botelho, *J. Math. Phys.* **30**, 2160 (1989).

14. L. C. L. Botelho, *Phys. Rev.* **49**D, 1975 (1994).

15. A. M. Polyakov, *Phys. Lett.* **103**B, 211 (1981).

16. L. Brink and J. Schwarz, *Nucl. Phys.* B**121**, 285 (1977).

17. L. C. L Botelho, *Phys. Rev.* D**33**, 1195 (1986).

18. B. Durhuus, *Quantum Theorem of String*, Nordita Lectures (1982).

19. C. G. Callan Jr., S. Coleman, J. Wess and B. Zumine, *Structure of Phenomenological Lagrangians — II, Phys. Rev.* **177**, 2247 (1969).

20. L. C. L. Boteho, Path integral Bosonization for a non-renormalizable axial four-dimensional. Fermion Model, *Phys. Rev.* D**39**(10), 3051–3054 (1989).

Appendix A

In this appendix, we have given new functional analytical proofs of the Bochner–Martin–Kolmogorov theorem of Sec. 5.2.

Bochner–Martin–Kolmogorov Theorem (Version I). *Let* $f : E \to R$ *be a given real function with domain being a vector space E and satisfying the following properties*:

(1) $f(0) = 1$.

(2) *The restriction of f to any finite-dimensional vector subspace of E is the Fourier Transform of a real continuous function of compact support.*

Then there is a measure $d\mu(h)$ on σ-algebra containing the Borelians if the space of linear functionals of E with the topology of pontual convergence denoted by E^{alg} such that for any $y \in E$

$$f(g) = \int_{E^{\text{alg}}} \exp(ih(g))d\mu(h). \qquad (A.1)$$

Proof. Let $\{\hat{e}_{\lambda \in A}\}$ be a Hamel (vectorial) basis of E, and $E^{(N)}$ a given subspace of E of finite-dimensional. By the hypothesis of the theorem, we have that the restriction of the functions to $E^{(N)}$ (generated by the elements of the Hamel basis $\{\hat{e}_{\lambda_1}, \dots, \hat{e}_{\lambda_N}\} = \{e_\lambda\}_{\lambda \in \Lambda_F}$) is given by the Fourier transform

$$f\left(\sum_{\ell=1}^{N} \sigma \lambda_\ell \hat{e}_{\lambda_\ell}\right) = \int_{\prod_{\lambda \in \Lambda_F} R^\lambda} (dP_{\lambda_1} \cdots dP_{\lambda_N})$$

$$\times \exp\left[\sum_{\ell=1}^{N} a_{\lambda_\ell} P_{\lambda_\ell}\right] \hat{g}(P_{\lambda_1}, \dots, P_{\lambda_N}), \qquad (A.2)$$

with $\hat{g}(P_{\lambda_1}, \dots, P_{\lambda_N}) \in C_c(\prod_{\lambda \in \Lambda_F} R^\lambda)$. $\qquad \square$

As a consequence of the above result, we consider the following well-defined family of linear positive functionals on the space of continuous function on the product space of the Alexandrov compactifications of R denoted by R^w:

$$L_{\lambda_F} \in \left[C\left(\prod_{\lambda \in \Lambda_F} (R^w)^\lambda; R\right)\right]^{\text{Dual}}, \qquad (A.3)$$

with

$$L_{\Lambda_F}[\hat{g}(P_{\lambda_1}, \dots, P_{\lambda_N})] = \int_{\prod_{\lambda \in \Lambda_F} (R^w)^\lambda} \int \hat{g}(P_{\lambda_1}, \dots, P_{\lambda_N})(dP_{\lambda_1} \cdots dP_{\lambda_N}).$$

$$(A.4)$$

Here, $\hat{g}(P_{\lambda_1}, \dots, P_{\lambda_N})$ still denotes the unique extension of Eq. (A.2) to the Alexandrov compactification R^w.

Now, we remark that the above family of linear continuous functionals have the following properties:

(1) The norm of L_{λ_F} is always the unity since

$$\|L_{\lambda_1}\| = \int_{\prod_{\lambda \in \Lambda_F} (R^w)^\lambda} \hat{g}(P_{\lambda_1}, \dots, P_{\lambda_N}) dP_{\lambda_1} \cdots dP_{\lambda_N} = 1. \qquad (A.5)$$

(2) If the index set Λ_F, contains Λ_F the restriction of the associated linear functional Λ_F, to the space $C(\prod_{\lambda \in \Lambda_f} (R^w)^\lambda, R)$ coincides with L_{Λ_F}.

Now, a simple application of the Stone–Weierstrass theorem shows us that the topological closure of the union of the subspace of functions of finite variable is the space $C(\prod_{\lambda \in A}(R^w)^\lambda, R)$, namely

$$\overline{\bigcup_{\Lambda_F \subset A} C\left(\prod_{\lambda \in \Lambda_F}(R^w)^\lambda, R\right)} = C\left(\prod_{\lambda \in A}(R^w)^\lambda, R\right), \qquad (A.6)$$

where the union is taken over all family of subsets of finite elements of the index set A.

As a consequence of the Remark 2 and Eq. (A.6) there is a unique extension of the family of linear functionals $\{L_{\Lambda_F}\}$ to the whole space $C(\prod_{\lambda \in A}(R^w)^\lambda, R)$ and denoted by L_∞. The Riesz Markov theorem gives us a unique measure $d\bar{\mu}(h)$ on $\prod_{\lambda \in A}(R^w)^\lambda$ representing the action of this functional on $C(\prod_{\lambda \in A}(R^w)^\lambda, R)$.

We have, thus, the following functional integral representation for the function $f(g)$:

$$f(g) = \int_{(\prod_{\lambda \in A}(R^w)^\lambda)} \exp(i\bar{h}(g)) d\mu(\bar{h}). \qquad (A.7)$$

Or equivalently (since $\bar{h}(g) = \sum_{i=1}^N p_i a_i$ for some $\{p_i\}_{i \in N} < \infty$), we have the result

$$f(g) = \int_{(\prod_{\lambda \in A} R^\lambda)} (\exp ih(g)) d\mu(h) \qquad (A.8)$$

which is the proposed theorem with $h \in (\prod_{\lambda \in \lambda_F} R^\lambda)$ being the element which has the image of $\bar{h}$ on the Alexandrov compactification $\prod_{\lambda \in \lambda_F}(R^w)^\lambda$.

The practical use of the Bochner–Martin–Kolmogorov theorem is difficult by the present day' non-existence of an algoritm generating explicitly a Hamel (vectorial) basis on function of spaces. However, if one is able to apply the theorems of Sec. 5.3, one can construct explicitly the functional measure by only considering the topological basis as in the Gaussian functional integral equation (5.32).

Bochner–Martin–Kolmogorov Theorem (Version II). We now have the same hypothesis and results of Theorem Version 1 but with the more general condition.

(3) *The restriction of f to any finite-dimensional vector subspace of E is the Fourier Transform of a real continuous function vanishing at "infinite."*

For the proof of the theorem under this more general mathematical condition, we will need two lemmas and some definitions.

Definition A.1. Let X be a normal space, locally compact and satisfying the following σ-compacity condition:

$$X = \bigcup_{n=0}^{\infty} K_k, \tag{A.9}$$

with

$$K_n \subset \mathrm{int}(K_{n+1}) \subset K_{n+1}. \tag{A.10}$$

We define the following space of continuous function "vanishing" at infinite:

$$\tilde{C}_0(X, R) = \left\{ f(x) \in C(X, R) \,\middle|\, \lim_{n \to \infty} \sup_{x \in (K_n)^c} |f(x)| = 0 \right\}. \tag{A.11}$$

We have, thus, the following lemma.

Lemma A.1. *The topological closure of the functions of compact support contains* $\tilde{C}_0(X, R)$ *in the topology of uniform convergence.*

Proof. Let $f(x) \in \tilde{C}_0(X, R)$ and $g_\mu \in C(X, R)$, the (Uryhson) functions associated with the closed disjoints sets $\bar{K}_n$ and (K_{n+1}^c). Now it is straightforward to see that $(f \cdot g_n)(x) \in C_i(X, R)$ and converges uniformly to $f(x)$ by definition (A.11).

At this point, we consider a linear positive continuous functional L on $\tilde{C}_0(X, R)$. Since the restriction of L to each subspace $C(K_n, R)$ satisfy the conditions of the Riesz–Markov theorem, there is a unique measure $\mu^{(n)}$ on K_n containing the Borelians on K_n and representing this linear functional restriction. We now use the hypothesis equation (A.10) to have a well-defined measure on a σ-algebra containing the Borelians of X

$$\bar{\mu}(A) = \limsup \mu^{(n)} \left(A \bigcap K_n \right). \tag{A.12}$$

for A in this σ-algebra and representing the functional L on $\tilde{C}_0(X, R)$

$$L(f) = \int_X f(x) d\bar{\mu}(x). \tag{A.13}$$

Note that the normal of the topological space X is a fundamental hypothesis used in this proof by means of the Uryhson lemma.

Unfortunately, the non-countable product space $\prod_{\lambda \in A} R^\lambda$ is not a normal topological space (the famous Stone counter example) and we cannot, thus, apply the above lemma to our vectorial case equation (A.8). However, we can overcome the use of the Stone–Weierstrass theorem in the proof of the Bochner–Martin–Kolmogorov theorem by considering directly a certain functional space instead of that given by Eq. (A.6).

Thus, we define the following space of infinite-dimensional functions vanishing at finite

$$C_0(R^\infty, R) \equiv C_0 \left(\prod_{\lambda \in A} R^\lambda, R \right) \overset{\text{def}}{=} \overline{\bigcup_{\Lambda_F \subset A} \tilde{C}_0 \left(\prod_{\lambda \in \lambda_F} R^\lambda, R \right)}, \qquad \text{(A.14)}$$

where the closure is taken in the topology of uniform convergence.

If we consider a given continuous linear functional L on $C_0(\prod_{\lambda \in A} R^\lambda, R)$ there is a unique measure μ^∞ on the union of the Borelians $\prod_{\lambda \in \Lambda_F} R^\lambda$ representing the action of L on $C_0(R^\infty, R)$.

Conversely, given a family of consistent measures $\{\mu_{\Lambda_F}\}$ on the finite-dimensional spaces $(\prod_{\lambda \in \Lambda_F} R^\lambda)$ satisfying the property of $\mu_{\Lambda_F}(\prod_{\lambda \in \Lambda_F} R^\lambda) = 1$, there is a unique measure on the cylinders $\prod_{\lambda \in A} R^\lambda$ associated with the functional L on $C_0(\prod_{\lambda \in A} R^\lambda, R)$.

By collecting the results of the above written lemmas we will get the proof of Eq. (A.8) in this more general case. $\qquad \square$

Appendix B. On the Support Evaluations of Gaussian Measures

Let us show explicitly by one example of ours of the quite complex behavior of cylindrical measures on infinite-dimensional spaces R^∞.

Firstly, we consider the family of Gaussian measures on $R^\infty = \{(x_n)_{1 \le n \le \infty}, x_n \in R\}$ with $\sigma_n \in \ell^2$.

$$d^{(\infty)} \mu(\{x_n\}) = \limsup_N \left\{ \prod_{n=1}^N \left(dx_n \frac{1}{\sqrt{\sigma_n \pi}} \right) e^{-\frac{x_n^2}{2\sigma_n^2}} \right\}. \qquad \text{(B.1)}$$

Let us introduce the measurable sets on R^∞

$$E_{(\alpha_n)} = \left\{ (x_n) \in R^\infty; \|x\|_{(x_n)}^2 = \sum_{n=1}^\infty \alpha_n^2 x_n^2 < \infty \right\} \quad \text{and} \quad \sum_{\ell=1}^\infty \alpha_n^2 \sigma_n^2 < \infty. \qquad \text{(B.2)}$$

Here, $\{\alpha_n\}$ is a given sequence suppose to belonging to ℓ^2 either.

Now, it is straightforward to evaluate the "mass" of the infinite-dimensional set $E_{(x_n)}$, namely

$$
\begin{aligned}
{}^{(\infty)}\mu(E_{(\alpha_n)}) &= \int_{R^\infty} d^\infty\mu(\{x_n\}) \left[\lim_{\varepsilon\to 0^+} e^{-\varepsilon\left(\sum_{n=1}^\infty \alpha_n^2 x_n^2\right)} \right] \\
&= \lim_{\varepsilon\to 0^+} \left\{ \lim_{0\le\ell\le n}\sup \left[\prod_{\ell=1}^n (1+2\varepsilon\,\alpha_n^2\sigma_n^2)^{-\frac12} \right] \right\} \quad\text{(B.3)}
\end{aligned}
$$

Note that

$$
\left(\prod_{\ell=1}^n (1+2\varepsilon\,\alpha_n^2\sigma_n^2)^{-\frac12} \right) \le \frac{1}{1+\sum_{\ell=1}^n \alpha_n^2\sigma_n^2}. \quad\text{(B.4)}
$$

As a consequence, one can exchange the order of the limits in Eq. (B.3) and arriving at the result

$$
\begin{aligned}
{}^{(\infty)}\mu(E_{(\alpha_n)}) &= \lim_{0\le\ell\le n}\sup \left\{ \lim_{\varepsilon\to 0^+} \left[\prod_{\ell=1}^n (1+2\varepsilon\,\alpha_n^2\sigma_n^2)^{-\frac12} \right] \right\} \\
&= \lim_{0\le\ell\le n}\sup \{1\} = 1. \quad\text{(B.5)}
\end{aligned}
$$

So we can conclude, on the basis of Eq. (B.5) that the support of the measure equation (B.1) is the set of $E_{(\alpha_n)}$ for any possible sequence $\{\alpha_n\} \in \ell^2$. Let us show that $(E_{(\alpha_n)})^C \cap E_{(\beta_n)} \ne \{\phi\}$, so that these sets are not coincident.

Let the sequences be

$$
\begin{aligned}
\sigma_n &= n^{-\sigma}, \\
\alpha_n &= n^{\sigma-1}, \quad\quad\text{(B.6)} \\
\beta_n &= n^{\sigma-\lambda},
\end{aligned}
$$

with $\gamma > 1$ and $\sigma > 0$.

We have that

$$
\sum \alpha_n^2\sigma_n^2 = \sum \frac{1}{n^2} = \frac{\pi^2}{6}, \quad\text{(B.7)}
$$

$$
\sum \beta_n^2\sigma_n^2 = \sum n^{-2\lambda} < \infty. \quad\text{(B.8)}
$$

So $E_{\{\alpha_n\}}$ and $E_{\{\beta_n\}}$ are non-empty sets on R^∞.

Let us consider the point $\{\bar{x}_n\} \in R^\infty$ and defined by the relation

$$\bar{x}_n^2 = n^{-2(\sigma-1)-\varepsilon}. \tag{B.9}$$

We have that

$$\sum (\bar{x}_n)^2 \alpha_n^2 = \sum n^{-2(\sigma-1)-\varepsilon} \cdot n^{n^2(\sigma-1)} = \sum n^{-\varepsilon}$$

and

$$\sum (\bar{x}_n)^2 \beta_n^2 = \sum n^{-2(\sigma-1)-\varepsilon} \cdot n^{2(\sigma-\lambda)} = \sum n^{2-\varepsilon-2\lambda}$$

If we choose $\varepsilon = 1$; $\gamma > 1$ ($\gamma = \frac{3}{2}!$), we obtain that the point $\{\bar{x}_n\}$ belongs to the set $E_{\{\beta_n\}}$ (since $\sum n^2 = \frac{\pi^2}{6}$), however it does not belongs to $E_{\{x_n\}}$ (since $\sum_{n=0}^{\infty} n^{-1} = +\infty$), although the support of the measure equation (B.1) is any set of the form $E_{\{\gamma_n\}}$ with $\{\gamma_n\} \in \ell^2$.

Appendix C. Some Calculations of the Q.C.D. Fermion Functional Determinant in Two-Dimensions and (Q.E.D.)$_2$ Solubility

Let us first define the functional determinant of a self-adjoint, positive definite operator A (without zero modes) by the proper-time method

$$\log \det_F(A) = -\lim_{\beta \to 0^+} \left\{ \int_\varepsilon^\infty \frac{dt}{t} \operatorname{Tr}_f(e^{-tA}) \right\}, \tag{C.1}$$

where the subscript f reminds us the functional nature of the objects under study and so its trace.

It is thus expected that the definition equation (C.1) has divergents counter terms as $\varepsilon \to 0^+$, since $\exp(-tA)$ is a class trace operator only for $t \geq \varepsilon$. Asymptotic expressions at the short-time limit $t \to 0^+$ are well-known in mathematical literature. However, this information is not useful in the first sight of Eq. (C.1) since one should know $\operatorname{Tr}_F(e^{-tA})$ for all t-values in $[\varepsilon, \infty)$.

A useful remark on the exact evaluation is in the case where the operator A is of the form $A = B + m^2 \mathbf{1}$ and one is mainly interested in the effective asymptotic limit of large mass $m^2 \to \infty$. In this particular case, one can use a saddle-point analysis of the expression in Eq. (C.1)

$$\lim_{m^2 \to \infty} [\log \det{}_F(B + m^2\mathbf{1})] = -\lim_{\varepsilon \to 0} \left\{ \int_\varepsilon^\infty \frac{dt}{t} e^{-m^2 t} \left[\lim_{t \in 0^+} \mathrm{Tr}_F(e^{-tB}) \right] \right\}.$$

(C.2)

Another very important case is covered by the (formal) Schwarz–Romanov theorem announced below.

Theorem C.1. *Let $A(\sigma)$ be a one-parameter family of positive-definite self-adjoints operators and satisfying the parameter derivative condition* ($0 \leq \sigma \leq 1$)

$$\frac{d}{d\sigma} A(\sigma) = fA(\sigma) + A(\sigma)g,$$

(C.3a)

where f and A are σ-independent objects (may be operators).

Then we have an explicit result

$$\log \left(\frac{\det{}_F(A(1))}{\det{}_F(A(0))} \right) = \left\{ \int_0^1 d\sigma \lim_{\varepsilon \to 0^+} \mathrm{Tr}_F[fe^{-\varepsilon(A^{(\sigma)})^2}] \right.$$

$$\left. + \int_0^1 d\sigma \lim_{\varepsilon \to 0^+} \mathrm{Tr}_F[ge^{-\varepsilon(A^{(\sigma)})^2}] \right\}.$$

(C.3b)

The proof of Eqs. (C.3a) and (C.3b) is based on the validity of the differential equation in relation to the σ-parameter

$$\frac{d}{d\sigma} [\log \det{}_F(A(\sigma))^2] = \lim_{\varepsilon \to 0^+} 2 \left\{ \mathrm{Tr}_F(fe^{-\varepsilon(A(\sigma))^2}) + \mathrm{Tr}_\varepsilon(ge^{-\varepsilon(A(\sigma))^2}) \right\}$$

(C.4)

which can be seen from the obvious calculations written down in the above equation

$$[\log \det{}_F(A(\sigma))^2]$$

$$= -\lim_{\varepsilon \to 0^+} \left\{ \int_\varepsilon^\infty dt\, \mathrm{Tr}_F \left[(fA(\sigma) + A(\sigma)g)A(\sigma) + A(\sigma)(fA(\sigma) \right.\right.$$

$$\left.\left. + A(\sigma)g)(-(A(\sigma))^{-2}) \frac{d}{dt} (\exp(-t(A(\sigma))^2)) \right] \right\}$$

$$= \text{Eq. (C.4)}.$$

(C.5)

Let us apply the above formulae in order to evaluate the functional determinant of the "Chrially transformed" self-adjoint Dirac operator in a two-dimensional space-time

$$D(\sigma) = \exp(\sigma\gamma^5\varphi^2(x)\lambda_a)|\partial|(\exp(\sigma\gamma^5\varphi^a(x)\lambda_a). \tag{C.6}$$

Here, the Chiral Phase $W[\psi]$ in Eq. (C.6) takes value in $SU(N)$ for instance.
One can see that

$$(\not{D}(\sigma))^2 = -(\partial_\mu + iG_\mu(\sigma))^2\mathbf{1} - \frac{1}{4}[\gamma_\mu, \gamma_\nu]\, F_{\mu\nu}(-iG_\mu(\sigma)) \tag{C.7a}$$

with the Gauge field

$$\gamma_\mu G_\mu = \gamma_\mu(W^{-1}\partial_\mu W). \tag{C.7b}$$

Hence, the asymptotics of the operator (C.7a) are easily evaluated (see Chap. 1, Appendix E)

$$\lim_{\varepsilon\to 0^+} \mathrm{Tr}_F\left(\exp(-\varepsilon(D(\sigma))^2)\right)$$

$$= \lim_{\varepsilon\to 0^+} \mathrm{Tr}\left\{\frac{1}{4\pi\varepsilon}\left\{1 + \frac{g(i\varepsilon_{\mu\nu}\gamma_5)}{2}F_{\mu\nu}(-iG_\mu(\sigma))\right\}\right\}, \tag{C.8a}$$

where we have used the Seeley expansions below for the square of the Dirac operator in the presence of a non-Abelian connection

$$A : C_0^\infty(R^2) \to C_0^\infty(R^2), \tag{C.8b}$$

$$A\varphi = (-\Delta) - V_1(\xi_1, \xi_2)\frac{\partial}{\partial\xi_1} - V_2(\xi_1, \xi_2)\frac{\partial}{\partial\xi_2} - V_0((\xi_1, \xi_2), \tag{C.8c}$$

$$\lim_{t\to 0^+}\left\{\mathrm{Tr}_{C_c^\infty(R^2)}(e^{-tA})\right\}$$

$$= \frac{1}{4\pi t} + \left\{\left(-\frac{1}{8\pi}\left(\frac{\partial}{\partial\xi_1}V_1 + \frac{\partial}{\partial\xi_2}V_0\right)\right) - \frac{1}{16\pi}(V_1^2 + V_2^2) - \frac{1}{4\pi}V_0\right\}$$

$$\times (\xi_1, \xi_2) + O(t), \tag{C.8d}$$

$$(\not{\partial} - ig\not{G}_\mu)^2 = (-\partial^2)_\xi + (2igG_\mu\partial_m u)_\xi$$

$$+ \left[ig(\partial_\mu G_\mu) + \frac{ig\sigma^{\mu\nu}}{2}F_{\mu\nu}(G) + g^2G_\mu^2\right]_\xi, \tag{C.8e}$$

which leads to the following exact integral (non-local) representation for the non-Abelian Dirac determinant

$$\log \left\{ \frac{\det_F(D(1))}{\det_F(\partial)} \right\}$$

$$= \frac{i}{2\pi} \int d^2x \, \text{Tr}_{SU(N)} \left\{ \varphi_a(x)\lambda^a \left[\int_0^1 d\sigma \mathcal{E}_{\mu\nu} F^{\mu\nu}(-iG_\mu(\sigma)) \right] \right\}. \tag{C.9}$$

A more invariant expression for Eq. (C.9) can be seen by considering the decomposition of the "$SU(N)$ gauge field" $G_\mu(\sigma)$ in terms of its vectorial and axial components:

$$(W^{-1}(\sigma)\partial_\mu W(\sigma)) = V_\mu(\sigma) + \gamma^5 A_\mu(\sigma) \tag{C.10}$$

or equivalently ($\gamma_\mu \gamma_5 = it\mathcal{E}_{\mu\nu}Y_\nu, \gamma_s = i\gamma_0\gamma_1, [\gamma_\mu, \gamma_\nu] = -2i\mathcal{E}_{\mu\nu}\gamma_5$):

$$G_\mu(\sigma) = V_\mu(\sigma) + i\mathcal{E}_{\mu\nu}A_\nu(\sigma). \tag{C.11}$$

At this point we point out the formula

$$F_{\mu\nu}(-iG_\mu(\sigma)) = \{(iD_\alpha^v(\sigma)A_\alpha(\sigma))\mathcal{E}^{\mu\nu} - [A_\mu(\sigma), A_\nu(\sigma)] + F_{\mu\nu}(V_\mu(\sigma))\}. \tag{C.12}$$

Here

$$D_\alpha^V A_\beta = \partial_\alpha A_\beta + [V_\alpha, A_\beta]. \tag{C.13}$$

Note that $A_\mu(\sigma)$ and $V_\mu(\sigma)$ are not independent fields since the chiral phase $W(\sigma)$ satisfies the integrability condition

$$F_{\mu\nu}(W\partial_\mu W) \equiv 0 \tag{C.14}$$

or equivalently

$$F_{\mu\nu}(V_\beta(\sigma)) = -[A_\mu(\sigma), A_\nu(\sigma)], \tag{C.15}$$

$$D_\mu^V A_\nu(\sigma) = D_\nu^V A_\mu(\sigma). \tag{C.16}$$

After substituting Eqs. (C.12)–(C.16) in Eq. (C.9), one can obtain the results

$$
\log\left\{\frac{\det_F(D(1))}{\det_F(\partial)}\right\} \left\{
\begin{aligned}
&= \frac{i}{2\pi}\left\{\int d^2x\,\mathrm{Tr}_{SU(N)}\left(\lambda^a\phi_a(\alpha)\int_0^1 d\sigma(2iD_\mu(\sigma)A_\mu(\sigma)\right.\right.\\
&\qquad\qquad\qquad\qquad\qquad\overbrace{}^{-[A_\mu(\sigma),A_\nu(\sigma)]}\\
&\qquad\left.\left.+\,\mathcal{E}_{\mu\nu}(\,\overbrace{F_{\mu\nu}(V_\alpha(\sigma))}\,)-[A_\mu(\sigma),A_\nu(\sigma)]\right),\right. \qquad\text{(C.17)}\\[6pt]
&= \frac{i}{2\pi}\left\{\int d^2x\,\mathrm{Tr}_{SU(N)}\right.\\
&\qquad\times\left.\left(\lambda^a\phi_a(x)\int_\theta^1 d\sigma(2iD_\mu(\sigma)A_\mu(\sigma))\right)\right\}\\
&\quad+\frac{i}{2\pi}\left\{\int d^2x\,\mathrm{Tr}_{SU(N)}\right.\\
&\qquad\times\left.\left(\lambda^a\phi_a(x)\int_0^1 d\sigma(-2[A_\mu(\sigma),A_\nu(\sigma)])\right)\right\}\\[6pt]
&= I_1(\phi)+I_2(\phi). \qquad\qquad\qquad\qquad\qquad\qquad\text{(C.18)}
\end{aligned}
\right.
$$

Let us now show that the term $I_1(\phi)$ is a mass term for the physical gauge field $A_\mu(\sigma=1)=A_\mu$.

First, we observe the result

$$
\mathrm{Tr}_{\text{Dirac}}\otimes\mathrm{Tr}_{SU(N)}\left\{\gamma^5 A_\mu(\sigma)\overbrace{\tfrac{d}{d\sigma}(W\partial_\mu W)(\sigma)}^{L_\mu(\sigma)}\right\}
$$

$$
=\mathrm{Tr}_{\text{Dirac}}\otimes\mathrm{Tr}_{SU(N)}\{\gamma^5\phi_a(x)\lambda^a(\gamma^5(\partial_\mu A_\mu(\sigma))-[\gamma^5 A_\mu(\sigma),L_\mu(\sigma)])(x)\}
$$

$$
=2\,\mathrm{Tr}_{SU(N)}\{\lambda_a\phi^a(x)(\partial_\mu A_\mu(\sigma)+[V_\mu(\sigma),A_\mu(\sigma)](x))\}
$$

$$
=2\,\mathrm{Tr}_{SU(N)}\{\lambda_a\phi_a(x)D_\mu(\sigma)A_\mu(\sigma)\}=I_1(\phi). \qquad\text{(C.19)}
$$

By the other side

$$
\mathrm{Tr}_{\text{Dirac}}\otimes\mathrm{Tr}_{SU(N)}\{\gamma^5 A_\mu(\sigma)L_\mu(\sigma)\}
$$

$$
=\mathrm{Tr}_{\text{Dirac}}\otimes\mathrm{Tr}_{SU(N)}\left\{\gamma^5 A_\mu(\sigma)\left(\frac{d}{d\sigma}V_\mu(\sigma)+\gamma_5\frac{d}{d\sigma}A_\mu(\sigma)\right)\right\}
$$

$$
=\mathrm{Tr}_{SU(N)}\left(\frac{d}{d\sigma}(A_\mu(\sigma)A_\mu(\sigma))\right). \qquad\text{(C.20)}
$$

At this point it is worth to see the appearance of a dynamical Higgs mechanism for Q.C.D. in two-dimensions.

Let us now analyze the second term $I_2(\phi)$ in Eq. (C.18)

$$\exp\left\{-\frac{i}{2\pi}\int d^2x\,\mathrm{Tr}_{SU(N)}\left(\int_0^1 d\sigma\mathcal{E}_{\mu\nu}(2\phi_a(x)\lambda^a[A_\mu(\sigma),A_\nu(\sigma)](x))\right)\right\}$$

$$=\exp\left\{-\frac{i}{2\pi}\int d^2x\,\mathrm{Tr}_{\mathrm{Dirac}\otimes SU(N)}\left(\int_0^1 d\sigma(\overbrace{\gamma_5\gamma_5}^{1}\mathcal{E}_{\mu\nu}\phi_a(x)\lambda^a\right.\right.$$

$$\times\left.\left.[\gamma_5 W^{-1}(\sigma)\partial_\mu W(\sigma)-\gamma_5 V_\mu(\sigma),\gamma_5 W^{-1}(\sigma)\partial_\nu W(\sigma)-\gamma_5 V_\nu(\sigma)]\right)\right\}.$$

$$(C.21)$$

Since we have the identity as a consequence of the fact that $[\sigma\gamma_5\phi_a(x)\lambda^a,$ $\gamma_5\phi_b(x)\lambda^b]=\sigma(\phi_a(x)\phi_b(x)[\lambda^a,\lambda^b]\equiv 0$;

$$W^{-1}(\sigma)\frac{\partial}{\partial\sigma}W(\sigma)=\gamma_5\phi_a(x)\lambda^a,\qquad(C.22)$$

we can see the appearance of a term of the form of a Wess–Zumino–Novikov topological functional for the Chiral group $SU(N)$, namely

$$I_2(\phi)=\exp\left\{-\frac{i}{2\pi}\int d^2x\,\mathrm{Tr}_{\mathrm{Color}\otimes\mathrm{Dirac}}\left(\mathcal{E}_{\mu\nu}\gamma_5 W^{-1}(\sigma)\frac{\partial}{\partial\sigma}W(\sigma)\right.\right.$$

$$\times\left.\left.[\gamma_5 W^{-1}(\sigma)\partial_\mu W(\sigma),\gamma_5 W^{-1}(\sigma)\partial_\nu W(\sigma)]\right)\right\}$$

$$+\text{terms }(\phi,V_\mu),\qquad(C.23)$$

which after the one-point compactification of the space-time to S^3 and considering only smooth phases ($\phi(x)\in C^\infty(S^3)$) one can see that the Wess–Zumino–Novikov functional is a homotopical class invariant. For the closed ball $S^3\times[0,1]=(\{\bar{x}\equiv(x^1,x^2,\sigma)\})$

$$\int_{S^3\times[0,1]}d^3\bar{x}\,\mathrm{Tr}_{SU(N)_{\mathrm{axial}}}\left\{(\gamma_5(W^{-1}\partial_\alpha W)(\bar{x}))(\gamma_5(W^{-1}\partial_\mu W)(\bar{x})\right.$$

$$\times\left.(\gamma_5 W^{-1}\partial_\nu W))(\bar{x})\right\}=\alpha\pi n\qquad(C.24)$$

with α an over all factor and $n\in\mathbb{Z}^+$.

Finally, let us call our readers attention that the Dirac operator in the presence of a non-Abelian $SU(N)$ gauge field $A_\mu(x) = A_\mu^a(x)\lambda_a$, can always be re-written in the "chiral phase" in the so-called Roskies gauge fixing

$$i\gamma^\mu(\partial_\mu - gG_\mu) = e^{i\gamma_s \widetilde{\phi}^a(x)}(i\gamma^\mu\partial_\mu)e^{i\gamma_s \widetilde{\phi}^a(x)} = \widetilde{W}[\phi](i\gamma^\mu\partial_\mu)\widetilde{W}[\phi]. \quad \text{(C.25)}$$

Here

$$\widetilde{W}[\phi] = e^{i\gamma_s \widetilde{\phi}^a(x)} \equiv \mathbb{P}_{\text{Dirac}}\left\{\mathbb{P}_{SU(N)}e^{i\gamma_s \int_{-\infty}^x d\xi^\mu (\varepsilon_{\mu\nu}G\nu)(x)}\right\}. \quad \text{(C.26)}$$

Since

$$(\gamma_\mu G_\mu)(x) = +\gamma_\mu(\widetilde{W})\partial_\mu(\widetilde{W})^{-1}(x). \quad \text{(C.27)}$$

It is worth now to use the formalism of invariant functional integration — Appendix in Chap. 1 — to change the quantization variables of the gauge field $A_\mu(x)$ to the $SU(N)$-axial phases $\widetilde{W}(\phi)$. This task is easily accomplished through the use of Riemannian functional metric on the manifold of the gauge connections

$$\begin{aligned}
dS^2 &= \int d^2x \, \text{Tr}_{SU(N)}(\delta G_\mu \delta G^\mu)(x) \\
&= \frac{1}{4}\int d^2x \, \text{Tr}_{SU(N)\otimes\text{Dirac}}[(\gamma_\mu\delta G_\mu)(\gamma^\mu\delta G^\mu)](x) \\
&= \det_F[\slashed{D}\slashed{D}^*]_{\text{adg}} \times \left\{\int d^2x \, \text{Tr}_{SU(N)_{\text{Axial}}}[(\delta\widetilde{W}W^{-1})(\delta\widetilde{W}W^{-1})]\right\},
\end{aligned}$$

$$\text{(C.28)}$$

since

$$\begin{aligned}
\gamma_\mu(\delta G_\mu) &= \gamma_\mu\left\{\partial_\mu(\delta\widetilde{W})\widetilde{W}^{-1} + \partial_\mu\widetilde{W}(-\widetilde{W}^{-1}(\delta\widetilde{W})\widetilde{W}^{-1})\right\} \\
&= \gamma_\mu(\partial_\mu - [G_\mu,])(\delta\widetilde{W}\widetilde{W}^{-1})
\end{aligned}$$

and

$$[\gamma_\mu, \delta\widetilde{W}\widetilde{W}^{-1}] = \delta\widetilde{W}\{\gamma_\mu, \widetilde{W}^{-1}\} + \{\gamma_\mu, \delta\widetilde{W}\}\widetilde{W}^{-1} \equiv 0 \quad \text{(C.29)}$$

and leading to new parametrization for the gauge field measure

$$D^F[G_\mu(x)] = [\det_{F,\text{adj}}(\not{D}\not{D}^*)]^{\frac{1}{2}} D^{\text{Haar}}[\widetilde{W}(x)]. \qquad (C.30)$$

Here, $\det_{F,\text{adj}}(\phi\phi^*)^{\frac{1}{2}}$ is the functional Dirac operator in the presence of the gauge field and in the adjoint $SU(N)$-representation. Its explicit evaluations are left to our readers.

The full gauge-invariant expression for the Fermion determinant is conjectured to be given on explicit integration of the gauge parameters considered now as dynamical variables in the gauge-fixed result. For instance, in the Abelian case and in the gauge fixed Roskies gauge equation (C.6) result, we have the Schwinger result

$$\det_F[i\gamma^\mu(\partial_\mu - ieA_\mu)]$$

$$= \frac{1}{2} \int D^F[W(x)] \exp\left\{ -\frac{e^2}{\pi} \int d^2x \frac{1}{2}(A_\mu - \partial_\mu W(x))^2 \right\}$$

$$= \int D^F[W(x)] \exp\left\{ -\frac{e^2}{2\pi} \int d^2x [A_\mu^2 + (\partial_\mu W)^2 + 2A_\mu \partial_\mu W] \right\}$$

$$= \exp\left\{ -\frac{e^2}{\pi} \int d^2x \left[A_\mu \left(\delta_{\mu\nu} - \frac{\partial_\mu \partial_\nu}{(-\partial^2)} \right) A_\nu \right](x) \right\}. \qquad (C.31)$$

Note that the Haar measure on the Abelian Group $U(1)$ is

$$\delta S_W^2 = \int d^2x [\delta(e^{iW} \partial_\mu e^{-iW}) \delta(e^{iW} \partial_\mu e^{-iW})](x)$$

$$= \int d^2x (\delta W(-\partial^2) \delta W)(x). \qquad (C.32)$$

Let us solve exactly the two-dimensional quantum electrodynamics.

Firstly, Eq. (C.10) takes the simple form in term of the chiral phase in the Roskies gauge

$$G_\mu(x) = (\varepsilon_{\mu\nu} \partial_\nu \phi)(x). \qquad (C.33)$$

Now, the somewhat cumbersome non-Abelian equations (C.3a) and (C.3b) has a straightforward form in the Abelian case

$$D^F[G_\mu] = \det_F(-\Delta) D^F[\phi] \qquad (C.34)$$

and thus, we have the exactly soluble expression for the $(Q.E.D.)_2$- generating functional (a non-gauge invariant object!)

$$Z[J_\mu, \eta, \bar{\eta}] = \frac{1}{2}\left\{ \int D^F[\phi]D^F[\chi]D^F(\bar{\chi}) \right.$$

$$\times \exp\left\{ -\frac{1}{2}\int d^2x\left[\phi\left(-\partial^4 + \frac{e^2}{\pi}\partial^2\right)\phi + \mathcal{E}_{\mu\nu}(\partial_\nu J_\mu)\phi\right](x) \right\}$$

$$\times \exp\left\{ -\frac{1}{2}\int d^2x(\chi, \bar{\chi})\begin{bmatrix} 0 & i\slashed{\partial} \\ i\slashed{\partial} & 0 \end{bmatrix}\begin{pmatrix} \chi \\ \bar{\chi} \end{pmatrix} \right\}$$

$$\left. \times \exp\left\{ -\int d^2x(\chi, \bar{\chi})\begin{bmatrix} e^{-ig\gamma_s\phi} & 0 \\ 0 & e^{-ig\gamma_s\phi} \end{bmatrix}\begin{pmatrix} \eta \\ \bar{\eta} \end{pmatrix} \right\} \right\}$$

For instance, correlations functions are exactly solved and possessing a "coherent state" factor given by the ϕ-average below (after "normal ordenation" at the coincident points) and explicitly given a proof of the confinement of the fermionic fields since one cannot assign LSZ-scattering fields configuration for them by the Coleman theorem since then grown as the factor $|x - y|^{\frac{1}{2g^2}}$ at large separation distance

$$\left\langle e^{-i(\gamma_s)_x\phi(x)}e^{-i(\gamma_s)_y\phi(y)} \right\rangle_\phi$$

$$= \langle \cos(\varphi(x))\cos(\varphi(y))\rangle_\phi \mathbf{1}_x \otimes \mathbf{1}_x + \gamma_s \otimes \mathbf{1}\langle\sin(\varphi(x))\cos(\varphi(y))\rangle_\phi$$

$$+ \gamma_s \otimes \mathbf{1}\langle\cos(\varphi(x))\sin(\varphi(y))\rangle_\phi - \gamma_s \otimes \gamma_s\langle\sin(\varphi(x))\sin(\varphi(s))\rangle_\phi$$

$$= e^{-g^2\left[\left(-\partial^2 + \frac{g^2}{\pi}\right)^{-1} - (-\partial^2)^{-1}\right](x,y)}$$

$$= \exp\left\{ +\frac{\pi}{g^2}\left(\frac{1}{2\pi}K_0\left(\frac{g}{\sqrt{\pi}}|x-y|\right) + \frac{1}{2\pi}\ell g|x-y|\right\}(\mathbf{1}_x \otimes \mathbf{1}_y). \quad (C.35)$$

The two-point function for the two-dimensional-electromgnetic field shows clearly the presence of a massive excitation (Fotons have acquired a mass term by dynamical means)

$$\langle G_\mu(x)G_\nu(g)\rangle_\phi = \int \frac{d^2k}{(2\pi)}(\mathcal{E}_{\mu\alpha}\mathcal{E}_{\nu\beta})(k^\alpha k^\beta)\frac{1e^{ik(x-s)}}{k^2(k^2 + \frac{e^2}{\pi})}$$

$$= \frac{1}{2\pi}\int d^2k \overbrace{[\mathcal{E}_{\mu\alpha}\mathcal{E}_{\nu\beta}]}^{\delta^{\mu\nu}}\frac{\delta^{\alpha\beta}k^2e^{ik|x-y|}}{k^2(k^2 + \frac{e^2}{\pi})}. \quad (C.36)$$

As an important point of this supplementary appendix, we wish to point out that the chirally transformed Dirac operator equation (C.6) in four-dimensions, still have formally an exactly integrability as expressed by the integral representation equation (C.3) (see the asymptotic expansion equations (4.38) in Chap. 4). It reads as follows

$$\log\left[\frac{\det(\not{D}(\sigma)^2)}{\det(\partial)^2}\right] = -\frac{i}{2\pi^2}\int d^4x\, \mathrm{Tr}_{SU(N)}\left\{\phi_a(x)\lambda^a\right.$$

$$\left.\times\left[\int_0^1 d\sigma F^c_{\alpha\beta}(-iG_\mu(\sigma))F^{c'}_{\mu\nu}(-iG_\mu(\sigma))\varepsilon^{\alpha\beta\mu\nu}\lambda_c\lambda_{c'}\right]\right\},$$

$$(C.37)$$

where

$$-i\gamma_\mu G_\mu(\sigma) = \exp\left(\sigma\gamma^5\phi^a(x)\lambda_a\right)(i\partial)\exp\left(\sigma\gamma^5\phi^a(x)\lambda_a\right) \qquad (C.38)$$

and we have the formulae[19]

$$-i\gamma_\mu G_\mu(\sigma) = V_\mu(\sigma) + \gamma^5 A_\mu(\sigma), \qquad (C.39a)$$

$$A_\mu(\sigma) = \Delta^{-1}_{\gamma_s\phi^a\lambda_a}\{\sin h(\Delta_{\gamma_s\phi_a\lambda^a}) \circ \partial_\mu(\gamma_5\phi^a\lambda_a)\}, \qquad (C.39b)$$

$$V_\mu(\sigma) = \Delta^{-1}_{\gamma_s\phi^a\lambda_a}\{(1-\cos h(\Delta_{\gamma_s\phi^a\lambda_a})) \circ \partial_\mu(\gamma_5\phi^a\lambda_a)\} \qquad (C.39c)$$

with the matrix operation

$$\Delta_X \circ Y = [X, Y] \qquad (C.39d)$$

and $\Delta_X^{(n)}$ denoting its n-power.

For a complete quantum field theoretic analysis of the above formulae in an Abelian (theoretical) axial model we refer our work to L. C. L. Botelho.[20]

Finally and just for completeness and pedagogical purposes, let us deduce the formal short-time expansion associated with the second-order positive differential elliptic operator in R^ν used in the previous cited reference

$$\mathcal{L} = -(\partial^2)_x + a_\mu(x)(\partial_\mu)_x + V(x). \qquad (C.40)$$

Its evolution kernel $k(x, y, t) = \langle x| - \exp(^{-t}\mathcal{L})|y\rangle$ satisfies the heat-kernel equation

$$\frac{\partial}{\partial t}K(x, y, t) = -\mathcal{L}_x K(x, y, t), \qquad (C.41)$$

$$K(x, y, 0) = \delta^{(\nu)}(x - y). \tag{C.42}$$

After substituting the asymptotic expansion below into Eq. (C.41) [with $K_0(x, y, t)$ denoting the free kernel ($a_\mu \equiv 0$ and $V \equiv 0$)]

$$K(x, y, t) \overset{t \to 0^+}{\simeq} K_0(x, y, t) \left[\sum_{n=0}^{\infty} t^n H_n(x, y) \right] \tag{C.43}$$

and by taking into account the obvious relationship for $t > 0$

$$(\partial_\mu K_0(x, y, t))|_{x=y} = 0. \tag{C.44}$$

We obtain the following recurrence relation for the coefficients $H_n(x, x)$:

$$(n + 1)H_{n+1}(x, x) = - \left\{ (-\partial_x^2)H_n(x, x) + a_\mu(x) \cdot \partial_\mu H_n(x, x) + V(x) \right\}. \tag{C.45}$$

For the axial Abelian case in R^4, we have the result:

$$(e^{g\gamma_s\phi}(i\partial)e^{g\gamma_s\phi})^2 = (-\partial^2)\mathbf{1}_{4\times4} + \left(\left(\frac{1}{2}g\gamma^s[\gamma^\mu, \gamma^\nu] \right) \partial_\mu\phi(x) \right) \partial_\nu$$
$$+ [-g\gamma_s\partial^2\phi + (g)^2(\partial_\mu\phi)^2]. \tag{C.46}$$

Note that in R^4

$$H_0(x, x) = \mathbf{1}_{4\times4},$$

$$H_1(x, x) = -V(x),$$

$$H_2(x, x) = \frac{1}{2}[-\partial^2 V + a_\mu\partial_\mu V + V^2](x). \tag{C.47}$$

Appendix D. Functional Determinants Evaluations on the Seeley Approach

In this technical appendix, we intend to highlight the mathematical evaluation of the functional determinants involved in the theory of random surface path-integral.[18]

Let us start by considering a differential elliptic self-adjoint operator of second-order acting on the space of infinitely differentiable functions of compact support in $R^2, C_c^\infty(R^2, \mathbb{C}^q)$ with values in $\dot{\mathbb{C}}^q$

$$A = \sum_{|\alpha|\leq 2} a_\alpha(x)D_x^\alpha \tag{D.1}$$

with $\alpha = (\alpha_1, \alpha_2)$ multi-indexes and

$$D_x^\alpha = \left(\frac{1}{i}\frac{\partial}{\partial x_1}\right)^{\alpha_1}\left(\frac{1}{i}\frac{\partial}{\partial x_2}\right)^{\alpha_2} \tag{D.2}$$

and $A_\alpha(x) \in C_c^\infty(R^2, \mathbb{C}^q)$.

By introducing the usual square-integrable inner product in $C_c^\infty(R^2, \mathbb{C}^q)$ and making the hypothesis that A is a positive definite operator, one may consider the (contractive) semi-group generated by A and defined by the spectral calculus

$$e^{-tA} = \frac{1}{2\pi i}\left\{\int_C d\lambda \frac{e^{-t\lambda}}{(\lambda \mathbf{1}_{q\times q} - A)}\right\} \tag{D.3}$$

with C being an (arbitrary) path containing the positive semi-weis $\lambda > 0$ (the spectrum of the operator A) with a counter-clockwise orientation.

According to Seeley, one must consider the symbol associated with the resolvent pseudo-differential operator $(\lambda \mathbf{1}_{q\times q} - A)^{-1}$ and defined by the relation below

$$\sigma(A - \lambda \mathbf{1}) = e^{ix\xi}(A - \lambda \mathbf{1})e^{ix\xi} = \sum_{|j|=0}^{2} A_j(x, \xi, \lambda). \tag{D.4}$$

with

$$A_j(x, \xi, \lambda) = \left(\sum_{|\alpha|=\alpha_1+\alpha_2=j} a_j(x)(\xi_1^{\alpha_1})(\xi_2^{\alpha_2})\right) \quad 0 \le j < 2, \tag{D.5}$$

$$A_2(x, \xi, \lambda) = -\lambda \mathbf{1}_{q\times q} + \left(\sum_{|\alpha|=\alpha_1+\alpha_2=j} a_j(x)(\xi_1^{\alpha_1})(\xi_2^{\alpha_2})\right) \quad j = 2. \tag{D.6}$$

This is basic for symbols calculations, the important scaling properties as written below

$$A_j(x, c\xi, c^2\lambda) = (c)^j A_j(x, \xi, \lambda). \tag{D.7}$$

It too is a fundamental result of the Seeley's theory of pseudo-differential operators that the resolvent operator $(A - \lambda \mathbf{1})^{-1}$ (the associated Green functions of the operator A) has an expansion of the form in a suitable functional space

$$\sigma((A - \lambda 1)^{-1}) = \sum_{j=0}^{\infty} C_{-2-j}(x, \xi, \lambda) \qquad (D.8)$$

and satisfies the relations below

$$\sigma(A - \lambda 1) \cdot \sigma((A - \lambda 1)^{-1}) = 1, \qquad (D.9)$$

$$\frac{1}{a!} \left\{ \sum_{|\alpha| \leq 2} \sum_{j=0}^{\infty} [(D_\xi^\alpha(\sigma(A - \lambda 1))(x, \xi) D_x^\alpha [C_{-2-j}(x, \xi, \lambda)] \right\} = 1. \qquad (D.10)$$

Recurrence relationships for the explicitly determination of the Seeley coefficients of the resolvent operator equation (D.8) can be obtained through the use of the scaling properties

$$\xi = p\xi'; \quad \lambda^{\frac{1}{2}} = p(\lambda')^{\frac{1}{2}}, \qquad (D.11)$$

$$C_{-2-j}(x, p\xi', (p(\lambda')^{\frac{1}{2}})^2) = p^{-(2+j)} C_{-2-j}(x, \xi', \lambda'). \qquad (D.12)$$

After using Eq. (D.11) in Eq. (D.10) and for each integer j, comparing the resultant power series in the variable $1/p(1 = 1 + 0(\frac{1}{p}) + \cdots + 0(\frac{1}{p})^n + \cdots)$, one can get the result as

$$C_{-2}(x, \xi) = (A_2(x, \xi)^{-1}) \qquad (D.13)$$

$$0 = a_2(x, \xi) C_{-2-j}(x, \xi) + \left\{ \frac{1}{\alpha!} \sum_{\substack{\ell < j \\ k - |\alpha| - 2 - \ell = -j}} D_\xi^\alpha a_k(x, \xi) \left((i D_x^\alpha C_{-2-\ell}(x, \xi, \lambda)) \right) \right\}$$

$$(D.14)$$

For the explicit operator in $C_c^\infty(R^2, R^q)$ is given below

$$A = - \left(g_{11} \frac{\partial^2}{\partial x_1^2} + g_{22} \frac{\phi^2}{\partial x_2^2} \right) 1_{q \times q}$$

$$- (A_1)_{q \times q} \frac{\partial}{\partial x_1} - (A_2)_{q \times q} \frac{\partial}{\partial x_2} - (A_0)_{q \times q} \qquad (D.15)$$

with all the coefficients in $C_c^\infty(R^2, R^q)$, one obtains the following results after calculations:

$$A_2(x, \xi, \lambda) = (g_{11}(x)\xi_1^2 + g_{22}\xi_2^2 - \lambda)1_{q\times q},$$
$$A_2(x, \xi, \lambda) = -iA_1(x)\xi_1 - iA_2(x)\xi_2, \quad\quad\quad (D.16)$$
$$A_0(x, \xi, \lambda) = -A_0(x),$$

and

$$C_{-2}(x, \xi) = (g_{11}(x)\xi_1^2 + g_{22}(x)\xi_2^2 - \lambda)^{-1},$$
$$C_{-3}(x, \xi) = i(A_1(x)\xi_1 + A_2(x)\xi_2)(C_{-2}(x, \xi))^2$$
$$-2ig_{11}(x)\xi_1\left[\left(\frac{\partial}{\partial x_1}g_{11}\right)(\xi_1)^2 + \left(\frac{\partial}{\partial x_2}g_{22}\right)(\xi_2)^2\right](C_{-2}(x, \xi))^3$$
$$-2ig_{22}(x)\xi_2\left[\left(\frac{\partial}{\partial x_2}g_{11}\right)(\xi_1)^2 + \left(\frac{\partial}{\partial x_2}g_{22}\right)(\xi_2)^2\right](C_{-2}(x, \xi))^3$$

$$(D.17)$$

By keeping in view evaluations of the heat kernel of our given differential operator, let us write its expansion in terms of the Seeley coefficients of (D.8):

$$Tr(e^{-tA}) = \sum_{j=0}^\infty \left(\frac{1}{2\pi}\right)^2 \left\{\int_{R^2} d^2x\sigma(e^{-tA})(x, \xi)\right\}$$
$$= \sum_{j=0}^\infty \left(\frac{1}{2\pi}\right)^2 \left(\frac{1}{2\pi i}\right)$$
$$\times\left[\int_{+\infty}^{-\infty} d(-is)e^{ist}\left(\int_{R^2\times R^2} d^2x d^2\xi C_{-2-j}(x, \xi, -is)\right)\right]$$
$$= \sum_{j=0}^\infty \left\{\frac{1}{t}\frac{1}{(2\pi)^3}\int_{R^2} d^2\xi\int_{R^2} d^2x\left[\int_{-\infty}^{+\infty} e^{is}C_{-2-j}(x, \xi, \frac{-is}{t})ds\right]\right\}$$
$$= \sum_{j=0}^\infty \left\{\frac{1}{t}\frac{1}{(2\pi)^3}\int_{R^2} d^2\xi\int_{R^2} d^2x e^{is}t^{(\frac{2+j}{2})}C_{-2-j}(x, t^{\frac{1}{2}}\xi, -is)\right\}$$
$$= \sum_{j=0}^\infty \left\{t^{\frac{(j-2)}{2}}(2\pi)^{-3}\int_{R^2} d^2\xi\int_{R^2} d^2x\int_{-\infty}^{+\infty} C_{-2-j}(x, \xi, -is)\right\}.$$

$$(D.18)$$

By applying Eq. (D.17) to the differential operator as given in Eq. (D.15), we obtain the short-time Seeley expansion as an asymptotic expansion in the variable t

$$Tr(e^{-tA}) \sim \frac{q}{4\pi t} \left(\int d^2x \sqrt{g_{11}g_{22}} \right) + \frac{q}{4\pi} \left(\int d^2x \sqrt{g_{11}g_{22}} \left(-\frac{1}{6}R \right) \right)$$

$$+ \int d^2x \left(-\frac{1}{2} \frac{1}{\sqrt{g_{11}g_{22}}} \mathrm{Tr} \left[\overbrace{\left(\frac{\partial}{\partial x_1} (\sqrt{g_{11}g_{22}} A_1) \right) + \left(\frac{\partial}{\partial x_2} (\sqrt{g_{11}g_{22}} A_2) \right)}^{(-\frac{1}{2} \mathrm{div}_{\mathrm{cov}} \vec{A})} \right] \right)$$

$$+ \int d^2x \left(-\frac{1}{4} \mathrm{Tr} \left[\frac{(A_1)^2}{g_{11}} + \frac{(A_2)^2}{g_{22}} + A_0 \right] \right) + 0(t). \tag{D.19}$$

By applying the above formula for the Polyakov's covariant path-integrals, let us first introduce the R^2 complex structure

$$z = x_1 + x_2, \quad \bar{z} = x_1 - ix_2, \tag{D.20}$$

$$\frac{\partial}{\partial z} = \frac{\partial}{\partial x_1} - i\frac{\partial}{\partial x_2}, \quad \frac{\partial}{\partial \bar{z}} = \frac{\partial}{\partial x_1} + i\frac{\partial}{\partial x_2}. \tag{D.21}$$

For each integer j, let us define two Hilbert spaces H_j and $\bar{H}_j$ as follows:

(a) H_j is defined as the vector space of all complex functions $f(z, \bar{z}) = f_1(x, y) + if_2(x, y)$ with the following tensorial behavior under the action of a conformal tranformation

$$z = z(w), \tag{D.22}$$

$$f(z, \bar{z}) = \left(\frac{\partial w}{\partial z} \right)^{-j} \tilde{f}(w, \bar{w}). \tag{D.23}$$

Let us introduce into H_j the following inner product

$$(q, f)_{H_j} = \int_{R^2} dz\bar{z}(\rho(z, \bar{z})^{j+1} \bar{g}(z, \bar{z}) f(z, \bar{z})), \tag{D.24}$$

with $\rho(\bar{z})$ is a positive continuous real-valued function of compact support in R^2 and associated with a conformal metric.

$$ds^2 = \rho(z, \bar{z}) \, dz \wedge d\bar{z}.$$

(b) $\bar{H}_j$ is defined same as above exposed with the following tensor

$$f(z, \bar{z}) = \left(\left(\frac{\overline{\partial w}}{\partial z}\right)\right)^{-j} \tilde{f}(w, \bar{w}). \qquad (D.25)$$

At this point, we can verify that the above written inner products are conformal invariant

$$(g, f)_{H_j} = \int_{R^2} dw d\bar{w} \left(\frac{\partial z}{\partial w}\right) \left(\frac{\partial \bar{z}}{\partial w}\right) \left(\left|\frac{\partial w}{\partial z}\right|^2 \tilde{\rho}(w, \bar{w})\right)^{j+1}$$

$$\times \left[\left(\frac{\partial}{\partial z} w\right)^{-j} \tilde{f}(w, \bar{w})\right] \left[\left(\frac{\overline{\partial w}}{\partial z}\right)^{-j} (\tilde{g}(w, \bar{w}))\right]$$

$$= \int_{R^2} dw d\bar{w} (\tilde{\rho}(w, \bar{w}))^{j+1} \tilde{f}(w, \bar{w}) (\overline{\tilde{g}(w, \bar{w})}). \qquad (D.26)$$

Let us now introduce the following weighted Cauchy–Riemann operators with a $U(1)$-real valued connection $A = (A_z, A_{\bar{z}})$ in $R^2 \equiv \mathbb{C}$.

$$
\begin{aligned}
\text{(a)} \quad & L_j = H_j \to \bar{H}_{-(j+1)} \\
& f \to (\rho(z, \bar{z}))^j (\partial_{\bar{z}} + A_{\bar{z}}) f, \\
\text{(b)} \quad & \bar{L}_j = \bar{H}_j \to H_{(j+1)} \\
& f \to (\rho(z, \bar{z}))^j (\partial_z + A_z) f
\end{aligned}
\qquad (D.27)
$$

together with the adjoint operators $(L_j \varphi, f)_{\bar{H}_{-(j+1)}} = \langle \varphi, L_j^* f \rangle_{H_j}$, namely:

$$L_j^* = -\bar{L}_{-(j+1)} : \bar{H}_{-(j+1)} \to H_j \quad (\text{and } \bar{L}_j^* = -L_{-(j+1)})$$
$$f \to -(\rho(z, \bar{z}))^{-(j+1)} (\partial_z + A_z)) f. \qquad (D.28)$$

For simplicity, we consider the case of $A_z = A_z \equiv 0$.

The second-order positive definite operators are given by

$$
\begin{aligned}
\mathcal{L}_j &= L_j^* L_j = -\bar{L}_{-(j+1)} L_j : H_j \to H_j, \\
\bar{\mathcal{L}} &= (\bar{L}_j)^* \bar{L}_j : -L_{-(j+1)} \bar{L}_j : \bar{H}_{-(j+1)} \to \bar{H}_{-(j+1)}.
\end{aligned}
\qquad (D.29)
$$

Possesses the explicitly expressions (for $\rho(z, \bar{z}) = e^{\varphi(z, \bar{z})}$)

$$
\begin{aligned}
\mathcal{L}_j &= -e^{-(j+1)\varphi(z, \bar{z})} \partial_z e^{j\varphi(z, \bar{z})} \partial_{\bar{z}}, \\
\bar{\mathcal{L}}_j &= -e^{-(j+1)\varphi(z, \bar{z})} \partial_{\bar{z}} e^{j\varphi(z, \bar{z})} \partial_z.
\end{aligned}
\qquad (D.30)
$$

They have the following Seeley expansion:

$$\lim_{t \to 0^+} \mathrm{Tr}_{C_c^\infty(R^2)}(e^{-t\mathcal{L}_j}) = \int dz\,d\bar{z}\left(\frac{\rho(z,\bar{z})}{2\pi t} - \frac{(1+3j)}{j2\pi}\Delta\ell g\rho(z,\bar{z})\right),$$

(D.31)

$$\lim_{t \to 0^+} \mathrm{Tr}_{C_c^\infty(R^2)}(e^{-t\bar{\mathcal{L}}_j}) = \int dz\,d\bar{z}\left(\frac{\rho(z,\bar{z})}{2\pi t} + \frac{(2+3j)}{j2\pi}\Delta\ell g\rho(z,\bar{z})\right).$$

(D.32)

The above expressions are from the Seeley asymptotic expansion

$$\lim_{t \to 0^+} \mathrm{Tr}_{C_c^\infty(R^2)}(e^{-tA})$$

$$\sim \int d^2x\left\{\left(\frac{\sqrt{g}}{4\pi}\mathrm{Tr}(1)_{2\times 2}\right)\left(\frac{1}{t}\right) - \left(\frac{1}{24\pi}\sqrt{g}R\right)\mathrm{Tr}(1)_{2\times 2}\right.$$

$$\left. + \frac{1}{4\pi}\sqrt{g}B_0\right\} + O(t),$$

(D.33)

where A is the elliptic second-order self-adjoint differential operator in the presence of a Riemann metric $ds^2 = g_{\mu\nu}dx^\mu dx^\nu$ (in a tensorial notation in the space $L^2(R^2, \sqrt{g}dx^1dx^2)$:

$$A = \left(-\frac{1}{\sqrt{g}}(\partial_\mu 1_{2\times 2} + B_\mu)\sqrt{g}g^{\mu\nu}(\partial_\nu 1_{2\times 2} + B_\nu)\right) - (B_0)$$

(D.34)

with $B_\mu(x^\nu)$ denoting $C_c^\infty(R^2, R^2)$ functions.

After all these preliminaries discussions, we pass to the problem of evaluating functional determinant (without zero modes)

$$\ell g \det \mathcal{L}_j = \lim_{\varepsilon \to 0^+} -\left\{\int_\varepsilon^\infty \frac{dt}{t}\mathrm{Tr}_{C_c^\infty(R^2)}(e^{-t\mathcal{L}_j})\right\}.$$

(D.35)

It is straightforward to verify that the following chain of equations related to the functional variations of the conformal structure hold true [Herewith $\mathrm{Tr} \equiv \mathrm{Tr}_{C_c^\infty(R^2)}$]

$$\delta\ell g\det\mathcal{L}_j = \lim_{\varepsilon \to 0^+}\left\{\int_\varepsilon^\infty dt\,\mathrm{Tr}\left(\overbrace{\frac{\delta\mathcal{L}_j}{\delta\varphi}}^{\delta\mathcal{L}_j}\delta\,\varphi e^{-t\mathcal{L}_j}\right)\right\}$$

$$= \lim_{\varepsilon \to 0^+}\left\{\int_\varepsilon^\infty dt\,\mathrm{Tr}\left[(-(j+1)\delta\varphi\mathcal{L}_j - j\bar{L}_{(j+1)}\delta\varphi L_j)e^{-t\mathcal{L}_j}\right]\right\}$$

$$= \lim_{\varepsilon \to 0^+}\left\{\int_\varepsilon^\infty dt\,\mathrm{Tr}\left[-(j+1)\delta\varphi\mathcal{L}_j e^{-t\mathcal{L}_j} + j\delta\varphi\bar{\mathcal{L}}_{-(j+1)}e^{-t\bar{\mathcal{L}}_{-(j+1)}}\right]\right\},$$

(D.36)

where we have used the functional identity

$$\text{Tr}(-j\bar{L}_{-(j+1)}\delta\varphi L_j e^{-t\mathcal{L}_j})$$

$$= \text{Tr}\left[-j\bar{L}_{-(j+1)}(\bar{L}_{-(j+1)})^{-1}(\delta\varphi)L_j\bar{L}_{-(j+1)}e^{-t\bar{\mathcal{L}}_{-(j+1)}}\right]$$

$$= \text{Tr}\left[-j \cdot \mathbf{1}\delta\varphi(-\bar{\mathcal{L}}_{-(j+1)})e^{-t\bar{\mathcal{L}}_{-(j+1)}}\right] \tag{D.37}$$

since

$$e^{-t\bar{\mathcal{L}}_{-(j+1)}} = (\bar{L}_{-(j+1)})^{-1}e^{-t\mathcal{L}_j}\bar{L}_{-(j+1)} \tag{D.38}$$

is a consequence of the operatorial relationship

$$\bar{\mathcal{L}}_{-(j+1)} = (\bar{L}_{-(j+1)})^{-1}\mathcal{L}_j\bar{L}_{-(j+1)}. \tag{D.39}$$

As a consequence, we obtain, the following results:

$$\delta(\ell g \det \mathcal{L}_j)$$

$$= -(j+1)\lim_{\varepsilon\to 0^+}\text{Tr}(\delta\varphi e^{-\varepsilon\mathcal{L}_j}) + j\lim_{\varepsilon\to 0^+}\text{Tr}(\delta\varphi e^{-\varepsilon\bar{\mathcal{L}}_{-(j+1)}}) \tag{D.40}$$

$$= -(j+1)\left[\int_{R^2} d^2x\delta\varphi(x)\left\{\frac{1}{2\pi\varepsilon}e^{\varphi(x)} - \frac{(1+3j)}{12\pi}\Delta\varphi(x)\right\}\right]\Bigg|_{\varepsilon\to 0^+}$$

$$+ j\left[\int_{R^2} d^2x\delta\varphi(x)\left\{\frac{1}{2\pi\varepsilon}e^{\varphi(x)} + \frac{(2+3j)}{12\pi}\Delta\varphi(x)\right\}\right]\Bigg|_{\varepsilon\to 0^+}. \tag{D.41}$$

Combining together, we obtain our final "basic-brick" formulae of the quantum geometric path-integrals for random surfaces:

$$\delta\ell g \det \mathcal{L}_j = \lim_{\varepsilon\to 0^+}\left(-\frac{1}{2\pi\varepsilon}\right)\left[\int_{R^2} d^2x\delta\varphi(x)e^{\varphi(x)}\right]$$

$$+ \frac{(1+6j(j+1))}{12\pi}\left[\int_{R^2} d^2x\delta\varphi(x)\Delta\varphi(x)\right] \tag{D.42}$$

which produces the "brick" result (Polyakov)

$$\ell g \det \mathcal{L}_j = \left[\lim_{\varepsilon\to 0^+}\left(-\frac{1}{2\pi\varepsilon}\right)\int_{R^2} d^2x e^{\varphi(x)}\right]$$

$$\times \left[-\frac{1+6j(j+1))}{12\pi}\int_{R^2} d^2x\left\{\frac{1}{2}(\partial_a\varphi)^2(x)\right\}\right] \tag{D.43}$$

The result in the presence of gauge fields can be obtained through Bosonization techniques and only will lead to the additional term to be added to Eq. (D.43) in its right-hand side

$$\exp\left[-\frac{1}{2\pi}\int_{R^2} d^2x e^{\varphi(x)}\left\{A_1(x)A^1(x) + A_2(x)A^2(x)\right\}\right]$$

$$= \exp\left[-\frac{1}{2\pi}\int (dz \wedge d\bar{z})(A_z \cdot A_{\bar{z}})\right]. \tag{D.44}$$

Chapter 6

Basic Integral Representations in Mathematical Analysis of Euclidean Functional Integrals*

In this complementary chapter to Chap. 5, we expose additional rigorous mathematical concepts and theorems behind Euclidean functional integrals as proposed in Chap. 5 and used thoroughly in other chapters.

In Sec. 6.1, we present a pure topological proof of the basic measure theory (Riesz–Markov theorem), mathematical concept basic to construct rigorously functional integrals. In Sec. 6.2, we present analogous results on the mathematical structure of the L. Schwartz distributions.

In Sec. 6.3, we present the important Kakutani theorem on the equivalence of Gaussian measures in Hilbert spaces, the mathematical basis for the rigorous framework for Jacobian transformations in Euclidean path-integrals.

6.1. On the Riesz-Markov Theorem

> The words set and function are not as simple as they may seem. They are potent words. They are like seeds, which are primitive in appearance but have the capacity for vast and intricate developments — G.F. Simmons.

The Riesz representation theorem

Theorem 6.1. *Let X be a topological compact Hausdorff space. Let L be a positive linear functional on $C(X)$. There exists a unique positive measure $d_L\mu$ and an associated σ-algebra on X which represents L in the sense that*

$$L(f) = \int_X f(x)d_L u(x). \tag{6.1}$$

Let us begin our proof by introducing the smallest ring containing the compact subsets of X.

---

*Author's original results.

Another mathematical structure we need is the following Banach space. Let $_pC(X)$ be the vector space formed by all linear combinations of the elements of $C_0(X)$ and the characteristic functions of the compact sets of X. We introduce the sup norm on this vector space and take its completion still denoted by $_pC(X)$ (which is a Banach space). It is a straightforward consequence of the Hahn–Banach theorem that the given positive linear functional L has a unique extension to $_pC(X)$ still denoted by L in what follows.

We define now an equivalence relation on the algebra of sets introduced above through the relationship

$$\forall \beta, \alpha \in A \quad \text{and} \quad \alpha \sim \beta \Leftrightarrow L(\chi_{\overline{\alpha \Delta \beta}}) = 0, \tag{6.2}$$

here $\overline{\alpha \Delta \beta}$ denotes the topological closure of the difference set $\alpha \Delta \beta = (\alpha - \beta) \cup (\beta - \alpha) = (\alpha \cap \beta') \cup (\beta \cap \alpha')$. On this coset algebra of sets $A/\sim$, denoted by $E_{\text{Baire}}(X)$, we introduce a metrical structure by means of the metric set function

$$d_L(A, B) = L((\chi_{\overline{A \Delta B}})). \tag{6.3}$$

By considering the topological completion of the metric space $(E_{\text{Baire}}(X), d_L)$, we obtain our proposed σ-algebra on X and a measure defined by the simple metrical relation

$$\mu_L(\alpha) = d_L(\alpha, \phi). \tag{6.4}$$

At this point, it is evident that the extension theorem of Caratheodory is a simple rephrasing of Eq. (6.4), since for a given μ_L-measurable set $\Omega \in (\overline{E}_{\text{Baire}}(X), \mu_L)$ and $\varepsilon > 0$, there exists a finite family of disjoint compact sets on X: $\{K_e\}_{e=1,\ldots,N(c)}$ such that

$$d\left(\Omega, \bigcup_{\ell=1}^{N(\varepsilon)} K_\ell\right) \le \varepsilon \Leftrightarrow \left(\sum_{\ell=1}^{N(\varepsilon)} \mu_L(K_e)\right) - \varepsilon$$

$$\le \mu_L(\Omega) \le \varepsilon + \left(\sum_{\ell=1}^{N(\varepsilon)} \mu_L(K_e)\right) \tag{6.5}$$

Let us introduce a large Banach space $C_{\text{bounded}}(\overline{E}_{\text{Baire}}(X), R)$ with the usual sup norm. It is straightforward to see that the given functional $L \in (_pC(X))^*$ has a unique extension $\widetilde{L}$ to this new space of continuous function on $\bar{E}_{\text{Baire}}(X)$ (which is straightforwardly identified with the measurable functions on $(\bar{E}_{\text{Baire}}(X), u_L)$!). Since the characteristic functions of compact sets are elements of $C_{\text{bounded}}(\bar{E}_{\text{Baire}}(X), R)$ any $f \in C(x)$ is the limit on the topology of $C(\bar{E}_{\text{Baire}}(X), R)$ of the simple functions (monotone nondecreasing) sequence

$$f(X) = \lim_{n \to \infty} \left\{ \sum_{j=1}^{n2^n} \frac{j-1}{2^n} \chi_{E_{n,j}}(x) + n\chi_{F_n}(x) \right\} \equiv \lim_{n \to \infty} S_n(f). \qquad (6.6)$$

Here, the (compact!) sets $E_{n,j}$ and F_n in X are defined by

$$E_{n,j} = f^{-1}\left(\left[\frac{j-1}{2^n}, \frac{j}{2^n}\right]\right), \qquad (6.7a)$$

$$F_n = f^{-1}\left([n, \|f\|_{C(x)}]\right). \qquad (6.7b)$$

Now the assertive expressed by Eq. (6.1) is a simple result of the definition of integration

$$L(f) = \tilde{L}(f) = \tilde{L}\left(\lim_{n \to \infty} S_n(f)\right) = \lim_{n \to \infty} L(S_n(f))$$

$$= \lim_{n \to \infty} \left(\sum_{j=1}^{n2^n} \frac{j-1}{2^n} \mu_L(E_{n,j}) + n\mu_L(F_n) \right)$$

$$= \lim_{n \to \infty} \left(\int_X f_n(x) d_L\mu(x) \right) \overset{\text{def}}{\equiv} \int_X f(x) d_L\mu(x), \qquad (6.8)$$

which proves the Riesz–Markov theorem.

As a last point of this section let us give a criterion for the existence of invariant sets in relation to a given (measurable) transformation

$$T \colon (\bar{E}_{\text{Baire}}(X), d_L) \to (\bar{E}_{\text{Baire}}(X), d_L). \qquad (6.9)$$

If the measurable transformation is a contraction (or some of its power!) between the above complete metric spaces, namely, if $c < 1$ and an integer $\rho \in \mathbb{Z}^+$ such that

$$d_L(T^\rho A, T^\rho B) = \int_X d_L\mu(x)\chi_{(T^\rho A \triangle T^\rho B)}(x)$$

$$\leq e\left(\int_X d_L\mu(x)\chi_{(A \triangle B)}(x) \right), \qquad (6.10)$$

then there exists a point-fixed set $\bar{A}$, such that

$$T(\bar{A}) = \bar{A}, \tag{6.11}$$

in the sense that

$$d_L(T(\bar{A}), \bar{A}) = \int_X d_L \mu(x) X_{(T(\bar{A}) \triangle \bar{A})}(x) = 0. \tag{6.12}$$

We now show a concrete version of the Riesz representation theorem.

Theorem 6.2. *Let L be a continuous linear functional on $C(\bar{\Omega})$, the space of the continuous function defined in a compact set $\bar{\Omega} \subset R^4$ and satisfying the following property*

$$L\left(e^{i(\sum_{j=1}^{N} p_j x_j)} \chi_{\bar{\Omega}}(x)\right) = f_{\Omega}(p_1, \ldots, p_n) = f_{\Omega}(p) \in L^1(R^N) \tag{6.13}$$

Then, there exists a (unique) function $\phi_L(x) \in C(\bar{\Omega})$ representing the action of the functional Eq. (6.13) on $C(\bar{\Omega})$ by the integral representation

$$L(g(x)) = \int_{\bar{\Omega}} d^N x \phi_L(x) g(x). \tag{6.14}$$

Proof. Let us firstly consider the Fourier transform of the function $f_{\Omega}(p)$. Namely

$$\phi_L(x) = \left(\frac{1}{\sqrt{2\pi}}\right)^n \int_{F^N} d^N p \cdot e^{iP \cdot x} f_{\Omega}(p). \tag{6.15}$$

Obviously $\phi_L(x) \in C_0(R^N)$.

Due to supposed continuity of the functional L, one can show that $f_{\Omega}(p) \in C^{\infty}(R^N)$, and we have the differentiability relation ($M = (\ell_1, \ldots, \ell_N)$) (exercise)

$$\frac{\partial^{|M|}}{\partial p_1^{\ell_1} \cdots \partial p_n^{\ell_N}}(f_{\Omega}(p)) = L\left\{(ix_1)^{\ell_1} \cdots (ix_n)^{\ell_N}\left(\exp i\left(\sum_{j=1}^{N} p_j x_j\right)\right)\right\}, \tag{6.16}$$

which means that the inversion Fourier transform theorem holds true

$$L((x_1)^{\ell_1} \cdots (x_n)^{\ell_n}) = \left(\frac{1}{\sqrt{2\pi}}\right)^N \int_{R^N} d^N x e^{-ipx} \{(x_1)^{\ell_1} \cdots (x_n)^{\ell_n} \phi_L(x)\}|_{p=0}$$

$$= \left(\frac{1}{\sqrt{2\pi}}\right)^N \left\{\int_{R^N} d^N x (x_1)^{\ell_1} \cdots (x_n)^{\ell} \phi_L(x)\right\}. \tag{6.17}$$

By the Weierstrass theorem, we have, finally, our envisaged result on $C(\bar{\Omega})$

$$L(g(x)) = \int_{R^N} d^N x g(x) \phi_L(x) \chi_{\bar{\Omega}}(x) = \int_{\bar{\Omega}} d^N x g(x) \phi_L(x). \qquad (6.18)$$

$\square$

6.2. The L. Schwartz Representation Theorem on $C^\infty(\Omega)$ (Distribution Theory)

The Quantum and Random World is an application of Cantor Set Theory in its developments — Luiz Botelho.

After having exposed the fundamental abstract result of Riesz–Markov on the structure of the elements of the dual space of continuous linear functionals in $C(X)$, with X denoting a general compact topological space, we pass on to the problem of describing continuous functionals on the vector space $C^\infty(\Omega)$, with Ω denoting a open set of R^N.

Let us, thus, start by considering a sequence of compact sets K_n, with the interior property $(K_{n+1}) \supset K_n$ and such that $\Omega = \bigcup_{n=1}^\infty K_n$, together with the complete metric space $C^\infty(K_n)$, defined by the vector space of infinitely differentiable functions in Ω, with support in K_n with the Frechet metric

$$d(f,g) = \sum_{m=0}^\infty \frac{2^{-m} \|f - g\|_m}{1 + \|f - g\|_m}, \quad \text{where } \|f\|_m \colon \sup_{x \in \Omega} \sup_{|p| \le m} |D^p f(x)|.$$

The basic contribution of L. Schwartz is to consider the topology of the inductive limit on $C^\infty(\Omega)$ as writing formally as topological spaces $C^\infty_{\text{ind}}(\Omega) = \bigcup_{n=0}^\infty C^\infty(K_n)$ rigorously means that the topology in $C^\infty(\Omega)$ is the weakest topology which makes all the canonical injections

$$D_N \colon C^\infty(K_n) \to C^\infty(\Omega), \qquad (6.19)$$

continuous applications.

A topological basis for the origin of $C^\infty(\Omega)$ is formed by all those convex and barreleds sets $U \subset C^\infty(\Omega)$ such that $U \cap C^\infty(K_n)$ is always a neighborhood of the origin in $C^\infty(K_n)$. The main result and reason for introducing such inductive topology in $C^\infty(\Omega)$ is that it leads to the fundamental result that $C^\infty_{\text{ind}}(\Omega)$ is a sequentially complete topological vector space.

At this point, it is worth to call to the readers' attention that the usual nondistributional topological definition of $C^\infty(\Omega)$ as $\bigcup_{m=0}^\infty C^m(\Omega)$ (always used in others approach of generalized functions) is stronger than the L. Schwartz inductive topology introduced above.

We always rewrite $C_{\text{ind}}^{\infty}(\Omega)$ in the well-known L. Schwartz notation as $D(\Omega)$, the Schwartz test function space. The description of the notion of convergence in $D(\Omega)$ is straightforward, since we have the sequential completeness topological property. Namely, a sequence $\varphi_n(x) \in D(\Omega)$ converges in $D(\Omega)$ if there exists a set K_n such that $\varphi_n \to \bar{\varphi}$ in $C^{\infty}(K_n)$.

Another basic result as a consequence of the introduction of inductive limit topology in $C^{\infty}(\Omega)$ is the straightforward description of the dual space of $D(\Omega)$, denoted by $D'(\Omega)$ and named as the L. Schwartz distribution space in Ω

$$D'(\Omega) = \left(\bigcup_n C^{\infty}(K_n) \right)^* = \bigcup_n (C^{\infty}(K_n))^*. \qquad (6.20)$$

Note that the structural description of $(C^{\infty}(K_n))^*$ is expected to be closely related to that of $C(K_n)^*$ (the Riesz–Markov theorems). In fact, we have L. Schwartz generalization of the functional integral representation of Riesz–Markov theorem.

Theorem 6.3 (Laurent Schwartz). *Any given continuous linear functional $L \in D'(\Omega)$ may be represented by a sequence of complex Borel measures $d_n u(x)$ in K_n, a sequence of multi-indexes $\{P_n\} = \{p_n^1, \dots, P_n^N\}$ through the integral representation*

$$L(\varphi) = \sum_{n=0}^{\infty} \left(\int_{K_n} d\mu_n(x)(D^{P_n}\varphi)(x) \right), \qquad (6.21)$$

where

$$(D^p \varphi)(x) = \frac{\partial^{(P_n^1 + \cdots + P_n^N)}}{\partial x_1^{P_n^1} \cdots \partial x_N^{P_n^N}} \varphi(x_1, \dots, x_N). \qquad (6.22)$$

Proof. Let $L \in D'(\Omega)$, but with compact support $K_s \subset \Omega$. By the inductive limit topology (exercise 1), there exists a constant $c_s > 0$, and an integer $m_s > 0$, such that for any $\varphi \in D(\Omega)$, we have the estimate

$$|L(\varphi)| \leq c_s \sup_{x \in K_s} \left(\sup_{|p| \leq m_s} |D^p \varphi(x)| \right). \qquad (6.23)$$

Note that the triple (C, K_s, m) is not unique. We now consider the following elliptic operator $\mathcal{L} = (\frac{\partial}{\partial x_1}^{p_1}) \cdots (\frac{\partial}{\partial x_n})^{p_n}$ $(p = p_1 + \cdots + p_n)$. Since

$\mathcal{L}_p$ is an injective application of $C^\infty(\Omega)$ into $C^\infty(\Omega)$, and this T restricts the dense subspace $\mathcal{L}_p[D(\Omega)]$ (range of $\mathcal{L}^p$ in $D(\Omega)$) which satisfies the obvious estimate for any

$$f \in C(K_s): L(\mathcal{L}_p^{-1}f) \le c_s \cdot \sup_{x \in K_s} |f(x)| \qquad (6.24)$$

we can apply the Riesz–Markov Theorem 6.1 to the composed functional $L \circ \mathcal{L}_p^{-1}$ in $C(K_s)$

$$L(\mathcal{L}_p^{-1}f) = \int_{K_s} d_s u(x) f(x), \qquad (6.25)$$

or equivalently (exercise) for any $\varphi \in D(\Omega)$, we have the functional integral representation

$$L(\varphi) = \int_{K_s} d_s \mu(x)(D^{p_s}\varphi)(x). \qquad (6.26)$$

$\square$

In general, we just consider a unity partition subordinate to a given open cover of Ω ($1 = \sum_n h_n(x)$, $K_n \subset \operatorname{supp} h_n \subset K_{n+1}$, K_n compact set of Ω and $UK_u = \Omega$ and $h_n(x) = 1$, for $x \in K_n$):

$$L = \sum_{n=1}^{\infty} h_n(x)L = \sum_{n=1}^{\infty} L_n. \qquad (6.27)$$

At this point, we introduce the weak-* topology in $D'(\Omega)$ through a sequential criterion. A sequence of distributions $L_n \in D'(\Omega)$ converges weak-star if the sequence of measures in Eq. (6.25) converges in the weak-star topology of $C(K_s)^*$.

After the proof of L. Schwartz representation theorem, let us introduce the operation of derivation in the distributional sense. First, let us recall some definitions in the functional analysis of vector topological spaces. Let E and F be two vector spaces with topologies compatible with its vectorial structure and $U: E \to F$ a linear continuous application between them. For any $y' \in F^*$ (dual of F), we can associate the element x' of E' through the definition ($^tU: F' \to E'$):

$$(x') = x'(x) = y'(U(x)) = (^tU(y)). \qquad (6.28)$$

It can be showed that if U is continuous, the ^tU remains continuous if E and F are Frechet spaces like $C^\infty(K_s)$.

As a consequence of the above remarks, the usual derivative operator is a linear continuous application between $D(\Omega)$. Namely

$$D: D(\Omega) \to D(\Omega).$$

By the duality equation (6.26) mentioned above, one has a natural derivative application in $D'(\Omega)$

$$(-DL) \stackrel{\text{def}}{\equiv} ({}^t D)(L)(f) \stackrel{\text{def}}{\equiv} L(Df), \tag{6.29}$$

besides being always a continuous operation in $D'(\Omega)$ if $L_n \stackrel{D'(\Omega)}{\longrightarrow} \Leftrightarrow {}^t DL_n \stackrel{D'(\Omega)}{\longrightarrow} {}^t DL$.

At this point, we call our reader to show that the sequence of functions $f_n(x) = \frac{1}{n}\sin(nx)$ seen as kernels of distributions in $D'(R)$ obviously converges to the zero distribution in $D'(R)$:

$$\lim_{n\to\infty} \int_R dx \left(\frac{1}{n}\sin nx \right) \varphi(x) = 0. \tag{6.30}$$

As a consequence of the above remark, we have the validity of the result called the Riemann–Lebesgue lemma

$$\lim_{n\to\infty} \int_R dx \cos(nx)\varphi(x) = 0. \tag{6.31}$$

Namely

$$\frac{d}{dx}\left(\frac{\sin nx}{n} \right)^{D'(R)} = \cos(nx) \stackrel{D'(R)}{\longrightarrow} 0. \tag{6.32}$$

A further finer structural analysis can be implemented to Eq. (6.21) by means of an application of the Radon–Nikodym theorem to the pair of complex Borel measures $(du_s(x), d^N x)$ on the Borelians of Ω.

$$L(\varphi) = \sum_{s=0}^{\infty} \left(\int_{\Omega} (D^{p_s}\varphi)(x) \overbrace{\left(\frac{du_s(x)}{d^n x} \right)}^{h_s(x)} d^n x \right)$$

$$+ \sum_{s=0}^{\infty} \left(\int_{\Omega} (D^{p_s}\varphi)(x) d\nu_r^{\text{sing}}(x) \right), \tag{6.33}$$

where $h_s(x) \in L^1(\Omega, d^N x)$, and $dV_n^{\mathrm{sing}}(x)$ is a singular measure (in relation to the Lebesgue measure $d^N x$ in Ω) with support at points (Dirac delta functions) and on sets of Lebesgue zero measure

$$dV_s^{\mathrm{sing}}(x) = \sum_{\ell=0}^{\infty} a_{\ell,s} \delta(x - x_{\ell,s}) + dV_s^{(\mathrm{continuous \ singular})}(x). \qquad (6.34)$$

Another important distributional space is the (topological) dual space of test functions with polynomial decreasing $S(R^N)$, a very basic object in wave fields quantum path-integral (see Chap. 19)

$$S(R^N) = \left\{ u \in C^\infty(R^N) \mid ||\varphi||_{n,r} = \sup_{x \in R^N} |x^n D^m \varphi(x)| < (\infty) \right\}. \qquad (6.35)$$

We have the following structural theorem, analogous to Theorem 6.1 of L. Schwartz.

Theorem 6.4. *Given a functional L in $(S(R^n))'$, we can always represent L by a Borel complex measure $du(x)$ in R^N by means of ($x^p = x_1^{p_1} \cdots x_N^{p_n}$, etc.)*

$$L(\varphi) = \int_{R^N} d\mu(x)(x^p D^q \varphi)(x). \qquad (6.36)$$

The proof of the above integral representation for distributions in $S^q(R^N)$ is based on the fact that for a given $L \in S^q(R^N)$, these are multi-indexes (p, q), such that is a positive constant c with

$$|L(\varphi)| \leq c||\varphi||_{p,q}. \qquad (6.37)$$

Note that the elliptic operator $\mathcal{L}_p = x^p D^q = x_1^{p_1} \cdots x_n^{p_n} (\frac{\partial}{\partial x_1})^{q_1} \cdots (\frac{\partial}{\partial x_N})^{q_N}$ is a subjective application of $S(R^N)$ into $S(R^N)$, and $\mathcal{L}_p(S(R^N))$ is dense in $C_0(R^N)$ (continuous functions vanishing at ∞), the great usefulness of $S^1(R^N)$ in quantum field theoretic path-integrals is related by the fact that the usual Fourier transform is a vectorial/topological isomorphism in $S(R^N)$. By duality, one straightforwardly define the Fourier transforms in $S^1(R^N)$ which remains as a topological isomorphism in the distributional space $S^1(R^N)$

$$\mathcal{F}: S(R^N) \to S(R^N), \qquad (6.38)$$

$$^t\mathcal{F}: S^q(R^N) \to S^q(R^N), \qquad (6.39)$$

$$^t\mathcal{F}(L)(\varphi) = L(\mathcal{F}(\varphi)). \qquad (6.40)$$

The above equations are important results in the applications of distribution theory of L. Schwartz given by the following result. Let A be a continuous linear application between a locally convex topological vector space E with value in the topological dual of another locally convex topological vector space F'. Then, the bilinear form in $E \times F$ defined by the relation $B(f, g) = (Af)(g)$ is continuous in $E \times F'$, when one introduces the weak topology on F'. As a consequence, every continuous bilinear form on $S(R^N)$ is of linear superposition of forms written below for a pair of multi-indexes (m, n)

$$B(f, g) = \int_{R^N \times R^N} F(x, y)(D^m f(x))(D^n g(x)) d^n x d^n y. \qquad (6.41)$$

Here, $F(x, y)$ is a continuous function of polynomial grow in R^N

$$\left(\exists p \in \mathbb{Z}^+ \Big|_{\substack{|x| \to \infty \\ |y| \to \infty}} \lim F(x, y)(|x|^2 + |y|^2)^{-p} = 0 \right).$$

6.3. Equivalence of Gaussian Measures in Hilbert Spaces and Functional Jacobians

In this section, we present the mathematical analysis of the Jacobian change of variable in Gaussian functional integrals in Hilbert spaces through the formalism of the Kakutani theorem.

Let A^{-1} and B^{-1} be positive definite trace class operators in a given Hilbert space $(H, \langle, \rangle)$ and operators inverse of the operators A and B.

The spectral representations for theses operators

$$A\varphi_n = \lambda_n \varphi_n, \qquad (6.42a)$$

$$B\sigma_n = \alpha_n \sigma_n, \qquad (6.42b)$$

define Gaussian measures $d_{A^{-1}} u(\varphi)$ and $d_{B^{-1}} u(\varphi)$ in the Borelian algebra of the cylindrical sets in H and are defined by

$$d_{A^{-1}}\mu(\varphi) = \limsup_N \left\{ \prod_{n=1}^N d\langle \varphi, \varphi_n \rangle \exp\left\{ -\frac{\lambda_n}{2} \langle \varphi, \varphi_n \rangle^2 \right\} \left(\sqrt{\frac{\lambda_n}{2\pi}} \right) \right\},$$
$$(6.43a)$$

$$d_{B^{-1}}\nu(\varphi) = \limsup_N \left\{ \prod_{n=1}^N d\langle \varphi, \sigma_n \rangle \exp\left\{ -\frac{\alpha_n}{2} \langle \varphi, \sigma_n \rangle^2 \right\} \left(\sqrt{\frac{\alpha_n}{2\pi}} \right) \right\}.$$
$$(6.43b)$$

We have, thus, the Kakutani theorem and the equivalence of the above measures.

Theorem 6.5 (Kakutani). *The two measures Eqs. (6.43a) and (6.43b) are mutually equivalent or singular. In the first case, we have the criterion that*

$$\sum_n \left(\frac{\lambda_n^2 - \alpha_n^2}{\lambda_n \alpha_n} \right) < \infty, \qquad (6.44)$$

and the Radon–Nykodin derivative of the above measures is given by

$$\frac{d_{A^{-1}}\mu(\varphi)}{d_{B^{-1}}\nu(\varphi)} = \lim_{N \to \infty} \left\{ \prod^N \left[\frac{\lambda_n}{\alpha_n} \exp\left(-\frac{1}{2}(\lambda_n - \alpha_n) \left(\sum_n (\varphi, \sigma_n)^2 \right) \right) \right] \right\}. \qquad (6.45)$$

Note that in the case of the Radon–Nykodim derivative

$$\frac{d_{A^{-1}}\mu(\varphi)}{d_{B^{-1}}\nu(\varphi)} = \det(AB^{-1}) \exp\left\{ -\frac{1}{2}\langle \varphi, (A - B)\varphi \rangle \right\}. \qquad (6.46)$$

On the basis of Eqs. (6.45) and (6.46), one can show the Wienner result about translation invariant of Gaussian measures. Let $T_h \colon H \to H$ be the translation operator in H. Let us consider the translated Gauss measure

$$d_{A^{-1}}\mu(T_h\varphi) = d_{A^{-1}}\mu(\varphi + h) = \left(\prod_{n=1}^{\infty} d\langle \varphi | \varphi_n \rangle e^{-\frac{1}{2}\lambda_n [\langle \varphi, \varphi_n \rangle + \langle h, \varphi_m \rangle]^2} \right\}. \qquad (6.47)$$

By the Kakutani theorem, the translated measure $d_{A^{-1}}\mu(T_h\varphi)$ is equivalent to the measure $d_{A^{-1}}\mu(\varphi)$ if and only if

$$\sum_n \frac{|\langle h, \varphi_n \rangle|^2}{\frac{1}{\lambda_n}} = \langle Ah, h \rangle < \infty, \qquad (6.48)$$

or equivalently, the translational-invariance of the measure is insured if h belongs to the domain of the operator A.

At this point, we remark that $\text{Dom}(A)$ is a set of zero measure for $(H, d_{A^{-1}}\mu(\varphi))$. (Finite action configurations smooth field configurations, makes a set of zero functional measure.) Let us give a simple proof of such important result in practical calculations with path-integrals.

First, let us rewrite the finite action set of path integrated configurations $(= \mathrm{Dom}(A))$ in the following form (for $\varepsilon > 0$):

$$\chi_{\mathrm{Dom}(A)}(\varphi) = \lim_{\alpha \to 0^+} \lim_{N \to \infty} \exp\{-\alpha\langle p_N\varphi, AP_N\varphi\rangle\}$$

$$= \begin{cases} 1, & \text{if } \langle \varphi, A\varphi \rangle < \infty, \\ 0, & \text{otherwise}, \end{cases} \tag{6.49}$$

where the orthogonal projections $P_n \to 1$ in the strong sense.

Let as evaluate formally in the "physical way" its functional measure content

$$M_{A^{-1}}(\chi_{\mathrm{Dom}(A)}(\varphi)) = \lim_{\alpha \to 0} \left\{ \lim_{N \to \infty} \left[\int_H d_{A^{-1}}\mu(\varphi) e^{-\alpha\langle P_N\varphi, AP_N\varphi\rangle} \right] \right\}$$

$$\times \lim_{\alpha \to 0^+} \left(\lim_{N \to \infty} \exp\left(-\frac{N}{2}\ell g(1+\alpha) \right) \right)$$

$$= \lim_{\alpha \to 0} \lim_{N \to \infty} (e^{-\frac{N}{2}\alpha}) = e^{-\infty} = 0. \tag{6.50}$$

At this point, one can see that the usual Schwinger procedure to deduce functional equations for the quantum field generating functional does not make sense in the Euclidean framework of path-integrals since the measure is not translational invariant. Namely, in the usual Feynmann notation

$$\int_H D^F[\varphi]\frac{\delta}{\delta\varphi} \left\{ e^{-\frac{1}{2}\langle\varphi, A\varphi\rangle} e^{-V(\varphi)} \right\} \neq 0. \tag{6.51}$$

As one can see from the above exposed result that the Minlos theorem is a powerth "tool" in the functional integration theory in infinite dimension. In this context, we have the converse of the Minlos theorem in Hilbert spaces, the so called Sazonov theorem. Let us give a simple proof of this result.

Theorem 6.6 (Sazonov). *Given a Gaussian measure $d\mu(\varphi)$ in a separable Hilbert space H such that*

$$\int_H d\mu(\varphi)||\varphi||^2 < \infty. \tag{6.52}$$

Then, the functional Fourier transforms

$$Z(f) = \int_H d\mu(\varphi)\exp(i\langle f, \varphi\rangle_H), \tag{6.53}$$

is a continuous functional in relation to the Sazonov Hilbert–Schmidt inner product

$$\langle \bar{\varphi}, \bar{\psi} \rangle_{S_z} = \int_H d\mu(\varphi) \langle \varphi, \bar{\varphi} \rangle \langle \varphi, \bar{\psi} \rangle. \tag{6.54}$$

Proof. First, let us remark that the above-defined Sazonov norm is of Hilbert–Schmidt class by construction, since for any orthonormal basis (e_n) in $(H, \langle, \rangle)$, we have the relationship

$$\sum_{n=1}^{N} (e_n, e_n)_{S_z} = \sum_{n=1}^{N} \left\{ \int_H d\mu(\varphi) \langle \varphi, e_n \rangle \langle \varphi, e_n \rangle \right\}$$

$$\leq \int_H d\mu(\varphi) \|\varphi\|^2 < \infty. \tag{6.55}$$

It is obvious to see that if one has a sequence $\sigma_n \in H$ and $\sigma_n \xrightarrow{\|\ \|} 0$, then we have that

$$(\sigma_n, \sigma_n)_{S_z} = \int_H d\mu(\varphi) (\langle \varphi, \sigma_n \rangle)^2 \to 0. \tag{6.56}$$

Consequently by the Lebesgue theorem $\langle \varphi, \sigma_n \rangle \to 0$ almost everywhere in $(H, d\mu(\varphi))$. Obviously $(1 - \exp i(\langle \varphi, \sigma_n \rangle)) \to 0$ almost everywhere. As an easy estimate, we have

$$|Z(\sigma_n) - Z(0)| = \left| \int_H d\mu(\varphi)(1 - e^{i\langle \varphi, \sigma_n \rangle}) \right| \to 0. \tag{6.57}$$

Result showing the (sequential) continuity of the functional $Z(j)$ in the new separable Hilbert space $(H, \langle, \rangle_{S_z})$. Another important comment related to the Radom–Nykodin derivative Eq. (6.45) is the case one has the B-operator as a "formal" perturbation of the operator A. Namely

$$A = B + gC \quad \text{and} \quad [B, C] = 0. \tag{6.58}$$

In this situation, we have the relationship

$$\frac{d\mu(\varphi)}{B+gC} / d_B \nu(\varphi) = \det[1 + gCB^{-1}] \exp \left\{ -\frac{1}{2} \langle \varphi, C\varphi \rangle \right\}. \tag{6.59}$$

By using the proper time definition for evaluating the functional determinant in Eq. (6.59)

$$\det[1 + gCB^{-1}] = \exp \operatorname{Tr} \ell g[1 + gCB^{-1}]$$

$$= \exp \left[\int_0^g d\lambda \operatorname{Tr}(CB^{-1}(1 + \lambda CB^{-1})^{-1}) \right]$$

$$\stackrel{\text{def}}{=} \exp \operatorname{Tr} \left[\int_0^g R(\lambda) d\lambda \right], \tag{6.60}$$

where the resolvent operator $R(\lambda)$ satisfies the operatorial equation, and we suppose that $CB^{-1} \in \mathcal{J}_1(H)$ due to the obvious estimate

$$\left| \prod_{n=1}^{\infty} (1 + \lambda_n) \right| = e^{\left[\sum_{n=1}^{\infty} \ell n(1 + \lambda_n) \right]} \leq e^{\sum_{n=1}^{\infty} \lambda_n} = e^{\operatorname{Tr}(CB^{-1})}$$

$$R(\lambda) = CB^{-1} - \lambda(CB^{-1})R(\lambda). \tag{6.61}$$

Sometimes perturbative evaluations can be implemented for Eq. (6.61). Let us exemplify in this "toy model"

$$\langle x|A|x' \rangle = -\frac{1}{2} \frac{d^2}{d^2 x} \delta(x - x^2) + gV(x, x') = B + gC. \tag{6.62}$$

We can see that

$$[B, C] = 0 \Leftrightarrow \frac{d^2}{d^2 x'} V(x, x') = \frac{d^2}{d^2 x} V(x, x'). \tag{6.63}$$

So the resolvent $R(\lambda) \equiv R(x, x', \lambda)$ satisfies the integral equation below

$$R(x, w, \lambda) \left(-\frac{1}{2} \frac{d^2}{dw^2} \right) \delta(w - x')$$

$$= gV(x, x') - \lambda \int_{-\infty}^{+\infty} V(x, y)R(y, x', \lambda)dy. \tag{6.64}$$

As a geometrical comment, in this chapter, from a set of informal lecture notes for graduate students' seminars, we have a geometric stochastic criterion for differentiability of the samples (class of functions) in the space

$C^2(L^2(\Omega), d_{A^{-1}}\mu(\varphi))$. Let us start by considering the estimate of the Newton differential quotient

$$\int_{L^2(\Omega)} d_{A^{-1}}\mu(\varphi) \left| \lim_{h \to 0} \frac{\varphi(x+h) - \varphi(x)}{h} \right|$$

$$\leq \left(\int_{L^2(\Omega)} d_{A^{-1}}\mu(\varphi) \frac{[\phi(x+h)\phi(x+h) + \phi(x)\phi(x) - 2\phi(x+h)\phi(x)]}{h^2} \right)^{1/2}$$

$$\leq \frac{1}{h}[A^{-1}(x+h, x+h) + A^{-1}(x, x) - (A(x+h, x) + A(x, x+h))]^{1/2}, \tag{6.65}$$

which insures the differentiability in the average of the path-integral samples if we have the differentiability property for the correlation function

$$\left. \frac{\partial^2 A^{-1}(x,y)}{\partial x \partial y} \right|_{x=y} < \infty.$$

□

6.4. On the Weak Poisson Problem in Infinite Dimension

In this section, we intend to present a weak solution to the Poisson problem in infinite dimensions (see Chap. 1) by means of the simple basic Hilbert space techniques.

Let us start our analysis by considering a given separable Hilbert space H and an associated Gaussian cylindrical measure $d_Q u(x)$ on the Borel cylinder of H, with Q denoting a class trace positive-definite operator ($Q \in \mathfrak{f}_1^+(H, H)$). We intend now to show the existence of a real-valued weak solution to the following (functional) Hilbert space-valued Poisson problem in the Hilbert space $L^2(H, d_Q u)$

$$- \operatorname{Tr}_H[Q D^2 U(x)] + \langle x, DU \rangle_H = f(x)(\chi_{B_R(0)}(x)) \tag{6.66}$$

Here, $\chi_{B_R(0)}(x)$ is the characteristic function of the ball of radius R in H ($B_R(0) = \{x \in H \mid \langle x, x \rangle \leq R\}$, $u(x,t) \in C^2(H, R) \cap L^2(H, d_Q u)$) and D^2 the second-order (operator) Frechet derivative in H (for instance $\operatorname{Tr}\left[Q D^2(e^{i\langle P, x \rangle_H})\right] = -(\langle p, Qp \rangle_H)e^{i\langle p, x \rangle_H}$).

By using the following formula of integration by parts in infinite dimensions ($Q\beta_k = \lambda_k\beta_k$)

$$\int_H d_Q\mu(x)\langle D\varphi, D\psi\rangle_H(x) = -\int_H (\text{Tr}[D^2\psi(x)], \varphi(x))d_Q\mu(x)$$

$$+ \int_H \langle x, (QD\psi)(x)\rangle_H\varphi(x)d_Q\mu(x), \quad (6.67)$$

one can rewrite Eq. (6.66) into a functional integral form where it is possible now to consider a rigorous mathematical meaning for weak solutions in $L^2(H, d_Q\mu)$

$$\int_H d_A\mu(x)\langle BDu, BDg\rangle = \int_H d_A\mu(x)(f(x)\chi_{B_R}(x))g(x), \quad (6.68)$$

where $B \in \mathfrak{f}_2^+(H)$, and $BB^* = Q$ is the usual decomposition of a trace class (positive-definite) operator in terms of Hilbert–Schmidt root square operator B.

Since Q is positive-definite, there exists a positive constant c such that for any $x \in H$, we have the lower bound

$$\langle Qx, x\rangle_H \geq c\langle x, x\rangle. \quad (6.69)$$

Let us now introduce the following Hilbert–Sobolev space associated to the cylindrical measure $d_Q\mu(x)$

$$H^1(L^2(H, d_Q\mu)) = \{f \in C^1(H, R) \cap L^2(H, d_Q\mu)$$

$$\left|\langle f, f\rangle_{H^1} \equiv \int_H d_Q\mu(x)\langle BDf, BDf\rangle_H < \infty\right|. \quad (6.70)$$

We now show that the real linear function associated to our class of cut-off datum $f(x)\chi_{B_R}(x)$ as expressed by the right-hand side of Eq. (6.132) is continuous in the topology associated to Eq. (6.70). Since the above space is separable, one can verify the continuity through a sequential argument. Let us thus consider a sequence of elements $\{g_n\}$ in $H^1(L^2(H, d_Q\mu))$ converging to a given $\bar{g} \in H^1(L^2(H, D_Q\mu))$.

We have the estimate below

$$\left|\int_H d_Q\mu(x)(f(x)\chi_{B_R}(x))(g_n - g)(x)\right|$$

$$\leq \|f\|_{L^2(H, d_Q\mu)}^{1/2}\left(\int_H d_Q\mu(x)(g_n - g_n)^2(x)\chi_{B_R}(x)\right)^{1/2}. \quad (6.71)$$

At this point, we observe the validity of the estimate:

$$\left(\int_H d_Q\mu(x) q^2(x) \chi_{B_R}(x) \right)$$

$$\leq \left(\frac{4 \cdot [(\sum_k \lambda_k)^2]}{R^2 - 2\operatorname{Tr}_H(Q)} \right) \left(\int_H d_Q\mu(x) \langle Dg, Dg \rangle(x) \right). \qquad (6.72)$$

So, if one chooses R in a such way that $\operatorname{Tr}_H(Q) < \frac{R^2}{2}$ and observing that (see Eq. (6.69))

$$\int_H d_Q\mu(x) \langle Dg, Dg \rangle(x) \leq \frac{1}{c} \left\{ \int d_Q\mu(x) \langle BDg, BDg \rangle(x) \right\}, \qquad (6.73)$$

one can see naturally that the functional linear below

$$L_f(g) = \int_H d_Q\mu(x)(f(x)\chi_{B_R}(x))g(x), \qquad (6.74)$$

belongs to dual space of our Sobolev space (Eq. (6.70)).

We now apply the Riesz theorem used in Sec. 6.4 to find an element in $H^1(L^2(H, d_Q\mu))$ being the weak solution of the functional equation (6.68). Namely

$$L_f(g) = \int_H d_Q\mu(x) \langle BDg, BDu \rangle(x). \qquad (6.75)$$

One explicit construction of this weak-solution $U(s)$ is given by the series in $H^1(L^2(H, d_A\mu))$

$$U = \sum_{n=1}^{\infty} L_f(e_n)e_n = \sum_{n=1}^{\infty} \left(\int_H d_Q\mu(y) f(y) \overline{e_n(y)} \right) \cdot e_n(x), \qquad (6.76)$$

where e_n is any given orthonormal basis in $H^1(L^2(H, d_Q\mu))$ (not in $L^2(H, d_Q\mu)$!). Such basis are easily constructed by a simple application of the Grham–Schmidt orthogonalization process to any set of linearly independent elements $\{E_n(x)\}$ in $C^1(H, R) \cap L^2(d_Q\mu)$ like the Hermite functionals. Namely

$$e_1(x) = \frac{E_1(x)}{\|E_1\|_{H^1}}$$

$$\vdots$$

$$e_n(x) = \frac{E_n(x) - \sum_{n=1}^{n-1} \langle E_n, e_k \rangle_{H^1} \cdot e_n(x)}{\|E_n - \sum_{n=1}^{n-1} \langle E_n, e_k \rangle_{H^1} \cdot e_k\|_{H^1}}. \qquad (6.77)$$

The reader is invited to apply such procedure to the following Poisson problem in loop space theory for quantum fields as follows:

$$H = L^2([0, 2\pi])$$

$$Q(\sigma, \sigma') = \left(-\frac{d^2}{dx^2} + m^2\right)^{-1}(\sigma, \sigma'), \tag{6.78a}$$

Functional Laplacean equation

$$\int_0^{2\pi} d\sigma \int_0^{2\pi} d\sigma' \left[\frac{\delta}{\delta f(\sigma')}\left\{Q(\sigma, \sigma')\frac{\delta}{\delta f(\sigma)}\right\}\psi[f(\sigma)]\right]$$

$$= F_R[f(\sigma)]\chi_{\|f\|_{H_0^1} \leq R}(f(\sigma)). \tag{6.78b}$$

The functional data $F_R(f(\sigma))$ can be expressed as a functional (Minlos) Fourier transform of any analytical function $H(p)$ in H

$$F_R[f] = \int_H d_Q\mu(p) \cdot H(p) \exp(i\langle p, f\rangle_h). \tag{6.78c}$$

6.5. The Path-Integral Triviality Argument

We start our analysis by considering the chiral non-abelian $SU(N_c)$ thirring model Lagrangean on the Euclidean space-time of finite volume $\Omega \subset R^4$.

$$L(\psi, \bar{\psi}) = \frac{1}{2}[\bar{\psi}^a(i\gamma_\mu\overrightarrow{\partial_\mu}\psi^a) + (\bar{\psi}^a i\gamma_\mu\overleftarrow{\partial_\mu})\psi^a]$$

$$+ \left(\frac{g^2}{2}(\bar{\psi}_b\gamma^\mu\gamma^5(\lambda^A)_{bc}\psi_c)^2\right). \tag{6.79}$$

Here, $(\psi^a, \bar{\psi}^a)$ are the euclidean four-dimensional chiral fermion fields belonging to a fermionic fundamental representation of the $SU(N_c)$ nonabelian group with Dirichlet boundary condition imposed at the finite-volume region Ω. In the framework of path-integrals, the generating functional of the Green's functions of the quantum field theory associated with the Lagrangean equation (6.79) is given by ($\slashed{\partial} = i\gamma_\mu\partial_\mu$)

$$Z[\eta_a, \bar{\eta}_a] = \frac{1}{Z(0,0)}\int \prod_{a=1}^{N^2-N} D[\psi_a]D[\bar{\psi}_a]$$

$$\times \exp\left\{-\frac{1}{2}\int_\Omega d^4x(\psi_a, \bar{\psi}_a)\begin{bmatrix} 0 & \slashed{\partial} \\ \slashed{\partial}^* & 0 \end{bmatrix}\begin{pmatrix} \psi_a \\ \bar{\psi}_a \end{pmatrix}\right\}$$

$$\times \exp\left\{-\frac{g^2}{2}\int_\Omega d^4x(\bar{\psi}_b\gamma^5\gamma^\mu(\lambda^A)_{bc}\psi_c)^2(x)\right\}$$

$$\times \exp\left\{-i\int_\Omega d^4x(\bar{\psi}_a\eta_a + \bar{\eta}_a\psi_a)(x)\right\}. \tag{6.80}$$

In order to proceed with a bosonization analysis of the fermion field theory described by the above path-integral, it appears to be convenient to write the interaction Lagrangian in a form closely parallel to the usual fermion–vector coupling in gauge theories by making use of an auxiliary nonabelian vector field $A_\mu^a(x)$, but with a purely imaginary coupling with the axial vectorial fermion current (at the Euclidean world).

$$Z[\eta_a, \bar{\eta}_a]$$

$$= \frac{1}{Z(0,0)}\int \prod_{a=1}^{N^2-N} D[\psi_a(x)]D[\bar{\psi}_a(x)] \int \prod_{a=1}^{N^2-N}\prod_{\mu=0}^{3} D[A_\mu^a(x)]$$

$$\times \exp\left\{-\frac{1}{2}\int_\Omega d^4x(\psi_a, \bar{\psi}_a)\begin{bmatrix} 0 & \slashed{\partial}+ig\gamma_5\,\slashed{A} \\ (\slashed{\partial}+ig\gamma_5\,\slashed{A})^* & 0 \end{bmatrix}\begin{pmatrix}\psi_a \\ \bar{\psi}_a\end{pmatrix}(x)\right\}$$

$$\times \exp\left\{-\frac{1}{2}\int_\Omega d^4x(A_\mu^a A_\mu^a)(x)\right\}\exp\left\{-i\int_\Omega d^4x(\bar{\psi}_a\eta_a + \bar{\eta}_a\psi_a)(x)\right\}. \tag{6.81}$$

At this point, we present our idea to bosonize (solve) exactly the above fermion path-integral. The main point is to use the long-time ago suggestion that at the strong coupling $g_{\text{bare}} \to 0^+$ and at a large number of colors (the t'Hooft limit[2]), one should expect a great reduction of the (continuum) vector dynamical degrees of freedom to a manifold of constant gauge fields living on the infinite-dimensional Lie algebra of $SU(\infty)$. In this approximate leading t' Hooft limit of large number of colors, we can evaluate exactly the fermion path-integral by noting the Dirac kinetic operator in the presence of constant $SU(N)$ gauge fields written in the following suitable form

$$\exp\left\{-\frac{1}{2}\int_\Omega d^4x(\psi_a\bar{\psi}_a)\begin{bmatrix} 0 & V(\varphi)\,\slashed{\partial}V(\varphi) \\ V(\varphi)^* & \slashed{\partial}^*V^*(\varphi) & 0 \end{bmatrix}\begin{pmatrix}\psi_a \\ \bar{\psi}_a\end{pmatrix}(x)\right\}, \tag{6.82}$$

where the chiral phase-factor given by

$$V(\varphi) = \exp[-g\gamma_5(A_\mu^a \cdot x_\mu)\lambda_a], \tag{6.83}$$

with the chiral $SU(N)$ valued phase defined by the constant gauge field configuration

$$\varphi(x^\mu) = \varphi^a \lambda_a = (A_\mu^a \cdot x_\mu) \lambda_a. \tag{6.84}$$

Note that due to the attractive coupling on the axial current-axial current interaction of our thirring model (Eq. (6.80)), the axial vector coupling is made through an imaginary-complex coupling constant ig.

Now we can follow exactly as in the well-known chiral path-integral bosonization scheme[3] to solve exactly the quark field path-integral (Eq. (6.82)) by means of the chiral change of variables

$$\psi(x) = \exp\{-g\gamma_5\varphi(x)\}\chi(x), \tag{6.85}$$

$$\bar\psi(x) = \bar\chi(x)\exp\{-g\gamma_5\varphi(x)\}. \tag{6.86}$$

After implementing the variable change equations (6.7) and (6.8), the fermion sector of the generating functional takes to a form where the independent Euclidean fermion fields are decoupled from the interacting–intermediating non-Abelian constant vector field, A_μ^a, namely

$$Z[\eta_a \bar\eta_a] = \frac{1}{Z(0,0)} \int \prod_{a=1}^{N^2-N} D[\chi_a(x)] D[\bar\chi_a(x)]$$

$$\times \int_{-\infty}^{+\infty} \prod_{a=1}^{N^2-N} d[A_\mu^a] \times \exp\left\{-\frac{1}{2}(\text{vol}(\Omega))(A_\mu^a)^2\right\}$$

$$\times \det{}_F^{+1}\left[(\not\partial + ig\gamma_5 \not A)(\not\partial + ig\gamma_5 \not A)^*\right]$$

$$\times \exp\left\{-\frac{1}{2}\int_\Omega d^4x(\chi_a, \bar\chi_a)\begin{bmatrix} 0 & \not\partial \\ \not\partial^* & 0 \end{bmatrix}\begin{pmatrix} \chi_a \\ \bar\chi_a \end{pmatrix}(x)\right\}$$

$$\times \exp\left\{-i\int_\Omega d^4x(\bar\chi_a e^{-g\gamma_5\varphi(x)}\eta_a + \bar\eta_a e^{-g\gamma_5\varphi(x)}\chi_a)(x)\right\}. \tag{6.87}$$

Let us now evaluate exactly the fermionic functional determinant on Eq. (6.9) which is exactly the Jacobian functional associated to the chiral fermion field reparameterizations (Eqs. (6.7) and (6.8)).

In order to compute this fermionic determinant, $\ell n \det{}_F^{+1}[(\not\partial + ig\not A)(\not\partial + ig\not A)^*]$, we use the well-known theorem of Schwarz–Romanov[4] by introducing a σ-parameter ($0 \le \sigma \le 1$) dependent family of interpolating Dirac operators

$$\not D^{(\sigma)} = (\not\partial + ig\not A^{(\sigma)}) = \exp\{-g\sigma\gamma_5\varphi(x)\}(\not\partial)\exp\{-g\sigma\gamma_5\varphi(x)\}. \tag{6.88}$$

Since we have the straightforward relationship

$$\frac{d}{d\sigma} \not{D}^{(\sigma)} = (-g\gamma_5\varphi) \not{D}^{(\sigma)} + \not{D}^{(\sigma)}(-g\gamma_5\varphi), \tag{6.89}$$

and the usual proper time definition for the involved functional determinants on our study, namely

$$\log \det_F^{+1}(\not{D}^{(\sigma)} \not{D}^{(\sigma)^*}) = \lim_{\varepsilon \to 0^+} \int_\varepsilon^\infty \frac{ds}{s} \operatorname{Tr}_F(e^{-s(\not{D}^{(\sigma)}\not{D}^{(\sigma)^*})}). \tag{6.90}$$

As a consequence, we get the somewhat expected result that the Fermion functional determinant on the presence of constant gauge external fields coincides with the free one, namely

$$\frac{\det_F[(\not{\partial} + ig \not{A})(\not{\partial} + ig \not{A})^*]}{\det_F[(\not{\partial})(\not{\partial})^*]} = 1, \tag{6.91}$$

since $W[A_\mu^a] \equiv 0$ on basis of Eq. (6.16).

Let us return to our "Bosonized" generating functional (Eq. (6.9)) (after substituting the above results on it)

$$Z[\eta_a, \bar{\eta}_a] = \frac{1}{Z(0,0)} \int \prod_{a=1}^{N^2-N} D[\chi_a(x)]D[\bar{\chi}_a(x)]$$

$$\times \int_{-\infty}^{+\infty} \prod_{a=1}^{N^2-N} d[A_\mu^a] \exp\left\{ +\frac{1}{2} \operatorname{vol}(\Omega) \operatorname{Tr}_{SU(N)}(A_\mu)^2 \right\}$$

$$\times \exp\left\{ -\frac{1}{2} \int_\Omega d^4x(\chi_a, \bar{\chi}_a) \begin{bmatrix} 0 & \not{\partial} \\ \not{\partial}^* & 0 \end{bmatrix} \begin{pmatrix} \chi_a \\ \bar{\chi}_a \end{pmatrix}(x) \right\}$$

$$\times \exp\left\{ -i \int_\Omega d^4x(\bar{\chi}_a e^{-g\gamma_5(A_\mu^a \lambda_a)x^\mu} \eta_a + \bar{\eta}_a e^{-g\gamma_5(A_\mu^a \lambda_a)x^\mu} \chi_a)(x) \right\}. \tag{6.92}$$

Let us argue in favor of the theory's triviality by analyzing the long-distance behavior associated to the $SU(N)$ gauge-invariant fermionic composite operator $B(x) = \psi_a(x)\bar{\psi}_a(x)$. It is straightforward to obtain its exact expression from the bosonized path-integral (Eq. (6.18))

$$\langle B(x)B(y) \rangle = \langle (\chi_a(x)\bar{\chi}_a(x))(\chi_a(y)\bar{\chi}_a(y)) \rangle^{(0)} G((x-y)), \tag{6.93}$$

here, the reduced model's Gluonic factor is given exactly in its structural-analytical form by the path-integral (without bothering us with the γ_5-Dirac indexes)

$$G((x-y)) \sim \frac{1}{G(0)} \int_{-\infty}^{+\infty} \prod_{a=1}^{N^2-N} d[A_\mu^a] \exp\left\{+\frac{1}{2} \text{vol}(\Omega) \text{Tr}_{SU(N)}(A_\mu)^2\right\}$$

$$\times \text{Tr}_{SU(N_c)} \mathbb{P}\left\{\exp -g \oint_{C_{xy}} A_\alpha dx_\alpha\right\}, \tag{6.94}$$

with C_{xy} a planar closed contour containing the points x and y and possessing an area S given roughly by the factor $S = (x-y)^2$.

The notation $\langle \ \rangle^{(0)}$ means that the Fermionic average is defined solely by the fermion-free action as given in the decoupled form (Eq. (6.18)).

Let us pass to the important step of evaluating the Wilson phase factor average (Eq. (6.20)) on the coupling regime of $g_{\text{bare}} \to 0^+$ and formally at the limit of t'Hooft of large number of colors $N \to \infty$. As the first step to implement such evaluation, let us consider our loop C_{xy} as a closed contour lying on the plane $\mu = 0$, $\nu = 1$ bounding the planar region S.

We now observe that the ordered phase-factor for constant gauge fields can be exactly evaluated by means of a triangularization of the planar region S, i.e.

$$S = \bigcup_{l=1}^{M} \triangle_{\mu\nu}^{(i)}. \tag{6.95}$$

Here, each counter-clock oriented triangle $\triangle_{\mu\nu}^{(i)}$ is adjacent to next one $\triangle_{\mu\nu}^{(i)} \cap \triangle_{\mu\nu}^{(i+1)}$ common side with the opposite orientations.

At this point, we note that

$$\mathbb{P}\left\{e^{-g\int_{\triangle_{\mu\nu}^{(i)}} A_\alpha \cdot dx_\alpha}\right\} \cong e^{-gA_\alpha \cdot \ell_\alpha^{(1)}} \cdot e^{-gA_\alpha \cdot \ell_\alpha^{(2)}} \cdot e^{-gA_\alpha \cdot \ell_\alpha^{(3)}}, \tag{6.96}$$

where $\{\ell_\alpha^{(i)}\}_{1=1,2,3}$ are the triangle sides satisfying the (vector) identity $\ell_\alpha^{(1)} + \ell_\alpha^{(2)} + \ell_\alpha^{(3)} \equiv 0$.

Since we have that

$$\mathbb{P}\left\{e^{-g\oint_{(x)} A_\alpha dx_\alpha}\right\} = \lim_{n\to\infty} \prod_{i=1}^{n} \mathbb{P}\left\{e^{-g\int_{\triangle_{\mu\nu}^{(i)}} A_\alpha dx_\alpha}\right\}, \tag{6.97}$$

and by using the Campbel Hausdorff formulae to sum up the product limit (Eq. (6.23)) at $g^2 \to 0$ limit with X and Y denoting general elements of the $SU(N)$ — Lie algebra:

$$e^x \cdot e^y = e^{X+Y+\frac{1}{2}[X,Y]} + 0(g^2), \tag{6.98}$$

one arrives at the non-Abelian stokes theorem for constant gauge fields

$$\mathbb{P}\Big\{ e^{-g \int_{C_{xy}} A_\alpha dx_\alpha} \Big\} = \mathbb{P}\Big\{ e^{-g \iint_S F_{01} d\sigma^{01}(s)} \Big\}$$

$$= \mathbb{P}\Big\{ e^{+(g)^2 [A_0, A_1] \cdot S} \Big\}. \tag{6.99}$$

As a consequence, we have the effective approximate result (formally exact at $N \to \infty$) to be used in our analysis that follows

$$\text{Tr}_{SU(N)} \, \mathbb{P}\Big\{ e^{-g \int_{C_{xy}} A_\alpha dx_\alpha} \Big\}$$

$$\sim \exp\Big\{ +\frac{(g^2 s)^2}{2} (\text{Tr}_{SU(N)}[A_0, A_1])^2 \Big\} + O\left(\frac{1}{N}\right). \tag{6.100}$$

Let us now substitute Eq. (6.26) into Eq. (6.20) and taking into account the natural two-dimensional degrees of freedom reduction on the average Eq. (6.20)

$$G((x - y)) = \frac{1}{\tilde{G}(0)} \int_{-\infty}^{+\infty} \prod_{a=1}^{N^2 - N} d[A_1^a] d[A_0^a]$$

$$\times \exp\Big\{ +\frac{1}{2} \text{ vol } (\Omega)[\text{Tr}_{SU(N)}(A_0^2 + A_1^2)] \Big\}$$

$$\times \exp\Big\{ +\frac{(g^2 s)^2}{2} (\text{Tr}_{SU(N)}[A_0, A_1])^2 \Big\}, \tag{6.101}$$

where $\tilde{G}(0)$ is the normalization factor given explicitly by

$$\tilde{G}(0) = \int_{-\infty}^{+\infty} \prod_{a=1}^{N^2 - N} d[A_1^a] d[A_0^a] \exp\Big\{ -\frac{1}{2} \text{ vol}(\Omega)[(A_0^a)^2 + (A_1^a)^2] \Big\}. \tag{6.102}$$

By looking closely at Eqs. (6.27) and (6.28), one can see that the behavior at large N of the Wilson phase factor average is asymptotic to the value of the usual integral ($\text{Tr}_{SU(N)}(\lambda_a \lambda_b = -\delta_{ab})$).

$$G((x - y))_{N \gg 1} \sim \Big\{ \Big[\int_{-\infty}^{+\infty} da \exp\Big\{ -\frac{1}{2} \text{ vol}(\Omega)a^2 \Big\} \exp\Big\{ -\frac{(g^2 s)^2}{2} a^4 \Big\}$$

$$\times \Big(\int_{-\infty}^{+\infty} da \exp\Big\{ -\frac{1}{2} \text{ vol}(\Omega)a^2 \Big\} \Big)^{-1} \Big]^{N^2 - N} \Big\}. \tag{6.103}$$

By using the well-known result (see Ref. 5 — pp. 307, 323, Eq. (3))

$$\int_0^\infty \exp(-\beta^2 x^4 - 2\gamma^2 x^2)dx = 2^{-\frac{3}{2}}\left(\frac{\gamma}{\beta}\right)e^{\frac{\gamma^4}{2\beta^2}}K_{\frac{1}{4}}\left(\frac{\gamma^4}{2\beta^2}\right),\qquad (6.104)$$

we obtain the closed result at finite volume

$$G((x-y))_{N\gg 1} \sim \left\{\left(\frac{\sqrt{\mathrm{vol}(\Omega)}N}{2\cdot\left(\frac{g^2 SN}{\sqrt{2}}\right)}\right)\right.$$

$$\left.\times e^{+\frac{(\mathrm{vol}(\Omega))^2}{32\left(\frac{g^2 SN}{\sqrt{2}}\right)^2}} K_{\frac{1}{4}}\left(\frac{(\mathrm{vol}(\Omega))^2 N^2}{16g^4 N^2 S^2}\right)\left(\frac{\sqrt{\pi}}{2\cdot\left(\frac{\mathrm{vol}(\Omega)}{2}\right)^{\frac{1}{2}}}\right)^{-1}\right\}^{N^2-N}.$$

$$(6.105)$$

Let us now give a (formal) argument for the theory's triviality at infinite volume $\mathrm{vol}(\Omega)\to\infty$. Let us, first define the infinite-volume theory's limit by means of the following limit

$$\mathrm{vol}(\Omega) = S^2,\qquad (6.106)$$

and consider the asymptotic limit of the correlation function at $|x-y|\to\infty$ ($S\to\infty$).

By using the standard asymptotic limit of the Bessel function

$$\lim_{z\to\infty} K_{\frac{1}{4}}(z) \sim e^{-z}\sqrt{\frac{\pi}{2z}},\qquad (6.107)$$

one obtains the result ($\lim_{N\to\infty} g^2 N = g^2_\infty < \infty$) in four dimensions

$$G((x-y))_{\substack{N\gg 1\\|x-y|\to\infty}} \sim \lim_{S\to\infty}\left\{\frac{N}{S}\cdot e^{\frac{N^2 S^4}{16 S^2}}\sqrt{\frac{16\pi}{N^2 S^2 2}}e^{-\frac{S^2 N^2}{16}}\right\}^{N^2-N}$$

$$\sim \frac{1}{|x-y|^{4(N^2-N)}}.\qquad (6.108)$$

So, we can see that for big N, there exists a fast decay of Eq. (6.19) without any bound on the power decay. However, in the usual L.S.Z. framework for quantum fields, it would be expected a nondecay of such factor as in the two-dimensional case (see Eq. (6.33) for $\mathrm{vol}(\Omega) = S$), meaning physically that one can observe fermionic scattering free states at large separation. Naively at $N \to \infty$ where we expect the full validity of our analysis, one could expect on the basis of the behavior of Eq. (6.33), the vanishing of the above analyzed fermionic correlation function Eq. (6.19) as faster as any power of $|x - y|$ for large $|x - y|$. This result would imply that $g^2_{\text{bare}} \sim 0^+$ may be zero from the very beginning and making a sound support to the fact that the chiral thirring model — at $g^2 \sim 0^+$ and for large number of colors — may still remain a trivial theory, a result not expected at all in view of previous claims on the subject that resummation may turn nonrenormalizable field theories in nontrivial renormalizable useful ones.[1]

Finally as a last remark on our formulae (Eqs. (6.31)–(6.33)), let us point out that a mathematical rigorous sense to consider them is by taking the Eguchi–Kwaia continuum space-time Ω, a set formed of n hyper-four-dimensional cubes of side a — the expected size of the nonperturbative vacuum domain of our theory — and the surface S being formed, for instance, by n squares on the Ω plane section contained on the plane $\mu = 0$, $\nu = 1$. As a consequence of the construction exposed above, we can see that the large behavior is given exactly by $(N = N_c)$

$$G(na)_{N \gg 1} \overset{n \to \infty}{\sim} \left\{ \frac{N}{\bar{g}^2_\infty \cdot na^2} e^{-\frac{(N^2 n^2 a^8)}{32 \cdot \left(\frac{\bar{g}^2_\infty na^2}{\sqrt{2}}\right)^2}} K_{\frac{1}{4}} \left(\frac{N^2(n^2 a^8)}{16(g^2_\infty)^2 n^2 a^4} \right) \right\}^{N^2 - N}$$

$$\sim \left(\frac{1}{na^4} \right)^{N^2 - N}. \tag{6.109}$$

6.6. The Loop Space Argument for the Thirring Model Triviality

In order to argue the triviality phenomenon of the $SU(N)$ non-Abelian thirring model of Sec. 6.1 for finite N, let us consider the generating functional equation (6.3) for vanished fermionic sources $\eta_a = \bar{\eta}_a = 0$, the so-called vacuum energy theory's content

$$Z(0,0) = \int \prod_{a=1}^{N^2-N} \prod_{\mu=0}^{3} D[A_\mu^a(x)] e^{-\frac{1}{2} \int_\Omega d^4 x (A_\mu^a A_\mu^a)(x)}$$

$$\times \det{}_F [(\not{\partial} + ig\gamma_5 \not{A})(\not{\partial} + ig\gamma_5 \not{A})^*]. \tag{6.110}$$

At this point of our analysis, let us write the functional determinant on Eq. (6.35) as a functional on the space of closed bosonic paths $\{X_\mu(\sigma), 0 \leq \sigma \leq T, X_\mu(0) = X_\mu(T) = x_\mu\}$, namely[6]

$$\ell g \det{}_F[(\partial\!\!\!/ + ig\gamma_5 A\!\!\!/)(\partial\!\!\!/ + ig\gamma_5 A\!\!\!/)^*]$$

$$= \sum_{C_{xy}} \left\{ \mathbb{P}_{SU(N)} \cdot \mathbb{P}_{\text{Dirac}} \exp\left[-g \oint_{C_{xy}} A_\mu(X_\beta(\sigma)dX_\mu(\sigma) \right.\right.$$

$$\left.\left. + \frac{i}{2}[\gamma^\alpha, \gamma^\beta] \oint_{C_{xy}} F_{\alpha\beta}(X_\beta(\sigma))ds \right] \right\}. \tag{6.111}$$

The sum over the closed loops C_{xy} with fixed end-point x_μ is given by the proper-time bosonic path-integral

$$\sum_{C_{xy}} = \int_0^\infty \frac{dT}{T} \int d^4 x_\mu \int_{\chi_\mu(0)=x_\mu=\chi_\mu(T)} D^F[X(\sigma)] \exp\left\{ -\frac{1}{2}\int_0^T \dot{X}^2(\sigma)d\sigma \right\}. \tag{6.112}$$

Note the symbols of the path ordination $\mathbb{P}$ of both, Dirac and color indexes on the loop space phase factors on the expression equation (6.36).

By using the Mandelstam area derivative operator $\delta/\delta\sigma_{\gamma\rho}(X(\sigma))$,[6] one can rewrite Eq. (6.36) into the suitable form as an operation in the loop space with Dirac matries bordering the loop C_{xx}, i.e.

$$\ell g \det{}_F[(\partial\!\!\!/ + ig\gamma_5 A\!\!\!/)(\partial\!\!\!/ + ig\gamma_5 A\!\!\!/)^*]$$

$$\times \sum_{C_{xy}} \mathbb{P}_{\text{Dirac}} \left\{ \oint_{C_{xx}} d\sigma \frac{i}{2}[\gamma^\alpha, \gamma^\beta](\sigma) \frac{\delta}{\delta\sigma_{\alpha\beta}(X(\sigma))} \right.$$

$$\left. \times \mathbb{P}_{SU(N)} \left[\exp\left(-g \oint_{C_{xx}} A_\mu(X_\beta(\sigma)dX_\mu(\sigma)) \right) \right] \right\}. \tag{6.113}$$

In order to show the triviality of functional fermionic determinant when averaging over the (white-noise!) auxiliary non-Abelian fields as in Eq. (6.35), we can use a cummulant expansion, which in a generic form reads off as

$$\langle e^f \rangle_{A_\mu} = \exp\left\{ \langle f \rangle_{A_\mu} + \frac{1}{2}(\langle f^2 \rangle_{A_\mu} - \langle f \rangle_{A_\mu}^2) + \cdots \right\}. \tag{6.114}$$

So let us evaluate explicitly the first-order cummulant

$$\sum_{C_{xy}} \mathbb{P}_{\text{Dirac}} \left\{ \oint_{C_{xy}} ds \frac{i}{2} [\gamma^\alpha(\sigma), \gamma^\beta(\sigma)] \frac{\delta}{\delta \sigma_{\alpha\beta}(X(\sigma))} \right.$$

$$\left. \times \left\langle \mathbb{P}_{SU(N)} \left[\exp \left(-g \oint_{C_{xx}} A_\mu(X_\beta(\sigma)) dX_\mu(\sigma) \right) \right] \right\rangle_{A_\mu} \right\},$$

(6.115)

with the average $\langle \ \rangle_{A_\mu}$ defined by the path-integral equation (6.35).

By using the Grassmanian zero-dimensional representation to write explicitly the $SU(N)$ path-order as a Grassmanian path-integral[6]

$$\mathbb{P}_{SU(N)} \left[\exp \left(-g \oint_{C_{xx}} A_\mu(X_\beta(\sigma)) dX_\mu(\sigma) \right) \right]$$

$$= \int \prod_{a=1}^{N^2-N} D^F[\theta_a(\sigma)] D^F[\theta_a^*(\sigma)] \left(\sum_{a=1}^{N^2-N} \theta_a(0) \theta_a^*(T) \right)$$

$$\times \exp \left(\frac{i}{2} \int_0^T d\sigma \sum_{a=1}^{N^2-N} \left(\theta_a(\sigma) \frac{\vec{d}}{d\sigma} \theta_a^*(\sigma) + \theta_a^*(\sigma) \frac{d}{d\sigma} \theta_a(\sigma) \right) \right)$$

$$\times \exp \left(g \int_0^T d\sigma (A_\mu^a(X^\beta(\sigma))(\theta_b(\lambda_a)_{bc}\theta_c^*)(\sigma) dX^\mu(\sigma)) \right), \quad (6.116)$$

one can easily see that the average over the $A_\mu(x)$ fields is straightforward and produced as a result of the following self-avoiding loop action

$$\left\langle \mathbb{P}_{SU(N)} \left[\exp \left(-g \oint_{C_{xx}} A_\mu(X_\beta(\sigma)) dX_\mu(\sigma) \right) \right] \right\rangle_{A_\mu}$$

$$= \int \prod_{a=1}^{N^2-N} D^F[\theta_a(\sigma)] D^F[\theta_a^*(\sigma)] \left(\sum_{a=1}^{N^2-N} \theta_a(0) \theta_a^*(T) \right)$$

$$\times \exp \left(\frac{i}{2} \int_0^T d\sigma \sum_{a=1}^{N^2-N} \left(\theta_a(\sigma) \frac{\vec{d}}{d\sigma} \theta_a^*(\sigma) + \theta_a^*(\sigma) \frac{\vec{d}}{d\sigma} \theta_a(\sigma) \right) \right)$$

$$\times \exp \left\{ \frac{g^2}{2} \int_0^T d\sigma \int_0^T d\sigma' [(\theta_b(\lambda_a)_{bc}\theta_c^*)(\sigma)(\theta_b(\lambda_a)_{bc}\theta_c^*)(\sigma')] \right.$$

$$\left. \times \delta^{(D)}(X_\mu(\sigma) - X_\mu(\sigma')) dX_\mu(\sigma) dX_\mu(\sigma') \right\}. \quad (6.117)$$

At this point one can use the famous probabilistic-topological Parisi argument[7] well-used to understand the $\lambda \varphi^4$ triviality at the four-dimensional space-time to see that due to the fact that Hausdorff dimension of our Brownian loops $\{X_\mu(\sigma)\}$ is two, and the topological rule for continuous manifold holds true in the present situation, one obtains that for ambient space greater than (or equal to) four, the Hausdorff dimension of the closed-path intersection set of the argument of the delta function on Eq. (6.42) is empty. So, we have as a consequence

$$\left\langle \mathbb{P}_{SU(N)} \left[\exp \left(-g \oint_{C_{xx}} A_\mu(X_\beta(\sigma)) dX_\mu(\sigma) \right) \right] \right\rangle_{A_\mu} = 1, \qquad (6.118)$$

which in turn leads to the thirring model's triviality for space-time R^D with $D \geq 4$.

References

1. J. Zinn-Justin, *Quantum Field Theory and Critical Phenomena* (Clarendon, Oxford, 2001); G. Broda, *Phys. Rev. Lett.* **63**, 2709 (1989); L. C. L. Botelho, *Int. J. Mod. Phys. A* **15**(5), 755 (2000).
2. A. Di Giacomo, H. G. Dosch, V. S. Shevchenko and Yu A. Simonov, *Phys. Rep.* **372**, 319 (2002).
3. L. C. L. Botelho, *Phys. Rev.* **32D**(6), 1503 (1985); *Europhys. Lett.* **11**(4), 313 (1990).
4. L. C. L. Botelho, *Phys. Rev.* **30D**(10), 2242 (1984); V. N. Romanov and A. S. Schwarz, *Theoreticheskaya i Matematichesnaya, Fizika* **41**(2), 190 (1979).
5. I. Gradshteym and I. M. Ryzhik, *Tables of Integrals* (Academic Press, New York, 1980).
6. L. C. L. Botelho, *J. Math. Phys.* **30**, 2160 (1989).
7. B. Simon, *Functional Integration and Quantum Physics* (Academic Press, 1979).
8. K. Fujikawa, *Phys. Rev. D* **21**, 2848 (1980).
9. R. Roskies and F. A. Schaposnik, *Phys. Rev. D* **23**, 558 (1981).
10. K. Furuya, R. E. Gamboa Saravi and F. A. Schaposnik, *Nucl. Phys.* **B208**, 159 (1982).
11. A. V. Kulikov, *Theor. Math. Phys. (U.S.S.R.)* **54**, 205 (1983).
12. K. D. Rothe and I. O. Stamatescu, *Ann. Phys. (N.Y.)* **95**, 202 (1975).
13. L. C. L. Botelho and M. A. Rego Monteiro, *Phys. Rev. D* **30**, 2242 (1984).
14. J. D. Bjorken and S. D. Drell, *Relativistic Quantum Fields* (McGraw-Hill, New York, 1965).
15. J. Goldstone and F. Wilczek, *Phys. Rev. Lett.* **47**, 988 (1981).

16. K. Fujikawa, *Phys. Rev. D* **21**, 2848 (1980); R. Roskies and F. Schaposnik, *ibid.* **23**, 518 (1981).

17. L. Bothelho, *Phys. Rev. D* **31**, 1503 (1985); A. Das and C. R. Hagen, *ibid.* **32**, 2024 (1985).

18. L. Botelho, *Phys. Rev. D* **33**, 1195 (2986).

19. O. Alvarez, *Nucl. Phys.* **B238**, 61 (1984).

20. V. Romanov and A. Schwartz, *Teor. Mat. Fiz.* **41**, 190 (1979).

21. K. Osterwalder and R. Schrader, *Helv. Phys. Acata* **46**, 277 (1973).

22. P. B. Gilkey, in *Proceeding of Symposium on Pure Mathematics*, Stanford, California, 1973, eds. S. S. Chern and R. Osserman (American Mathematical Society, Poidence, RI, 1975), Vol. 127, Pt. II, p. 265.

23. B. Schroer, *J. Math. Phys.* **5**, 1361 (1964); *High Energy Physics and Elementary Particles* (IAEC, Vienna, 1965).

24. K. Bardakci and B. Schroer, *J. Math. Phys.* **7**, 16 (1966).

25. A. Jaffe, *Phys. Rev.* **158**, 1454 (1966).

Appendix A. Path-Integral Solution for a Two-Dimensional Model with Axial-Vector-Current-Pseudoscalar Derivative Interaction

Analysis of quantum field models in two space-time dimensions has proved to be a useful theoretical laboratory to understand phenomena like dynamical mass generation, topological excitations, and confinement; all features expected to be present in a realistic four-dimensional theory of strong interactions. Recently, a powerful nonperturbative technique has been used to analyze several two-dimensional fermionic models in the (Euclidean) path-integral approach. This technique is based on a chiral change of variables in the functional fermionic measure.[8−11] It is the purpose of this appendix to solve exactly another fermionic model by means of this nonperturbative technique. The model to be studied describes the interaction of a massive pseudoscalar field with massless fermion fields in terms of a derivative coupling and was analyzed previously in Ref. 12 by using the operator approach.

We start our study by considering the Euclidean Lagrangian of the model:

$$\mathsf{L}_1(\psi, \bar{\psi}, \phi)(x)$$
$$= \left[-i\bar{\psi}\gamma_\mu\partial_\mu\psi + \frac{1}{2}(\partial_\mu\phi)(\partial_\mu\phi) + \frac{1}{2}m^2\phi^2 + g\bar{\psi}\gamma^5\gamma_\mu\psi\partial_\mu\phi \right](x), \quad \text{(A.1)}$$

where $\psi = (\psi_1, \psi_2)$ denotes a massless fermion, ϕ a massive pseudoscalar field, and g is the coupling constant. The Lagrangian (A.1) is invariant under the global Abelian and Abelian-chiral groups

$$\psi \to e^{i\alpha}\psi, \quad \psi \to e^{i\gamma_5\beta}\psi, \quad (\alpha, \beta) \in \mathbb{R},$$

with the formally Noether conserved currents

$$\partial_\mu(\bar{\psi}\gamma^5\gamma_\mu\psi) = 0, \quad \partial_\mu(\bar{\psi}\gamma^\mu\psi) = 0.$$

The Hermitian γ matrices we are using satisfy the (Euclidean) relations

$$\{\gamma_\mu, \gamma_\nu\} = 2\delta_{\mu\nu}, \quad \gamma_\mu\gamma_5 = i\varepsilon_{\mu\nu}\gamma_\nu, \quad \gamma_5 = i\gamma_0\gamma_1,$$
$$\varepsilon_{01} = -\varepsilon_{10} = 1 \quad (\mu, \nu = 0, 2).$$

In the framework of path-integrals, the generating functional of the Green's functions of the quantum field theory associated with the Lagrangian (A.1) is given by

$$Z[J, n, \bar{n}] = \int D[\phi]D[\psi]D[\bar{\psi}]$$
$$\times \exp\left(-\int d^2x[\mathrm{L}_1(\psi, \bar{\psi}, \phi)(x) + \phi J + \bar{n}\psi + \bar{\psi}n](x)\right). \quad (A.2)$$

In order to perform a chiral change of variable in (A.2) it appears to be convenient to write the interaction Lagrangian in a form closely parallel to the usual fermion–vector coupling in gauge theories by making use of the identity

$$\exp\left(-\int d^2xg(\psi\gamma^5\gamma_\mu\psi\partial_\mu\phi)(x)\right)$$
$$= \int D[A_\mu]\delta(A_\mu + i\varepsilon_{\mu\alpha}\partial_\alpha\phi)\exp\left(\int d^2xg(\bar{\psi}\gamma_\mu A_\mu\psi)(x)\right), \quad (A.3)$$

where $A_\mu(x)$ is an auxiliary Abelian vector field. Substituting (B.3) into (B.2) we obtain a more suitable form for $Z[J, n, \bar{n}]$,

$$Z[J, n, \bar{n}] = \int D[\phi]D[\psi]D[\bar{\psi}]D[A_\mu]\delta(A_\mu + i\varepsilon_{\mu\alpha}\partial_\alpha\phi)$$
$$\times \exp\left(-\int d^2x[\mathrm{L}_1(\psi, \bar{\psi}, \phi, A_\mu) + J\phi + \bar{n}\psi + \bar{\psi}n]\right), \quad (A.4)$$

with

$$\mathsf{L}_2(\psi, \bar{\psi}, \phi, A_\mu) = -i\bar{\psi}\gamma_\mu(\partial_\mu - igA_\mu)\psi + \frac{1}{2}(\partial_\mu\phi)(\partial_\mu\phi) + \frac{1}{2}m^2\phi^2. \quad \text{(A.5)}$$

Now we can proceed as in the case of gauge theories[9,10] in order to decouple the fermion fields from the auxiliary vector field $A_\mu(x)$ by making the chiral change of variables

$$\psi(x) = e^{i\gamma_5\beta(x)}X(x),$$

$$\bar{\psi}(x) = \bar{\chi}(x)e^{i\gamma_5\beta(x)}, \quad \text{(A.6)}$$

$$A_\mu(x) = -\frac{i}{g}(\varepsilon_{\mu\nu}\partial_\nu\beta)(x).$$

We note the relation $\beta = g\Phi$ between the chiral phase β and the pseudoscalar field Φ due to the constraint $\delta(A_\mu + i\varepsilon_{\mu\alpha}\partial_\alpha\phi)$ in (A.4). After the change (A.6), the Lagrangian (A.5) takes a form where now the fermion fields $\chi(x)$, $\bar{\chi}(x)$ are free and decoupled from the auxiliary vector field $A_\mu(x)$,

$$\mathsf{L}_3(\chi, \bar{\chi}, \phi) = -i\bar{\chi}\gamma_\mu\partial_\mu\chi + \frac{1}{2}(\partial_\mu\phi)(\partial_\mu\phi) + \frac{1}{2}m^2\phi^2. \quad \text{(A.7)}$$

On the other hand, the fermionic measure $D\psi D\bar{\psi}$, defined in terms of the normalized eigenvectors of the Hermitian operator $[-i\gamma_\mu(\partial_\mu - igA_\mu)][-i\gamma_\mu(\partial_\mu - igA_\mu)]^*$, is not invariant under the chiral change and given the Jacobian

$$\exp\left(-\frac{g^2}{2\pi}\int d^2x(A_\mu)(x)\right)$$

$$= \{\text{Det}[-i\gamma_\mu(\partial_\mu - igA_\mu)](-i\gamma_\mu(\partial_\mu - igA_\mu)\}. \quad \text{(A.8)}$$

In terms of the new fermionic field $\chi(x)$, $\bar{\chi}(x)$ and taking into account the Jacobian (A.8), the generating functional reads

$$Z[J, n, \bar{n}] = \int D[\chi]D[\bar{\chi}]D[\phi]$$

$$\times \exp\left(-\int d^2x[-i\bar{\chi}\gamma_\mu\partial_\mu\chi + \frac{1}{2}(1 - g^2/\pi)(\partial_\mu\phi)(\partial_\mu\phi)\right.$$

$$\left. + \frac{1}{2}m^2\phi^2 + J\phi + \bar{n}e^{ig\gamma_5\phi}\chi + \bar{\chi}e^{ig\gamma_5\phi}\eta](x)\right). \quad \text{(A.9)}$$

We shall now use the effective generating function (A.9) to compute exactly the (bare) Green's functions of the model. The two-point Green's function of the pseudoscalar field $\phi(x)$ is straightforwardly obtained,

$$\langle\langle\phi(x)\phi(y)\rangle\rangle = \frac{1}{2\pi}K_0\left(\frac{m}{(1-g^2/\pi)^{1/2}}|x-y|\right), \qquad (A.10)$$

where $K_1(w)$ denotes the Hankel function of an imaginary argument of order zero. The two-point fermion Green's function is easily computed by noting that the fermion fields $\chi(x)$, $\bar{\chi}(x)$ are free and decoupled in (A.9). The result reads

$$\langle\langle\psi_\alpha(x)\bar{\psi}_\beta(y)\rangle\rangle = \frac{1}{2\pi}(\gamma_\mu)_{\alpha\beta}\frac{(x_\mu - y_\mu)}{|x-y|^2}$$
$$\times \exp\left[\frac{g^2}{2\pi(1-g^2/\pi)}K_0\left(\frac{m}{2\pi(1-g^2/\pi)^{1/2}}|x-y|\right)\right],$$
$$(A.11)$$

where we have used the dimensional regularization scheme to assign the value 1 to the "tadpole" contribution that appears in (A.11), i.e.

$$\exp\left(\frac{g^2}{1-g^2/\pi^2}K_0(0)\right) = 1.$$

It is also interesting to compute correlation functions involving fermions $\psi(x), \bar{\psi}(x)$ and the pseudoscalar field $\phi(x)$. For instance, we get the following expression for the vertex $\Gamma_{\alpha\beta}(x,y,z) = \langle\langle\psi_\alpha(x)\bar{\psi}_\beta(y)\phi(z)\rangle\rangle$:

$$\Gamma_{\alpha\beta}(x,y,z)$$
$$= \frac{1}{2\pi}(\gamma_\mu\gamma_5)_{\alpha\beta}\frac{(x_\mu - y_\mu)}{|x-y|^2}\frac{ig}{2\pi(1-g^2/\pi)^{1/2}}$$
$$\times \left[K_0\left(\frac{m}{(1-g^2/\pi)^{1/2}}|x-z|\right) - K_0\left(\frac{m}{(1-g^2/\pi)^{1/2}}|y-z|\right)\right]$$
$$\times \exp\left[\frac{g^2}{2\pi(1-g^2/\pi)}K_0\left(\frac{m}{(1-g^2/\pi)^{1/2}}|x-y|\right)\right]. \qquad (A.12)$$

It is instructive to point out that in a perturbative analysis of the model, the previous correlation functions correspond to the full sum of nonrenormalized Feynman diagrams involving all possible radiative corrections. The perturbative renormalization analysis can be implemented by applying the

. asymptotic Lehmann–Symanzik–Zimmermann conditions (or equivalently, the Dyson prescription) in the (bare) propagators and the vertex function of model.[12-14] This analysis results in that the pseudoscalar field ϕ gets (finite) wave-function and mass renormalizations, $\phi^{(R)} = (Z_\phi)^{-1}\phi$ and $m_{(R)} = m^2(Z_\phi)^2)$, respectively, with

$$Z_\phi = \frac{1}{(1 - g^2/\pi)^{1/2}},$$

and the coupling constant gets a multiplicative renormalization $g_{(R)} = gZ_\phi$. For instance, the renormalized two-point fermion Green's function is given by

$$\langle(\psi_\alpha(x)\bar{\psi}_\beta(y))\rangle^{(R)}$$
$$= \frac{1}{2\pi}(\gamma_\mu)_{\alpha\beta}\frac{(x_\mu - y_\mu)}{|x - y|^2}\exp\left(\frac{g^2_{(R)}}{2\pi}K_0(m((R)|x - y|)\right), \quad (A.13)$$

which has the following short- and long-distance behavior:

$$\langle(\psi_\alpha(x)\bar{\psi}_\beta(y))\rangle^{(R)}_{|x-y|\to 0} \cong (\gamma_\mu)_{\alpha\beta}|x - y|^{-(1+g^2_{(R)}/2w)}, \quad (A.14)$$

$$\langle(\psi_\alpha(x)\bar{\psi}_\beta(y))\rangle^{(R)}_{|x-y|\to 0} \cong \frac{1}{2\pi}(\gamma_\mu)_{\alpha\beta}\frac{(x_\mu - y_\mu)}{|x - y|^2}. \quad (A.15)$$

The ultraviolet behavior (A.14) implies that the fermion field carries an anomalous dimension $\gamma_\psi = g^2_{(R)}/4\pi$. We remark that the model displays the appearance of the axial anomaly as a consequence of the presence of the massive field π.[8,12] For instance, in terms of the bare field ϕ, we have

$$\partial_\mu(\bar{\psi}\gamma_\mu\gamma_5\psi) = \frac{g}{\pi}\Box\phi = \frac{g}{\pi}\frac{m^2}{1 - g^2/\pi}\phi. \quad (A.16)$$

Next, we shall consider the case of massive fermions in the model

$$Ł(\psi, \bar{\psi}, \phi) = \left[-i\bar{\psi}\gamma_\mu\partial_\mu\psi + \mu\bar{\psi}\psi + \frac{1}{2}(\partial_\mu\phi)(\partial_\mu\psi) + \frac{1}{2}m^2\phi^2 + g\bar{\psi}\gamma^5\gamma_\mu\psi\partial_\mu\psi\right].$$
$$(A.17)$$

In order to get an effective action as in (A.9), we make again the chiral change (A.6) where it becomes important to remark that the fermion measure $D\psi D\bar{\psi}$ is to be defined by the eigenvectors of the massless Hermitian Dirac operator $(D(A D(A)^*))$ and thus yields the same Jacobian.[15]

This fact is related to the fermion mass independence of the anomalous part of the divergence of chiral current $\partial_\mu(\bar{\psi}\gamma^\mu\gamma^5\psi)$ in fermion models interacting with gauge fields. Proceeding as above we obtain the new effective (bare) Lagrangian

$$\mathsf{L}(\chi, \bar{\chi}, \phi)(x)$$

$$= \left[-i\bar{\chi}\gamma_\mu\partial_\mu\chi + \mu\bar{\chi}e^{2ig\gamma_5\psi}\chi + \frac{1}{2}(1 - g^2/\pi)(\partial_\mu\phi)(\partial_\mu\phi) + \frac{1}{2}m^2\phi^2 \right](x),$$

$$(A.18)$$

where now the fermion fields $\chi(x)$, $\bar{\chi}(x)$ possess an interacting term $\mu\bar{\chi}[\cos(2g\phi) + i\gamma_5\sin(2g\phi)]\chi$ which contributes to the phenomenon of formation of fractionally fermionic solutions.[15]

As we have shown, chiral changes in path-integrals provide a quick, mathematically and conceptually simple way to solve and analyze the model studied in this appendix.

Note added. We could like to make some clarifying remarks on the analysis implemented above. First, in order to implement the regularization rule in Eqs. (A.3)–(A.11), we consider the associated perturbative power series in the coupling constant g dimensionally regularized which automatically takes into account the Wick normal-order operation in the correlation functions (Eqs. (A.9)–(A.11)). Second, we observe that the model should be redefined in the region $g^2 > \pi$ since its associated Gell–Mann–Low function has a nontrivial zero for $g^2 = \pi$ (the model contains a tachyonic excitation).[12]

Appendix B. Path-Integral Bosonization for an Abelian Nonrenormalizable Axial Four-Dimensional Fermion Model

B.1. Introduction

The study of two-dimensional fermion models in the framework of chiral anomalous path-integral has been shown to be a powerful nonperturbative technique to analyze the two-dimensional bosonization phenomenon.

It is the purpose of this appendix to implement this nonperturbative technique to solve exactly a nontrivial and nonrenormalizable four-dimensional axial fermion model which generalizes for four dimensions the two-dimensional model studied in the previous appendix.

B.2. The Model

Let us start our analysis by considering the (Euclidean) Lagrangian of the proposed Abelian axial model in R^4

$$L_1(\psi, \bar{\psi}, \phi) = \bar{\psi}\gamma_\mu(i\partial_\mu - g\gamma_5\partial_\mu\phi)\psi + \frac{1}{2}g^2(\partial_\mu\phi)^2 + V(\phi), \qquad (B.1)$$

where $\psi(x)$ denotes a massless four-dimensional fermion field, $\phi(x)$ a pseudoscalar field interacting with the fermion field through a pseudoscalar derivative interaction, and $V(\phi)$ is a ϕ self-interaction potential given by

$$V(\phi) = \frac{-g^4}{12\pi^4}\phi(\partial_\mu\phi)^2(-\partial^1\phi) + \frac{g^2}{4\pi^2}(-\partial^2\phi)(-\partial^2\phi). \qquad (B.2)$$

The presence of the above ϕ potential is necessary to afford the exact solubility of the model as we will show later (Eq. (B.18)).

The Hermitian γ matrices we are using satisfy the (Euclidean) relations

$$[\gamma_\mu, \gamma_\nu] = 2\delta_{\mu\nu}, \quad \gamma_5 = \gamma_0\gamma_1\gamma_2\gamma_3. \qquad (B.3)$$

The Lagrangian $L_1(\psi, \bar{\psi}, \phi)$ is invariant under the global Abelian and chiral Abelian groups

$$\psi \to e^{i\alpha}\psi, \quad \psi \to e^{i\gamma_5\beta}\psi, \quad (\alpha, \beta) \in \mathbb{R}, \qquad (B.4)$$

with the Noether conserved currents at the classical level:

$$\partial_\mu(\bar{\psi}\gamma^4\gamma^\mu\psi) = 0, \quad \partial_\mu(\bar{\psi}\gamma_\mu\psi) = 0. \qquad (B.5)$$

In the framework of path-integrals, the generating functional of the correlation functions of the mathematical model associated with the Lagrangian $L_1(\psi, \bar{\psi}, \phi)$ is given by

$$Z[J, \eta, \bar{\eta}] = \frac{1}{Z[0,0,0]} \int D[\phi]D[\psi]D[\bar{\psi}]$$

$$\times \exp\left[-\int d^4x[L_1(\psi, \bar{\psi}, \phi) + J\phi + \bar{\eta}\psi + \bar{\psi}\eta](x)\right]. \qquad (B.6)$$

In order to generalize for four-dimensions the chiral anomalous path-integral bosonization technique, we first rewrite the full Dirac operator in the following suitable form:

$$\mathcal{D}[\phi] = i\gamma_\mu(\partial_\mu + ig\gamma_5\partial_\mu\phi) = \exp(ig\gamma_5\phi)(i\gamma_\mu\partial_\mu)\exp(ig\gamma_5\phi). \qquad (B.7)$$

Now, we proceed as in the two-dimensional case by decoupling the fermion field from the pseudoscalar field $\phi(x)$ in the Lagrangian $L_1(\psi, \bar{\psi}, \phi)$ by making the chiral change of variables:

$$\psi(x) = \exp[ig\gamma_5\psi(x)]\chi(x),$$
$$\bar{\psi}(x) = \bar{\chi}(x)\exp[ig\gamma_5\phi(x)].$$
(B.8)

On the other hand, the fermion measure $D[\psi]D[\bar{\psi}]$, defined by the eigenvectors of the Dirac operator $\not{D}[\phi]\ \not{D}[\phi]^*$, is not invariant under the chiral change and yields a nontrivial Jacobian, as we can see from the relationship

$$\int D[\psi]D[\bar{\psi}]\exp\left[-\int d^4x(\bar{\psi}\ \not{D}[\phi]\psi)(x)\right]$$

$$= [\text{Det}(\not{D}[\psi]\ \not{D}[\psi]^*)]$$

$$= J[\phi]\int D[\chi]D[\bar{\chi}]\exp\left[-\int d^4x(\bar{\chi}i\gamma_\mu\partial_\mu(\chi)(x)\right].$$
(B.9)

Here, $J[\phi] = \text{Det}(\not{D}[\phi]\ \not{D}[\phi]^*)/\text{Det}(\not{D}[\phi=0]\ \not{D}[\phi=0]^*)$ is the explicit expression for this Jacobian.

It is instructive to point out that the model displays the appearance of the axial anomaly as a consequence of the nontriviality of $J[\phi]$, i.e. $\partial_\mu(\bar{\psi}\gamma_\mu\gamma_5\psi)(x) = \{(\delta/\delta\phi)J[\psi]\}(\chi)$.

So, to arrive at a complete bosonization of the model (Eq. (B.6)) we face the problem of evaluating $J[\phi]$.

Let us, thus, compute the four-dimensional fermion determinant $\det(\not{D}[\phi])$ exactly. In order to evaluate it, we introduce a one-parameter family of Dirac operators interpolating the free Dirac operator and the interacting one $\not{D}^{(\zeta)}[\phi]$, namely,

$$\not{D}^{(\zeta)}[\phi] = \exp(ig\gamma_5\zeta\phi)(i\gamma_\mu\partial_\mu)\exp(ig\gamma_5\zeta\phi),$$
(B.10)

with $\zeta \in [0,1]$.

At this point, we introduce the Hermitian continuation of the operators $\not{D}^{(\zeta)}[\phi]$ by making the analytic extension in the coupling constant $\bar{g} = ig$. This procedure has to be done in order to define the functional determinant by the proper-time method since only in this way $(\not{D}^\zeta[\phi])^2$ can be considered as a (positive) Hamiltonian $(\not{D}^\zeta[\phi])^2 \equiv (\not{D}^\zeta[\phi])(\not{D}^\zeta[\phi])^*$.

The justification of this analytic extension in the model coupling constant is due to the fact that typical interaction energy densities such as $\bar{\psi}\gamma^4\psi$, $\bar{\psi}\gamma_\mu\Delta^\mu\psi$, which are real in Minkowski space-time, become complex after continuation in Euclidean space-time.[21] As a consequence, the above analytic coupling extension must be done in the proper-time regularization for the Dirac functional determinant.

By using the property

$$\frac{d}{d\zeta} \, \mathcal{D}^{\zeta}[\phi] = \bar{g}\gamma_5\psi \, \mathcal{D}^{\zeta}[\phi] + \mathcal{D}^{\zeta}[\phi]\gamma_5\bar{g}\phi, \tag{B.11}$$

we can write the following differential equation for the functional determinant[19]:

$$\frac{d}{d\zeta}\ell n \, \text{Det} \, \mathcal{D}^{\zeta}[\phi]$$

$$= 2 \lim_{\varepsilon \to 0^+} \int d^3x \, \text{Tr}\langle x|(\bar{g}\gamma_5\psi)\exp\{-\varepsilon(\mathcal{D}^{\zeta}[\phi]^2\}|x\rangle, \tag{B.12}$$

where Tr denotes the trace over Dirac indices.

The diagonal part of $\exp\{-\varepsilon(\mathcal{D}^{\zeta}[\phi]^2\}$ has the asymptotic expansion[20,22]

$$\langle x|\exp\{-\varepsilon(\mathcal{D}^{\zeta}[\phi])^2\}|x\rangle \underset{\varepsilon \to 0^+}{\sim} \frac{1}{16\pi^2\varepsilon^2}\left[1 + \varepsilon H_1(\phi) + \frac{\varepsilon^2}{2!}H_2(\phi)\right], \tag{B.13}$$

with the Seeley coefficients given by (see the complements)

$$H_1(\phi) = \zeta\bar{g}\gamma_5\partial^2\phi - \bar{g}^2\zeta^2(\partial_\mu\phi)^2, \tag{B.14}$$

and

$$H_2(\phi) = 2\partial^2 H_1(\phi) + 2\zeta\bar{g}\gamma_4(\partial_\mu\phi)\partial_\mu H_1(\phi) + 2[H_1(\phi)]^2. \tag{B.15}$$

By substituting Eqs. (B.14) and (B.15) into Eq. (B.12), we obtain finally the result for the above-mentioned Jacobian:

$$J[\phi] = J_0[\phi, \varepsilon]J_1[\phi]. \tag{B.16}$$

Here, $H_0[\phi, \varepsilon]$ is the ultraviolet cut-off dependent Jacobian term

$$J_0[\phi, \varepsilon] = \exp\left[\frac{g^2}{4\pi^2\varepsilon}\int d^4x[\phi(-\partial^2)\phi](x)\right] \tag{B.17}$$

and $J_1(\phi)$ is the associated Jacobian finite part

$$J_1[\phi] = \exp\left[-\frac{g^2}{4\pi^2}\int d^4x(-\partial^2\phi)(-\partial^2\phi)(x)\right]$$

$$\times \exp\left[\frac{g^2}{12\pi^2}\int d^4x[\phi(\partial_\mu\phi)^2(-\phi^2\phi)](x)\right]. \tag{B.18}$$

From Eqs. (B.17) and (B.1), we can see that the (bare) coupling constant g^2 gets an additive (ultraviolet) renormalization. Besides the Jacobian term cancels with the chosen potential $V(\phi)$ in Eq. (B.2).

As a consequence of all these results, we have the following expression for $Z[J, \eta, \bar{\eta}]$ with the fermions decoupled:

$$Z[J, \eta, \bar{\eta}] = \frac{1}{Z[0,0,0]} \int D(\phi) D(\chi) D(\bar{\chi}) J[\phi](x)$$

$$\times \exp\left[-\int d^4x \left[\frac{1}{2}\bar{\chi}(\gamma_\mu \partial_\mu)\chi + \frac{1}{2}\bar{x}(i\gamma_\mu \partial_\mu)\chi + \frac{1}{2}g_R^2(\partial_\mu \phi)^2\right.\right.$$

$$\left.\left. + \bar{\eta}\exp(ig_R \gamma_5 \phi)\chi + \bar{\chi}\exp(ig_R \gamma_5 \phi)\eta\right]\right]. \qquad (B.19)$$

This expression is the main result of this appendix and should be compared with the two-dimensional analogous generating functional analyzed in Appendix A, Eq. (A.9). Now, we can see that the quantum model given by Eq. (6.6), although being nonrenormalizable by usual power counting and Feynman-diagrammatic analysis, it still has nontrivial and exactly soluble Green's functions. For instance, the two-point fermion correlation function is easily evaluated and produces the result

$$\langle \psi_\alpha(x_1)\bar{\psi}_\beta(x_2) \rangle = S^F_{\alpha\beta}(x_1 - x_2; m = 0)\exp[-\Delta_F(x_1, x_2.m = 0)]. \qquad (B.20)$$

Here, $\Delta(x_1 - x_2; m = 0)$ is the Euclidean Green's function of the massless free scalar propagator. We notice that correlation functions involving fermions $\psi(x)$, $\bar{\psi}(x)$ and the pseudoscalar field $\phi(x)$ are easily computed too (see Ref. 17).

As a conclusion of our appendix let us comment to what extent our proposed mathematical Euclidean axial model describes an operator quantum field theory in Minkowski space-time.

In the operator framework the fields $\psi(x)$ and $\phi(x)$ satisfy the following wave equations:

$$i\gamma_\mu \partial_\mu \psi = ig\gamma_\mu \gamma_5 (\partial_\mu \phi)\psi,$$

$$\Box \phi = \partial_\mu(\bar{\psi}\gamma_\mu \gamma_5 \psi) + \frac{\delta\Delta}{\delta\phi}. \qquad (B.21)$$

It is very difficult to solve exactly Eq. (B.12) in a pure operator framework because the model is axial anomalous $[\partial_\mu(\bar{\psi}\gamma_\mu \gamma_5 \psi) \neq 0]$.

However the path-integral study (Eq. (B.19)) shows that the operator solution of Eq. (B.21) (in terms of free (normal-ordered) fields) is given by

$$\psi(x) = \exp[i\gamma_5\phi(x)]\chi(x), \tag{B.22}$$

since it is possible to evaluate exactly the anomalous divergence of the above-mentioned axial-vector current and, thus, choose a suitable model potential $V(\phi)$ which leads to the above simple solution.

It is instructive to point out that the operator solution (Eq. (B.22)) coincides with the operator solution of the $U(1)$ vectorial model analyzed by Schroer in Ref. 1 (the only difference between the model's solutions being the γ_5 factor in the phase of Eq. (B.22)).

Consequently, we can follow Schroer's analysis to conclude that the Euclidean correlation function (Eq. (B.2)) defines Wightman functions which are distributions over certain class of analytic test functions.[23-25] But the model suffers the problem of the nonexistence of time-ordered Green's functions which means that the proposed axial model in Minkowski space-time does not satisfy the Einstein causality principle.[23]

Finally, we remark that the proposed model is to a certain extent less trivial than the vectorial model since for nondynamical $\phi(x)$ field the associated S matrix is nontrivial and is given in a regularized form by the result

$$S = T\left[\exp\left[i\int_{-\infty}^{+\infty}(\bar{\psi}\gamma_\mu\gamma_5\partial_\mu\phi\psi)(x)d^4x\right]\right]$$

$$= \exp\left[-i\int_{-\infty}^{+\infty}\phi(x)\left[\frac{\delta}{\delta\phi(x)}J_\varepsilon[\phi]\right](x)d^4x\right], \tag{B.23}$$

with $J_\varepsilon[\phi]$ expressed by Eq. (B.16).

B.3. Complement

We now briefly calculate the asymptotic term of the second-order positive differential elliptic operator:

$$(\mathcal{D}^\zeta[\phi])^2 = (-\partial^2) - (\bar{g}\gamma_5\zeta\partial_\mu\phi)\partial_\mu - \bar{g}\zeta\gamma_5\partial^2\phi + (\partial_\mu\phi)^2. \tag{B.24}$$

For this study, let us consider the more general second-order elliptic four-dimensional differential operator (non-necessarily) Hermitian in relation to the usual normal in $L^2(\mathbb{R}^D)$ (Ref. 22):

$$\mathrm{L}_x = -(\partial^2)_x + a_\mu(x)(\partial_\mu)_x + V(x). \tag{B.25}$$

$$\frac{\partial}{\partial \zeta} K(x, y. \zeta) = -\mathrm{L}_x K(x, y; \zeta),$$

$$\lim_{\eta \to 0^+} K(x, y; \zeta) = \delta^{(D)}(x - y). \tag{B.26}$$

The Greeen's function $K(x, y, \zeta)$ has the asymptotic expansion

$$\lim_{\zeta \to 0^+} K(x, y; \zeta) \sim K_0(x, y; \zeta) \left[\sum_{m=0}^{\infty} \zeta^m H_m(x, y) \right], \tag{B.27}$$

where $K_0(x, y, \zeta)$ is the evolution kernel for the n-dimensional Laplacian $(-\partial^2)$. By substituting Eq. (B.27) into Eq. (B.26) and taking the coinciding limit $x \to y$, we obtain the following recurrence relation for the coefficients $H_m(x, x)$:

$$\sum_{n=0}^{\infty} \frac{1}{n!} \zeta^n H_{n+1}(x, x) = - \left[\sum_{n=0}^{\infty} \frac{\zeta^n}{n!} (-\partial_x^2) H_n(x, x) + a_n(x) \sum_{n=0}^{\infty} \frac{\zeta^n}{n!} \partial_\mu^x H_n(x, x) \right.$$

$$\left. + V(x) \sum_{n=0}^{\infty} \frac{\zeta^n}{n!} H_n(x, x) \right]. \tag{B.28}$$

For $D = 4$, Eq. (A.5) yields the Seeley coefficients

$$H_0(x, x) = \mathbf{1}_{4 \times 4},$$

$$H_2(x) = -V(x),$$

$$H_2(x, x) = 2[-\partial_x^2 V(x) + a_\mu(x) \partial_\mu V(x) + V^2(x)]. \tag{B.29}$$

Now substituting the value $a_\mu(x) = [-\zeta \bar{g} \partial_\mu \phi] \mathbf{1}_{4 \times 4}$ and $V(x) = [(-\bar{g} \zeta \gamma_5 \partial_x^2 \phi + \partial_\mu \phi] \mathbf{1}_{4 \times 4}$ into Eq. (A.6), we obtain the result (Eqs. (B.14) and (B.15)) quoted in the main text of this appendix.

Chapter 7

Nonlinear Diffusion in R^D and Hilbert Spaces: A Path-Integral Study*

7.1. Introduction

The deterministic nonlinear diffusion equation is one of the most important topics in mathematical physics of the nonlinear evolution equation theory.[1-3] An important class of initial-value problem in turbulence has been modeled by nonlinear diffusion stirred by random sources.[4]

The purpose of this complementary chapter for Chap. 4, in mathematical methods for physics, is to provide a model of nonlinear diffusion were one can use and understand the compacity functional analytic arguments to produce the theorem of existence and uniqueness on weak solutions for deterministic stirring in $L^\infty([0,T] \times L^2(\Omega))$. We use these results to give a first step "proof" for the famous Rosen path-integral representation for the Hopf characteristic functional associated to the white-noise stirred nonlinear diffusion model. These studies are presented in Sec. 7.2.

In Sec. 7.3, we present a study of a linear diffusion equation in a Hilbert space, which is the basis of the famous loop wave equations in string and polymer surface theory.[3,4]

7.2. The Nonlinear Diffusion

Let us start our chapter by considering the following nonlinear diffusion equation in some strip $\Omega \times [0,T]$ with $\bar{\Omega}$ denoting a C^∞-compact domain of R^D.

$$\frac{\partial U(x,t)}{\partial t} = (+\Delta U)(x,t) + \Delta^{(\wedge)}F(U(x,t)) + f(x,t), \qquad (7.1)$$

---

*Author's original results.

with initial and Dirichlet boundary conditions as

$$U(x,0) = g(x) \in L^2(\Omega), \tag{7.2}$$

$$U(x,t)\,|_{\partial\Omega} \equiv 0 \quad (\text{for } t > 0). \tag{7.3}$$

We note that the nonlinearity of the diffusion-spatial term of the parabolic problem (Eq. (7.1)) takes into account the physical properties of nonlinear porous medium's diffusion saturation physical situation where this model is supposed to be applied[1] — by means of the hypothesis that the regularized Laplacean operator $\Delta^{(\wedge)}$ in the nonlinear term of the governing diffusion equation (7.1) has a cut-off in its spectral range. Additionally, we make the hypothesis that the nonlinear function $F(x)$ is a bounded real continuously differentiable function on the extended interval $(-\infty, \infty)$ with its derivative $F'(x)$ strictly positive there. The external source $f(x,t)$ is supposed to belong to the space $L^\infty([0,T] \times L^2(\Omega))$ or to be a white-noise external stirring of the form (Ref. 2, p. 61) when in the random case

$$f(\cdot,t) = \frac{d}{dt}\left\{ \sum_{n \in Z} \beta_n(t)\varphi_n(\cdot) \right\} = \frac{d}{dt}w(t). \tag{7.4}$$

Here, $\{\varphi_n\}$ denotes a complete orthonormal set on $L^2(\Omega)$, and $\beta_n(t)$, $n \in Z$ are independent Wiener processes.

Let us show the existence and uniqueness of weak solutions for the above diffusion problem stated by means of Galerking method for the case of deterministic $f(x,t) \in L^\infty([0,T] \times L^2(\Omega))$.

Let $\{\varphi_n(x)\}$ be spectral eigen-functions associated to the Laplacean Δ. Note that each $\varphi_n(x) \in H^2(\Omega) \cap H_0^1(\Omega)$.[3] We introduce now the (finite-dimensional) Galerkin approximants

$$U^{(n)}(x,t) = \sum_{i=1}^{n} U_i^{(n)}(t)\varphi_i(x),$$

$$f^{(n)}(x,t) = \sum_{i=1}^{n} (f(x,t),\varphi_i(x))_{L^2(\Omega)}\varphi_i(x), \tag{7.5}$$

subject to the initial-conditions

$$U^{(n)}(x,0) = \sum_{i=1}^{n} (g(x),\varphi_i)_{L^2(\Omega)}\varphi_i(x), \tag{7.6}$$

here $(,)_{L^2(\Omega)}$ denotes the usual inner product on $L^2(\Omega)$.

After substituting Eqs. (7.5) and (7.6) in Eq. (7.1), one gets the weak form of the nonlinear diffusion equation in the finite-dimension approximation as a mathematical well-defined systems of ordinary nonlinear differential equations, as a result of an application of the Peano existence-solution theorem.

$$\left(\frac{\partial U^{(n)}(x,t)}{\partial t}, \varphi_j(x) \right)_{L^2(\Omega)} + (-\Delta U^{(n)}(x,t), \varphi_j(x))_{L^2(\Omega)}$$

$$= (\nabla^{(\wedge)} \cdot [(F'(U^{(n)}(x,t))\nabla^{(\wedge)}U^{(n)}(x,t)], \varphi_j(x))_{L^2(\Omega)}$$

$$+ (f^{(n)}(x,t), \varphi_j(x))_{L^2(\Omega)}. \tag{7.7}$$

By multiplying the associated system (Eq.(7.7)) by $U^{(n)}$ we get the diffusion equation in the finite-dimensional Galerking sub-space in the integral form:

$$\frac{1}{2}\frac{d}{dt}\|U^{(n)}\|_{L^2(\Omega)}^2 + (-\Delta U^{(n)}, U^{(n)})_{L^2(\Omega)}$$

$$+ \int_\Omega d^3x (F'(U^{(n)})(\nabla^{(\wedge)}U^{(n)} \cdot \overline{\nabla^{(\wedge)}U^{(n)}})(x,t) = (f, U^{(n)})_{L^2(\Omega)}. \tag{7.8}$$

This result in turn yields a prior estimate for any positive integer p:

$$\frac{1}{2}\frac{d}{dt}(\|U^{(n)}\|_{L^2(\Omega)}^2) + \gamma(\Omega)\|U^{(n)}\|_{L^2(\Omega)}^2 + \|(F'(U^{(n)}))^{\frac{1}{2}}(\nabla U^{(n)})\|_{L^2(\Omega)}^2$$

$$\leq \frac{1}{2}\left\{ p\|f(x,t)\|_{L^2(\Omega)}^2 + \frac{1}{p}\|U^{(n)}\|_{L^2(\Omega)}^2 \right\}. \tag{7.9}$$

Here, $\gamma(\Omega)$ is the Garding–Poincaré constant on the inequality of the quadratic form associated to the Laplacean operator defined on the domain $H^2(\Omega) \cap H_0^1(\Omega)$.

$$\|U^{(n)}\|_{H^1(\Omega)}^2 = (-\Delta U^{(n)}, U^{(n)})_{L^2(\Omega)} \geq \gamma(\Omega)\|U^{(n)}\|_{L^2(\Omega)}^2. \tag{7.10}$$

By choosing the integer p big enough and applying the Gronwall lemma, we obtain that the set of function $\{U^{(n)}(x,t)\}$ forms a bounded set in $L^\infty([0,T], L^2(\Omega)) \cap L^\infty([0,T], H_0^1(\Omega))$ and in $L^2([0,T], L^2(\Omega))$. As a consequence of this boundedness property of the function set $\{U^{(n)}\}$, there is a sub-sequence weak-star convergent to a function $\bar{U}(t,x) \in L^\infty([0,T], L^2(\Omega))$, which is the candidate for our "weak" solution of Eq. (7.1).

Another important estimate is to consider again Eq. (7.9), but now considering the Sobolev space $H_0^1(\Omega)$ on this estimate (Eq. (7.9)), namely,

$$\frac{1}{2}(\|U^{(n)}(T)\|^2_{L^2(\Omega)} - \|U^{(n)}(0)\|^2) + \bar{C}_0 \int_0^T dt \|U^{(n)}\|^2_{H_0^1(\Omega)}$$

$$\leq \frac{1}{2}p\left(\int_0^T \|f\|^2_{L^2(\Omega)} dt\right) + \frac{1}{2p}\left(\int_0^T \|U^{(n)}\|^2_{L^2(\Omega)} dt\right) < \bar{M} < \infty,$$

$$(7.11)$$

since we have the coerciveness condition for the Laplacean operator

$$(-\Delta U^{(n)}, U^{(n)})_{L^2(\Omega)} \geq \bar{C}_0(U^{(n)}, U^{(n)})_{H_0^1(\Omega)}. \tag{7.12}$$

Note that $\|U^{(n)}(0)\|^2 \leq 2\|g(x)\|^2_{L^2(\Omega)}$ (see Eq. (7.8)), and $\{\|U^{(n)}(T)\|^2_{L^2(\Omega)}\}$ is a bounded set of real positive numbers.

As a consequence of *a prior* estimate of Eq. (7.11), one obtains that the previous sequence of functions $\{U^{(n)}\} \in L^\infty([0, T], H_0^1(\Omega) \cap H^2(\Omega))$ forms a bounded set on the vector-valued Hilbert space $L^2([0, T], H_0^1(\Omega))$ either.

Finally, one still has another *a prior* estimate after multiplying the Galerkin system (Eq. (7.7)) by the time-derivatives $\dot{U}^{(n)}$, namely

$$\int_0^T dt \left\| \frac{dU_n(t)}{dt} \right\|^2_{L^2(\Omega)} \leq \text{Real}(\Delta U_n(T), U_n(T))$$

$$- (U_n(0), U_n(0)) + \int_0^T dt \left(\Delta^{(\wedge)} F(U_n(t)), \frac{dU_n}{dt}\right)_{L^2(\Omega)}$$

$$\leq \frac{1}{2}p\left(\int_0^T \left\|\Delta^{(\wedge)} F(U_n(t))\right\|^2_{L^2(\Omega)} dt\right) + \frac{1}{2p}\left(\int_0^T dt \left\|\frac{dU_n}{dt}\right\|^2_{L^2(\Omega)}\right).$$

$$(7.13)$$

By noting that

$$\int_0^T \|\Delta^{(\wedge)} F(U_n(t))\|^2_{L^2(\Omega)} dt \leq \|\Delta^{(\wedge)}\|^2_{op} \left(\sup_{x \in [-\infty, \infty]} \{F(x)\}\right)^2$$

$$\times \int_0^T dt \|U_n(t)\|^2_{L^2(\Omega)} < \infty, \tag{7.14}$$

one obtains as a further result that the set of the derivatives $\{\frac{dU_n}{dt}\}$ is bounded in $L^2([0, T], L^2(\Omega))$ (so in $L^2([0, T], H^{-1}(\Omega))$).

At this point, we apply the famous Aubin–Lion theorem[3] to obtain the strong convergence on $L^2(\Omega)$ of the set of the Galerkin approximants $\{U_n(x,t)\}$ to our candidate $\bar{U}(x,t)$, since this set is a compact set in $L^2([0,T], L^2(\Omega))$ (see Appendix A).

By collecting all the above results we are lead to the strong convergence of the $L^2(\Omega)$-sequence of functions $F(U_n(x,t))$ to the $L^2(\Omega)$ function $F(\bar{U}(x,t))$.

We now assemble the above-obtained rigorous mathematical results to obtain $\bar{U}(x,t)$ as a weak solution of Eq. (7.1) for any test function $v(x,t) \in C_0^\infty([0,T]), H^2(\Omega) \cap H_0^1(\Omega))$

$$\lim_{n\to\infty} \int_0^T dt\left[\left(U^{(n)}, -\frac{dv}{dt}\right)_{L^2(\Omega)} + (-\Delta U^{(n)}, v)_{L^2(\Omega)}(F(U^{(n)}), -\Delta^{(\wedge)}v)_{L^2(\Omega)}\right]$$

$$= \lim_{n\to\infty} \int_0^T dt(f^{(n)}, v), \tag{7.15}$$

or in the weak-generalized sense mentioned above

$$\int_0^T dt\left(\bar{U}(x,t), -\frac{dv(x,t)}{dt}\right)_{L^2(\Omega)} + (\bar{U}(x,t), (-\Delta v)(x,t))_{L^2(\Omega)}$$

$$+ (F(\bar{U}(x,t), -(\Delta^{(\wedge)}v(x,t))_{L^2(\Omega)}$$

$$= \int_0^T dt(f(x,t), v(x,t))_{L^2(\Omega)}, \tag{7.16}$$

since $v(0,x) = v(T,x) \equiv 0$ by our proposed space of time-dependent test functions as $C_0^\infty([0,t], H^2(\Omega) \cap H_0^1(\Omega))$, suitable to be used on the Rosens path-integrals representations for stochastic systems (see Eqs. (7.22a) and (7.22b) in what follows).

The uniqueness of our solution $\bar{U}(x,t)$, comes from the following lemma.[4]

Lemma 7.1. *If $\bar{U}_{(1)}$ and $\bar{U}_{(2)}$ in $L^\infty([0,T] \times L^2(\Omega))$ are two functions satisfying the weak relationship below*

$$\int_0^T dt\left\{\left(\bar{U}_{(1)} - \bar{U}_{(2)}, -\frac{\partial v}{\partial t}\right)_{L^2(\Omega)} + (\bar{U}_{(1)} - \bar{U}_{(2)}, +\Delta v)_{L^2(\Omega)}\right.$$

$$\left.\times (F(\bar{U}_{(1)} - F(\bar{U}_{(2)}); +\Delta v)_{L^2(\Omega)}\right\} \equiv 0, \tag{7.17}$$

then $\bar{U}_{(1)} = \bar{U}_{(2)}$ a.e in $L^{\infty}([0,T] \times L^2(\Omega))$. The proof of Eq. (7.17) is easily obtained by considering the family of test functions on Eq. (7.16) of the following form $v_n(x,t) = g_{(\varepsilon)}(t)e^{+\alpha_n t}\varphi_n(x)$ with $-\Delta\varphi_n(x) = \alpha_n\varphi_n(x)$ and $g(t) = 1$ for $(\varepsilon, T - \varepsilon)$ with $\varepsilon > 0$ arbitrary. We can see that it reduces to the obvious identity $(\alpha_n > 0)$.

$$\int_{\varepsilon}^{T-\varepsilon} dt \exp(\alpha_n t)(F(\bar{U}_{(1)}) - F(\bar{U}_{(2)}), \varphi_n)_{L^2(\Omega)} \equiv 0, \qquad (7.18)$$

which means that $F(\bar{U}_{(1)}) = F(\bar{U}_{(2)})$ a.e on $(0,T) \times \Omega$ since ε is an arbitrary number. We have thus $\bar{U}_{(1)} = \bar{U}_{(2)}$ a.e, as $F(x)$ satisfies the lower bound estimate by our hypothesis on the kind of nonlinearity considered in our nonlinear diffusion equation (7.1).

$$|F(x) - F(y)| \geq \left(\begin{matrix} \inf(F'(x)) \\ -\infty < x < +\infty \end{matrix} \right) |x - y|. \qquad (7.19)$$

Let us now consider a path-integral solution of Eq. (7.1) (with $g(x) = 0$) for $f(x,t)$ denoting the white-noise stirring[4]

$$E(f(x,t)f(x',t')) = \lambda\delta^{(D)}(x - x')\delta(t - t'), \qquad (7.20)$$

where λ is the noise strength.

The first step is to write the generating process stochastic functional through the Rosen–Feynman path-integral identities[4]

$$Z[J(x,t)] = E_f\left[\exp\left\{ i \int_0^T dt \int_\Omega d^D x U(x,t,[f]) J(x,t) \right\} \right], \qquad (7.21a)$$

$$Z[J(x,t)] = E_f\left[\int D^F[U]\delta^{(F)}(\partial_t U - \Delta U - \Delta^{(\wedge)}(F(U)) - f) \right.$$
$$\left. \times \exp\left\{ i \int_0^T dt \int_\Omega d^D x U(x,t) J(x,t) \right\}, \right. \qquad (7.21b)$$

$$Z[J(x,t)] = E_f\left[\int D^F[U]D^F[\lambda] \exp\left\{ i \int_0^T dt \int_\Omega d^D x \lambda(x,t) \right.\right.$$
$$\left.\left. \times (\partial_t U - \Delta U - \Delta^{(\wedge)}(F(U) - f)) \right\} \right]$$
$$\times \exp\left\{ i \int_0^T dt \int_\Omega d^D x U(x,t) J(x,t) \right\}, \qquad (7.21c)$$

$$Z[J(x,t)] = \int D^F[U] \exp\left\{-\frac{1}{2\lambda}\int_0^T dt \int_\Omega d^D x\right.$$

$$\times \left.(\partial_t U - \Delta U - \Delta^{(\wedge)}(F(U))^2(x,t))\right\}$$

$$\times \exp\left\{i\int_0^T dt \int_\Omega d^D x\, U(x,t)J(x,t)\right\}. \tag{7.21d}$$

The important step made rigorous mathematically possible on the above (still formal) Rosen's path-integral representation by our previous rigorous mathematical analysis is the use of the delta functional identity on Eq. (7.21b) which is true only in the case of the existence and uniqueness of the solution of the diffusion equation in the weak sense at least for multiplier Lagrange fields $\lambda(x,t) \in C_0^\infty([0,T], H^2(\Omega) \cap H_0^1(\Omega))$.

As an important mathematical result to be pointed out is that in general case of a nonporous medium[4] in R^3, where one should model the diffusion nonlinearity by a complete Laplacean $\Delta F(U(x,t))$, one should observe that the set of (cut-off) solutions $\{\bar{U}^{(\wedge)}(x,t)\}$ of Eq. (7.1) still remains a bounded set on $L^\infty([0,T], L^2(\Omega))$. Since we have the *a priori* estimate uniform bound for the $U^{(n)}$-derivatives below in $D = 3$ (with $G'(x) = F(x)$). Namely

$$\left|\int_0^T dt \left\|\frac{dU^{(n)}}{dt}\right\|^2_{L^2(\Omega)}\right| \le \left|\int_0^T dt \left(\int_\Omega d^3 x (\Delta F(U^{(n)}(t))) \cdot \left(\frac{dU^{(n)}(t)}{dt}\right)\right)\right|$$

$$+ \left|\int_\Omega d^3 x f(x,t)\frac{d(U^{(n)}(x,t))}{dt}\right| \le \left|\int_0^T dt\, \text{Real}\left\{\frac{d}{dt}\int_\Omega d^3 x \Delta G(U^{(n)}(t))\right\}\right|$$

$$+ \frac{1}{2}\left\{\sup_{0\le t\le T} p\|f(x,t)\|^2_{L^2(\Omega)} + \frac{1}{p}\|\dot{U}^{(n)}\|^2_{L^2(\Omega)}\right\}$$

$$\le \left|\text{Real}\left(\int_\Omega d^3 x (\Delta G(U^{(n)}(T,x)) - \Delta G(U^{(n)}(0,x)))\right)\right|$$

$$+ \frac{1}{2}\sup_{0\le t\le T}\left\{p\|f(x,t)\|^2_{L^2(\Omega)} + \frac{1}{p}\|\dot{U}^{(n)}\|_{L^2(\Omega)}\right\}$$

$$\le \frac{1}{2}p\|f\|^2_{L^\infty((0,t),L^2(\Omega))} + \frac{1}{2p}\|\dot{U}^{(n)}(t)\|_{L^\infty((0,t),L^2(\Omega))} < \infty, \tag{7.22}$$

where $U^{(n)}(T,x)\big|_{\partial\Omega} = U^{(n)}(0,x)\big|_{\partial\Omega} = 0$ (see Eq. (7.3)). The uniform bound for the derivatives is achieved by choosing $\frac{1}{2p} < 1$.

As another point worth to call to attention is that the above considered function space is the dual of the Banach space $L^1([0,T], L^2(\Omega))$. So, one can extract from the above set of cut-off solutions a candidate $\bar{U}^{(\infty)}(x,t)$, in the weak-star topology of $L^\infty([0,T], L^2(\Omega))$ for the above cited case of cut-off removing $\wedge = +\infty$.[6] However, we will not proceed thoroughly in this straightforward technical question of cut-off removing our model of nonlinear diffusion in this chapter for general spaces R^D.

Finally, we remark that in the one-dimensional case $\Omega \in R^1$, one can further show by using the same compacity methods the existence and uniqueness of the diffusion equation added with the hydrodynamic advective term $\frac{1}{2}\frac{\partial}{\partial x}(U(x,t))^2$, which turns the diffusion equation (7.1) as a kind of nonlinear Burger equation on a porous medium.

It appears very important to remark that Galerking methods applied directly to the finite-dimensional stochastic equation (7.7) (see Eq. (7.4)) may be saving-time computer simulation candidates for the "turbulent" path-integral (Eqs. (6.22a)–(6.22d)) evaluations by approximate numerical methods (Ref. 2).

7.3. The Linear Diffusion in the Space $L^2(\Omega)$

Let us now present some mathematical results for the diffusion problem in Hilbert spaces formed by square-integrable functions $L^2(\Omega)$,[5] with the domain Ω denoting a compact set of R^D.

The diffusion equation in the infinite-dimensional space $L^2(\Omega)$ is given by the following functional differential equation (see Ref. 5 for mathematical notation).

$$\frac{\partial \psi[f(x);t]}{\partial t} = \frac{1}{2}\,\mathrm{Tr}_{L^2(\Omega)}([QD_f^2\psi[f(x,t)]) \qquad (7.23a)$$

$$\psi[f(x), t \to 0^+] = \Omega[f(x)]. \qquad (7.23b)$$

Here, $\psi[f(x), \cdot]$ is a time-dependent functional to be determined through the governing equation (7.23) and belonging to the space $L^2(L^2(\Omega), d_Q\mu(f))$ with $d_Q\mu(f)$ denoting the Gaussian measure on $L^2(\Omega)$ associated to Q — a fixed positive self-adjoint trace class operator $\oint_1(L^2(\Omega))$ — and D_f^2 is the second — Frechet derivative of the functional $\psi[f(x), t]$ which is given by a $f(x)$-dependent linear operator on $L^2(\Omega)$ with associated quadratic form $(D_f^2\psi[f(x), t] \cdot g(x), h(x))_{L^2(\Omega)}$.

By considering explicitly the spectral base of the operator Q on $L^2(\Omega)$

$$Q\varphi_n = \lambda_n\varphi_n, \tag{7.24}$$

The $L^2(\Omega)$-infinite-dimensional diffusion equation takes the usual form:

$$\Psi\left[\sum_n f_n\varphi_n, t\right] = \psi^{(\infty)}[(f_n), t], \tag{7.25a}$$

$$\Omega\left[\sum_n f_n\varphi_n\right] = \Omega^{(\infty)}[(f_n)], \tag{7.25b}$$

$$\frac{\partial\psi^{(\infty)}[(f_n), t]}{\partial t} = \sum_n [(\lambda_n\Delta_{f_n})\psi^{(\infty)}[(f_n), t]], \tag{7.25c}$$

$$\psi^{(\infty)}[(f_n), 0] = \Omega^{(\infty)}[(f_n)], \tag{7.25d}$$

or in the Physicist' functional derivative form (see Ref. 5).

$$\frac{\partial}{\partial t}\psi[f(x), t] = \frac{1}{2}\int_\Omega d^D x \int_\Omega d^D x' Q(x, x')\frac{\delta^2}{\delta f(x')\delta f(x)}\psi[f(x), t], \tag{7.26a}$$

$$\psi[f(x), 0] = \Omega[f(x)]. \tag{7.26b}$$

Here, the integral operator Kernel of the trace class operator is explicitly given by

$$Q(x, x') = \sum_n (\lambda_n\varphi_n(x)\varphi_n(x')). \tag{7.26c}$$

A solution of Eq. (7.26a) is easily written in terms of Gaussian path-integrals[5] which reads in the physicist's notations

$$\psi[f(x), t] = \int_{L^2(\Omega)} D^F[g(x)]\Omega[f(x) + g(x)] \times \det^{\frac{1}{2}}\left[\frac{1}{t}Q^{-1}\right]$$

$$\times \exp\left\{-\frac{1}{2t}\int_\Omega d^D x \int_\Omega d^D x' g(x)\cdot Q^{-1}(x, x')g(x')\right\}. \tag{7.27}$$

Rigorously, the correct functional measure on Eq. (7.27) is the normalized Gaussian measure with the following generating functional

$$Z[j(x)] = \int_{L^2(\Omega)} d_t Q\mu[g(x)]\exp\left\{i\int_\Omega j(x)g(x)d^D x\right\}$$

$$= \exp\left\{-\frac{t}{2}\int_\Omega d^D x \int_\Omega d^D x' j(x)Q^{+1}(x, x')j(x')\right\}. \tag{7.28}$$

At this point, it becomes important to remark that when writing the solution as a Gaussian path-integral average as done in Eq. (7.27), all the $L^2(\Omega)$ functions in the functional domain of our diffusion functional field $\psi[f(x), t]$ belongs to the functional domain of the quadratic form associated to the class trace operator Q the so-called reproducing kernel of the operator Q which is not the whole Hilbert space $L^2(\Omega)$ as naively indicated on Eq. (7.27), but the following subset of it:

$$\text{Dom}(\psi[\cdot, t]) = \{f(x) \in L^2(\Omega)|Q^{-\frac{1}{2}}f \in L^2(\Omega)\} \subsetneq L^2(\Omega). \qquad (7.29)$$

The above result gives a new generalization of the famous Cameron–Martin theorem that the usual Wienner measure (defined by the one-dimensional Laplacean with Dirichlet conditions on the interval end-points) is translation invariant, i.e. $d^{\text{Wien}}\mu[f + g] = d^{\text{Wien}}\mu[f] \times \left(\frac{d^{\text{Wien}}\mu[f+g]}{d^{\text{Wien}}\mu[f]}\right)$, if and only if the shift function $g(x)$ is absolutely continuous with derivative on $L^2([a, b])$. In other words, $g \in H_0^1([a, b]) = \text{Dom}\left\{\sqrt{-\frac{d^2}{d^2x}}\right\}$.

Another important point to call to the readers' attention is that one can write Eq. (7.27) in the usual form of diffusion in finite-dimensional case (see Appendix B):

$$\psi[f(x), t] = \int_{L^2(\Omega)} D^F[g(x)]\Omega[f(x) + \sqrt{t}g(x)] \times \left(\det^{\frac{1}{2}}\left[\frac{1}{t}Q\right]\right)$$

$$\times \exp\left\{-\frac{1}{2}\int_\Omega d^Dx \int_\Omega d^Dx' g(x)Q^{-1}(x, x')g(x)\right\}, \qquad (7.30)$$

At this point it is worth to call to the readers' attention that $d_{tQ}\mu$ and $d_Q\mu$ Gaussian measures are singular to each other by a direct application of Kakutani theorem for Gaussian infinite-dimensional measures for any time $t > 0$:

$$d_{tQ}\mu[g(x)]/d_Q\mu[g(x)] = +\infty. \qquad (7.31)$$

Let us apply the above results for the physical diffusion of polymer rings (closed strings) described by periodic loops $\vec{X}(\sigma) \in R^D, 0 \leq \sigma \leq A, \vec{X}(\sigma+A) = \vec{X}(\sigma)$ with a nonlocal diffusion coefficient $Q(\sigma, \sigma')$ (such that $\int_0^A d\sigma \int_0^A d\sigma' Q(\sigma, \sigma') = \text{Tr}[Q] < \infty$). The functional governing equation in loop space (formed by polymer rings) is given by

$$\frac{\partial \psi^{(\varepsilon)}[\vec{X}(\sigma); A]}{\partial A} = \int_0^A d\sigma \int_0^A d\sigma' Q_{ij}^{(\varepsilon)}(\sigma, \sigma')\frac{\delta^2}{\delta \vec{X}_i(\sigma)\delta \vec{X}_j(\sigma)}\psi^{(\varepsilon)}[\vec{X}(\sigma), A],$$

$$(7.32a)$$

$$\psi^{(\varepsilon)}[\vec{X}(\sigma);0] = \exp\left\{-\frac{\lambda}{2}\int_0^A d\sigma \int_0^A d\sigma' \vec{X}_i(\sigma)M_{ij}(\sigma,\sigma')\vec{X}_j(\sigma')\right\}.$$

$$\text{(7.32b)}$$

Here, the ring polymer surface probability distribution $\psi^{(\varepsilon)}[\vec{X}(\sigma),A]$ depends on the area parameter A, the area of the cylindrical polymer surface of our surface-polymer chain. Note that the presence of a parameter ε on the above objects takes into account the local (the integral operator kernel) case $Q(\sigma,\sigma') = \delta(\sigma - \sigma')$ as a limiting case of the rigorously mathematical well-defined (class trace) situation on the end of the observable evaluations

$$Q_{ij}^{(\varepsilon)}(\sigma,\sigma') = \frac{1}{\pi\varepsilon}\left[\exp\left(-\frac{(\sigma-\sigma')^2}{\varepsilon^2}\right)\right].$$

$$\text{(7.33)}$$

The solution of Eq. (7.32a) is straightforwardly written in the case of a self-adjoint kernel M on $L^2([0,A])$;

$$\exp\left\{-\frac{\lambda}{2}\int_0^A d\sigma \int_0^A d\sigma' \vec{X}_i(\sigma)M_{ij}(\sigma,\sigma')\vec{X}_j(\sigma')\right\}\det^{-\frac{1}{2}}[1 + A\lambda M(Q^{(\varepsilon)})]$$

$$\times \exp\left\{+\frac{\lambda^2}{2}\int_0^A d\sigma \int_0^A d\sigma' (\vec{M}X)_i(\sigma)\left(\left(\lambda M + (Q^{(\varepsilon)})^{-1}\cdot\frac{1}{A}\right)^{-1}(\vec{M}X)_j(\sigma')\right)\right\}.$$

$$\text{(7.34)}$$

The functional determinant can be reduced to the evaluation of an integral equation

$$\det^{-\frac{1}{2}}[1 + A\lambda M(Q^{(\varepsilon)})]$$

$$= \exp\left\{-\frac{1}{2}\text{Tr}_{L^2(\Omega)}\lg(1 + \lambda AM(Q^{(\varepsilon)})\right\}$$

$$= \exp\left\{-\frac{1}{2}\text{Tr}_{L^2(\Omega)}\int_0^\lambda d\lambda'[(Q^{(\varepsilon)})M)(1 + \lambda'A(Q^{(\varepsilon)})M)^{-1}]\right\}$$

$$= \exp\left\{-\frac{1}{2}\text{Tr}_{L^2(\Omega)}\int_0^\lambda d\lambda' R(\lambda')\right\}.$$

$$\text{(7.35)}$$

Here, the kernel operator $R(\lambda')$ satisfies the integral equation (accessible for numerical analysis)

$$R(\lambda')(1 + \lambda'A(Q^{(\varepsilon)})M = (Q^{(\varepsilon)})M.$$

$$\text{(7.36)}$$

Which in the local case of $\varepsilon \to 0^+$, when considered in the final result (Eqs. (7.34) and (7.35)), produces the explicitly candidate solutions for our polymer-surface probability distribution with M a class trace operator on the loop space, $L^2([0, A])$:

$$\psi[\vec{X}(\sigma), A] = \exp\left\{-\frac{A}{2}\operatorname{Tr}_{L^2(\Omega)}\int_0^\lambda d\lambda'[M(Q^{(\varepsilon)^{-1}} + \lambda'AM)^{-1}]\right\}$$

$$\times \exp\left\{-\frac{\lambda}{2}\int_0^A d\sigma \int_0^A d\sigma' X_i(\sigma) \cdot M_{ij}(\sigma, \sigma')\vec{X}_j(\sigma')\right\}$$

$$\times \exp\left\{+\frac{\lambda^2}{2}\int_0^A d\sigma \int_0^A d\sigma'(\vec{MX})_i(\sigma)\right.$$

$$\left. \times \left(\lambda M + (Q^{(\varepsilon)})^{-1} \cdot \frac{1}{A}\right)^{-1}(\sigma, \sigma')(\vec{MX})_j(\sigma')\right\}.$$

$$(7.37)$$

It is worth to call to the readers' attention that if $A \in \mathcal{J}_1$ and B is a bounded operator — so $A \cdot B$ is a class trace operator. The functional determinant $\det[1 + AB]$ is a well-defined object as a direct result of the obvious estimate, the result which was used to arrive at Eq. (7.37).

$$\lim_{N \to \infty} \prod_{n=0}^N (1 + \lambda_n) \leq \exp\left(\sum_{n=0}^N \lambda_n\right) = \exp(\operatorname{Tr} AB).$$

As a last comment on the linear infinite-dimensional diffusion problem (Eq. (7.23)), let us sketch a (rigorous) proof that Eq. (7.27) is the unique solution of Eq.(7.23). First, let us consider the initial condition on Eq. (7.23) as belonging to the space of all mappings $G{:}L^2(\Omega) \to R$ that are twice Fréchet differentiable on $L^2(\Omega)$ with uniformly continuous and bounded second derivative $D_f^2 G$ (a bounded operator of $\mathcal{L}(L^2(\Omega))$ with norm $\bar{C}$). This set of mappings will be denoted by $uC^2[L^2(\Omega), R]$. It is, thus, straightforward to see through an application of the mean-value theorem that the following estimate holds true

$$\sup_{f(x) \in Q^{\frac{1}{2}}L^2(\Omega)} |\psi[f(x), t] - G[f(x)]|$$

$$\leq \int_{L^2(\Omega)} |G(f(x) + g(x)) - G(g(x))|d_{tQ}\mu[g(x)]$$

$$\leq \int_{L^2(\Omega)} \left[\left\| DG(f(x), g(x))_{L^2(\Omega)} + \int_0^1 d\sigma(1-\sigma)(D^2 G[f(x) \right. \right.$$

$$\left. \left. + \sigma g(x)]g(x), g(x)_{L^2(\Omega)} \right\| \right] d_{tQ}\mu[g(x)]$$

$$\leq 0 + \bar{C} \int_0^1 d\sigma(1-\sigma) \int_{L^2(\Omega)} \|g(x)\|_{L^2(\Omega)}^2 d_{tQ}\mu[g(x)]$$

$$\leq \bar{C} \left(\int_{L^2(\Omega)} \|g(x)\|_{L^2(\Omega)}^2 d_{tQ}\mu[g(x)] \right)$$

$$\leq \bar{C} \operatorname{Tr}(tQ) = (\bar{C} \operatorname{Tr}(Q)t \to 0 \quad \text{as } t \to 0^+. \tag{7.38}$$

We have thus defined a strongly continuous semi-group on the Banach space $UC^2[L^2(\Omega), R]$ with infinitesimal generator given by the infinite-dimensional Laplacean $\operatorname{Tr}[QD^2]$ acting on the space $L^2(Q^{\frac{1}{2}}(L^2(\Omega)), R)$. By the general theory of semi-groups on Banach spaces, we obtain that Eq. (7.27) satisfies the infinite-dimensional diffusion initial-value problem Eq. (7.23), at least for initial conditions on the space $uC^2[L^2(\Omega), R]$. Since purely Gaussian functionals belong to $uC^2[L^2(\Omega), R]$ and they form a dense set on the space $L^2(L^2(\Omega), d_Q\mu)$, we get the proof of our result for general initial condition on $L^2(L^2(\Omega), d_Q\mu)$.

Finally, we point out that the general solution of the diffusion problem on Hilbert space with sources and sinks, namely

$$\frac{\partial}{\partial t}\psi[f(x), t] = \frac{1}{2}\operatorname{Tr}_{L^2(\Omega)}[QD_f^2\psi[f(x), t]] - V[f(x)]\psi[f(x), t] \tag{7.39}$$

with

$$\psi[f(x), t \to 0^+] = \Omega[f(x)], \tag{7.40}$$

possesses a generalized Feynman–Wiener–Kac Hilbert $L^2(\Omega)$ space-valued path-integral representation, which in the Feynman physicist formal notation reads as

$$\psi[h(x), T] = \int_{C([0,T], L^2(\Omega))} D^F[X(\sigma)]$$

$$\times \exp\left\{ -\frac{1}{2}\int_0^T d\sigma \left(\frac{dX}{d\sigma}, Q^{-1}\frac{dX}{d\sigma} \right)_{L^2(\Omega)}(\sigma) \right\}$$

$$\times \Omega \left[\left(\int_0^T X(\sigma) d\sigma \right) + X(0) \right]$$

$$\times \exp \left\{ - \int_0^T d\sigma V \left[\left(\int_0^T X(\sigma') d\sigma' \right) + X(0) \right] \right\}, \quad (7.41)$$

where the paths satisfy the end-point constraint $X(T) = h(x) \in L^2(\Omega); X(0) = f(x) \in L^2(\Omega)$.

References

1. I. Sneddon, *Fourier Transforms* (McGraw-Hill Book Company, INC, USA, 1951); P. A. Domenico and F. W. Schwartz, *Physical and Chemical Hydrogeology*, 2nd edn. (Wiley, 1998); P. L. Schdev, *Nonlinear Diffusive Waves* (Cambridge University Press, 1987); W. E. Kohler; B. S. White (ed.), *Mathematics of Random Media*, Lectures in Applied Mathematics, Vol. 27 (American Math. Soc., Providence, Rhode Island, 1980); R. Z. Sagdeev (ed.), Non-Linear and Turbulent Processes in Physics, Vols. 1–3 (Harwood Academic Publishers, 1987); Geoffrey Grimmett, *Percolation* (Springer-Verlag, 1989).
2. G. Da Prato and J. Zabczyk, *Ergodicity for Infinite Dimensional System*, London Math. Soc. Lect. Notes, Vol. 229 (Cambridge University Press, 1996); R. Teman, C. Foias, O. Manley and Y. Treve, *Physica* **6D**, 157 (1983).
3. R. Dortray and J.-L. Lion, *Analyse Mathématique et Calcul Numérique*, Masson, Paris, Vols. 1–9, A. Friedman, *Variational Principles and Free-Boundary Problems — Pure and Applied, Math.* (Wiley, 1988).
4. L. C. L. Botelho, *J. Phys. A: Math. Gen.* **34**, L31–L37 (2001); L. C. L. Botelho, *Il Nuovo Cimento* **117** (B1), 15 (2002); *J. Math. Phys.* **42**, 1682 (2001); L. C. L. Botelho, *Int. J. Mod. Phys. B* **13** (13), 1663 (1999); *Int. J. Mod. Phys.* **12**; *Mod. Phys. B* **14**(3), 331 (2002); A. S. Monin and A. M. Yaglon, *Statistical Fluid Mechanics*, Vol. 2 (Mit Press, Cambridge, 1971); G. Rosen, *J. Math. Phys.* **12**, 812 (1971); L. C. L. Botelho, *Mod. Phys. Lett* **13B**, 317 (1999); V. Gurarie and A. Migdal, *Phys. Rev. E* **54**, 4908 (1996); U. Frisch, *Turbulence* (Cambridge Press, Cambridge, 1996); W. D. Mc-Comb, *The Physics of Fluid Turbulence* (Oxford University, Oxford, 1990); L. C. L. Botelho, *Mod. Phys. Lett. B* **13**, 363 (1999); S. I. Denisov and W. Horsthemke, *Phys. Lett. A* **282** (6), 367 (2001); L. C. L. Botelho, *Int. J. Mod. Phys. B* **13**, 1663 (1999); L. C. L. Botelho, *Mod. Phys. Lett. B* **12**, 301 (1998); L. C. L. Botelho, *Mod. Phys. Lett. B* **12**, 569 (1998); L. C. L. Botelho, *Mod. Phys. Lett. B* **12**, 1191 (1998); L. C. L. Botelho, *Il Nuovo Cimento* **117** B(1), 15 (2002); *J. Math. Phys.* **42**, 1682 (2001).
5. G. da Prato and J. Zabczyk, Second Order Partial Differential Equations in Hilbert Spaces, Vol. 293 (London Math. Soc. Lect. Notes, Cambridge, UK, 2002).

6. J. Wloka, *Partial Differential Equations* (Cambridge University Press, 1987).
7. Soo Bong Chae, Holomorphy and Calculus in Normed Spaces, *Pure and Applied Math. Series* (Marcel Denner Inc, NY, USA, 1985).

Appendix A. The Aubin–Lion Theorem

Just for completeness in this mathematical appendix for our mathematical oriented readers, we intend to give a detailed proof of the basic result on compacity of sets in functional spaces of the form $L^2(\Omega)$ and throughout used in Sec. 2. We have, thus, the Aubin–Lion theorem[3] in the Gelfand triplet $H_0^1(\Omega) \hookrightarrow L^2(\Omega) \hookrightarrow H^{-1}(\Omega) = (H_0^1(\Omega))^*$

Aubin–Lion Theorem. *If $\{U_n(x,t)\}$ is a sequence of time-differentiable functions in a bounded set of $L^2([0,T], H_0^1(\Omega))$ such that its time derivatives forms a bounded set of $L^2([0,T], H_0^{-1}(\Omega))$, we have that $\{U_n(x,t)\}$ is a compact set on $L^2([0,T], L^2(\Omega))$.*

Proof. The basic fact we are going to use to give a mathematical proof of this theorem is the following identity (Ehrling's lemma). For any given $\varepsilon > 0$, there is a constant $C(\varepsilon)$ such that

$$\|U_n\|_{L^2(\Omega)} \leq \varepsilon \|U_n\|_{H_0^1(\Omega)} + C(\varepsilon)\|U_n\|_{H^{-1}(\Omega)}^2. \qquad (A.1)$$

As a consequence, we have the following estimate

$$\int_0^T \|U_n - U_m\|_{L^2([0,T]),L^2(\Omega)}^2$$

$$\leq \int_0^T dt (\varepsilon \|U_n - U_m\|_{H_0^1(\Omega)} + C(\varepsilon)\|U_n - U_m\|_{H^{-1}(\Omega)})^2$$

$$\leq \varepsilon^2 \left(\int_0^T dt \|U_n - U_m\|_{H_0^1(\Omega)}^2 \right) + (C(\varepsilon))^2 \left(\int_0^T dt \|U_n - U_m\|_{H^{-1}(\Omega)}^2 \right)$$

$$+ 2\varepsilon C(\varepsilon) \left(\int_0^T dt (\|U_n - U_m\|_{H_0^1(\Omega)} \times \|U_n - U_m\|_{H_0^{-1}(\Omega)}) \right)$$

$$\leq \varepsilon^2 \left(\int_0^T dt \|U_n - U_m\|_{H_0^1(\Omega)}^2 \right) + C(\varepsilon)^2 \left(\int_0^T dt \|U_n - U_m\|_{H^{-1}(\Omega)}^2 \right)$$

$$+ 2\varepsilon C(\varepsilon) \left(\int_0^T dt \|U_n - U_m\|_{H_0^1(\Omega)}^2 \right)^{\frac{1}{2}} + \left(\int_0^T dt \|U_n - U_m\|_{H^{-1}(\Omega)}^2 \right)^{\frac{1}{2}}$$

$$\leq 2\varepsilon^2 M + 2\varepsilon C(\varepsilon) M^{\frac{1}{2}} \left(\int_0^T dt \|U_n - U_m\|_{H^{-1}(\Omega)}^2 \right)^{\frac{1}{2}}$$

$$+ (C(\varepsilon)^2) \left(\int_0^T dt \|U_n - U_m\|_{H^{-1}(\Omega)}^2 \right). \tag{A.2}$$

At this point, we use the Arzela–Ascoli theorem to see that $\{U_n(x,t)\}$ is a compact set on the space $C([0,T], H^{-1}(\Omega))$ since we have the set equicontinuity:

$$\|U_n(t) - U_m(s)\|_{H^{-1}(\Omega)} \leq \int_s^t \|U_n'(\tau)\|_{H_0^{-1}(\Omega)} d\tau$$

$$\leq |t - s|^{(1-\frac{1}{2})} \left(\int_0^T \|U_n'(\tau)\|_{H^{-1}(\Omega)}^2 d\tau \right)^{\frac{1}{2}}$$

$$\leq \bar{M} |t - s|^{\frac{1}{2}}. \tag{A.3}$$

It is a crucial step now by remarking that $H_0^1(\Omega)$ is compactly immersed in $L^2(\Omega)$ (Rellich theorem). Let us note that for each t (almost everywhere in $[0,T]$), $U_n(x,t)$ is a bounded set on $H_0^1(\Omega)$ since $U_n(x,t)$ belongs to a bounded set $L^2([0,T], H_0^1(\Omega))$ by hypothesis. As a consequence, $\{U_{n_k}, (x,t)\}$ is a compact set on $L^2(\Omega)$ (Rellich theorem) and so in $H^{-1}(\Omega)$ almost everywhere in $[0,T]$ since $L^2(\Omega) \hookrightarrow H^{-1}(\Omega)$. By an application of the Arzela–Ascoli theorem, there is a sub-sequence $\{U_{nk}(x,t)\}$ of $\{U_n(x,t)\}$ (and still denoted by $\{U_n(x,t)\}$ such that it converges uniformly to a given function $\bar{U}(x,t) \in C([0,T], H^{-1}(\Omega))$. As a direct result of this fact we have that (for $T < \infty$!) for $(n,m) \to \infty$.

$$\left(\int_0^T \|U_n - U_m\|_{H^{-1}(\Omega)}^2 \right)^{\frac{1}{2}}$$

$$\leq (\sup |U_n - U_m|_{C([0,T],H^{-1}(\Omega))}) \left(\int_0^T 1 \cdot dt \right)^{\frac{1}{2}} \to 0. \tag{A.4}$$

Returning to our estimate (Eq. (A.2)), we see that this sub-sequence is a Cauchy sequence in $L^2([0,T], L^2(\Omega))$. As a consequence, for each fixed $t \in [0,T]$ (almost everywhere), $U_n(x,t)$ converges to $\bar{U}(x,t)$ in $L^2(\Omega)$. □

Appendix B. The Linear Diffusion Equation

Let us show mathematically the basic functional integral representation (Eq. (7.30)) for the $L^2(\Omega)$-space diffusion equation (7.23).

As a first step for such proof, let us call the readers' attention that one should consider the second-order (Laplacean) $D^2U(x,t)$ as a bounded operator in $L^2(\Omega)$ in order to the operatorial composition with the positive definite class trace operator Q still be a class trace operator as it is explicitly supposed in the right-hand side of Eq. (7.23).

We thus impose as the sub-space of initial condition the diffusion equation (7.23) for the (dense) vector sub-space of $C(L^2(\Omega), R)$ composed of all functionals of the form

$$ f(x) = \int_{L^2(\Omega)} d_Q\mu(p)F(p)\exp(i\langle p, x\rangle_{L^2(\Omega)}), \qquad (\text{B.1}) $$

with $F(p) \in L^2(L^2(\Omega), d_Q\mu)$.

By substituting the initial condition (Eq. (B.1)) into the integral representation equation (7.30) and by using the Fubbini–Toneli theorem to exchange the needed integrations order in the estimate below, we get:

$$ U(x,t) = \int_{L^2(\Omega)} f(x + \sqrt{t}\xi)d_Q\mu(\xi) $$

$$ = \int_{L^2} d_Q\mu(\xi)\left\{\int_{L^2} d_Q\mu(p)F(p)e^{i\langle p, x+\sqrt{t}\xi\rangle_{L^2}}\right\} $$

$$ = \int_{L^2} d_Q\mu(p)F(p)\cdot e^{i\langle p,x\rangle_{L^2}}e^{-\frac{1}{2}t\langle p, Qp\rangle_{L^2}}. \qquad (\text{B.2}) $$

Note that we have already proved that $U(x,t)$ is a bounded functional of $C(L^2(\Omega) \times [0,\infty]; R)$ on the basis of our hypothesis on the initial functional date (Eq. (B.1)).

At this point, we observe that the second-order Frechet derivatives of the functional, $\exp i\langle p, x\rangle_{L^2}$, are easily (explicitly) evaluated as[7]

$$ QD^2\left(e^{i\langle p,x\rangle_{L^2}}\right) = \left(\sum_{\ell=1}^{\infty} \lambda_\ell \frac{\partial^2}{\partial^2 x_\ell}\right)\left[e^{i(\sum_{n=1}^{\infty} p_n x_n)}\right] = -\left(\langle p, Qp\rangle_{L^2}\right)e^{i\langle p,x\rangle_{L^2}}. $$

$$ (\text{B.3}) $$

We have thus a straightforward proof of our claim cited above on the basis again of the chosen initial date sub-space

$$\mathrm{Tr}[QD^2 U(x,t)]$$

$$\leq \int_{L^2(\Omega)} d_Q \mu(p) |F(p)| \langle p, Qp \rangle_{L^2}$$

$$\leq \left(\int_{L^2(\Omega)} d_Q \mu(p) |F(p)|^2 \right)^{\frac{1}{2}} \left(\int_{L^2(\Omega)} d_Q \mu(p) |\langle p, Qp \rangle_{L^2(\Omega)}|^2 \right)^{\frac{1}{2}}$$

$$\leq (\mathrm{Tr}\, Q)^2 \|F\|^2_{L^2(L^2(\Omega), d_Q \mu)} < \infty. \tag{B.4}$$

Now, it is a simple application to verify that Eq. (B.2) satisfies the diffusion equation in $L^2(\Omega)$ (or in any other separable Hilbert space). Namely

$$\frac{\partial U(x,t)}{\partial t} = \int_{L^2(\Omega)} d_Q \mu(p) F(p) e^{i \langle p, x \rangle_{L^2}} \left\{ -\frac{1}{2} \langle p, Qp \rangle_{L^2(\Omega)} \right\}$$

$$\times e^{-\frac{t}{2}, \langle p, Qp \rangle_{L^2(\Omega)}}, \tag{B.5}$$

$$\mathrm{Tr}_{L^2(\Omega)}[QD^2 U(x,t)] = \int_{L^2(\Omega)} d_Q \mu(p) F(p) \{ D^2 e^{i \langle p, x \rangle} \} e^{-\frac{t}{2} \langle p, Qp \rangle_{L^2(\Omega)}}$$

$$= \int_{L^2(\Omega)} d_Q \mu(p) F(p) \{ -\langle p, Qp \rangle_{L^2} \} e^{i \langle p, x \rangle} e^{-\frac{t}{2} \langle p, Qp \rangle_{L^2(\Omega)}}, \tag{B.6}$$

with

$$U(x,0) = \int_{L^2(\Omega)} d_Q \mu(p) F(p) e^{i \langle p, x \rangle} \left\{ \lim_{t \to 0^+} e^{-\frac{t}{2} \langle p, Qp \rangle_{L^2(\Omega)}} \right\} = f(x). \tag{B.7}$$

Chapter 8

On the Ergodic Theorem*

8.1. Introduction

One of the most important phenomenon in numerical studies of the non-linear wave motion, especially in the two-dimensional case, is the existence of overwhelming majority of wave motions that wandering over all possible system's phase space and, given enough time, coming as close as desired (but not entirely coinciding) to any given initial condition.[1]

This phenomenon is signaling certainly that the famous recurrence theorem of Poincaré is true for infinite-dimensional continuum mechanical systems like that one represented by a bounded domain when subject to nonlinear vibrations.

A fundamental question appears in the context of this recurrence phenomenon and concerned to the existence of the time average for the associated nonlinear wave motion and naturally leading to the concept of an infinite-dimensional invariant measure for the nonlinear wave equation,[1] the mathematical phenomenon subjacente to the Poincaré recurrence theorem.

In this chapter, we intend to give applied mathematical arguments for the validity of the famous Ergodic theorem in a class of polynomial nonlinear and Lipschitz wave motions. Our approach is fully based on Hilbert space methods previously used to study dynamical systems in classical mechanics which in turn simplifies enormously the task of constructing explicitly the associated invariant measure on certain Sobolev spaces — the infinite-dimensional space of wave motion initial conditions. This study is presented in the main section of this chapter, namely Sec. 8.4.

In Sec. 8.2, we give a very detailed proof of the RAGE's theorem, a basic functional analysis rigorous method used in our proposed Hilbert

---

*Author's original results.

space generalization of the usual finite-dimensional Ergodic theorem and fully presented in Sec. 8.3.

In Sec. 8.5, we complement our studies by analyzing the important case of wave-diffusion under random stirring.

In Appendices A and B, we additionally present important technical details related to Sec. 8.4, and they contain material as important as those presented in the previous sections of our work.

8.2. On the Detailed Mathematical Proof of the RAGE Theorem

In this purely mathematical section, we intend to present a detailed mathematical proof of the RAGE theorem[2] used on the analytical proof of the Ergodic theorem on Sec. 8.3 for Hamiltonian systems of N-particles.

Let us, thus, start our analysis by considering a self-adjoint operator $\mathcal{L}$ on a Hilbert space $(H, (,))$, where $H_c(\mathcal{L})$ denotes the associated continuity sub-space obtained from the spectral theorem applied to $\mathcal{L}$. We have the following result.

Proposition (RAGE theorem). *Let $\psi \in H_c(\mathcal{L})$ and $\tilde{\psi} \in H$. We have the Ergodic limit*

$$\lim_{T \to \infty} \frac{1}{T} \int_0^T |(\tilde{\psi}, \exp(it\mathcal{L})\psi)|^2 dt = 0. \tag{8.1}$$

Proof. In order to show the validity of the above ergodic limit, let us rewrite Eq. (8.1) in terms of the spectral resolution of $\mathcal{L}$, namely

$$I = \frac{1}{T} \int_0^T |(\tilde{\psi}, \exp(it\mathcal{L})\psi)|^2 dt$$

$$= \frac{1}{T} \int_0^T \left\{ \left[\int_{-\infty}^{+\infty} e^{+it\lambda} d_\lambda(\tilde{\psi}, E_j(\lambda)\psi) \right] \left[\int_{-\infty}^{+\infty} e^{-it\mu} d_\mu(\tilde{\psi}, E_j(\mu)\psi) \right] \right.$$

$$\tag{8.2}$$

Here, we have used the usual spectral representation of $\mathcal{L}$

$$(g, \mathcal{L}h) = \int_{-\infty}^{+\infty} \lambda d_\lambda(g, E_j(\lambda)h), \tag{8.3}$$

with $g \in H$ and $h \in \text{Dom } (\mathcal{L})$.

Let us remark that the function $\exp(i(\lambda - \mu)t)$ is majorized by the function 1 which in turn is an integrable function on the domain $[0, T] \times R \times R$ with the product measure

$$\frac{dt}{T} \otimes d_\lambda(\tilde{\psi}, E_j(\lambda)\psi) \otimes d_\mu(\tilde{\psi}, E_j(\mu)\psi), \qquad (8.4)$$

since

$$\int_0^T \frac{dt}{T} \otimes d_\lambda(\tilde{\psi}, E_j(\lambda)\psi) \otimes d_\mu(\tilde{\psi}, E_j(\mu)\psi) = (\tilde{\psi}, \psi)^2 < \infty. \qquad (8.5)$$

At this point, we can safely apply the Fubbini theorem to interchange the order of integration in relation to the t-variable which leads to the partial result below

$$I = \int_{-\infty}^{+\infty} \int_{-\infty}^{+\infty} \left(\frac{e^{i(\lambda-\mu)T} - 1}{i(\lambda - \mu)T} \right) d_\lambda \langle \tilde{\psi}, E_j(\lambda)\psi \rangle \otimes d_\mu \langle \tilde{\psi}, E_j(\mu)\psi \rangle. \qquad (8.6)$$

Let us consider two cases. First, we take $\tilde{\psi} = \psi$. In this case, we have the bound

$$\left| \frac{e^{i(\lambda-\mu)T} - 1}{i(\lambda - \mu)T} \right| \leq \left| \frac{2\sin\left(\frac{(\lambda-\mu)}{2}T \right)}{(\lambda - \mu)T} \right| \leq 1. \qquad (8.7)$$

If we restrict ourselves to the case of $\lambda \neq \mu$, a direct application of the Lebesgue convergence theorem on the limit $T \to \infty$ yields that $I = 0$.

On the other hand, in the case of $\lambda = \mu$, we intend now to show that the set $\mu = \lambda$ is a set of zero measure with R^2 in respect to the measure $d_\lambda \langle \psi, E_j(\lambda)\psi \rangle \otimes d_\mu \langle \psi, E_j(\mu)\psi \rangle$. This can be seen by considering the real function on R

$$f(\lambda) = (\psi, E_j(\lambda)\psi). \qquad (8.8)$$

We observe that this function is continuous and nondecreasing since $\psi \in H_c(\mathcal{L})$ and range $E_j(\lambda_1) \subseteq$ range $E_j(\lambda_2)$ for $\lambda_1 \leq \lambda_2$. Now, $f(\lambda)$ is a uniform continuous function on the whole real line R, since for a given $\varepsilon > 0$, there is a constant $\tilde{\lambda}$ such that

$$(\psi, E_j(-\infty, -\tilde{\lambda})\psi) < \frac{\varepsilon}{2}, \qquad (8.9)$$

$$\|\psi\|^2 - (\psi, E_j(-\infty, \tilde{\lambda})\psi) < \frac{\varepsilon}{2}. \qquad (8.10)$$

Additionally, $f(\lambda)$ is uniform continuous on the closed interval $[-\tilde{\lambda}, \tilde{\lambda}]$, and $f(\lambda)$ cannot make variations greater than $\frac{\varepsilon}{2}$ on $(-\infty, -\tilde{\lambda}]$ and $[+\tilde{\lambda}, \infty)$, besides being a monotonic function on R. These arguments show the uniform continuity of $f(\lambda)$ on whole line R. Hence, we have that for a given $\varepsilon > 0$, there exists a $\delta > 0$ such that

$$|(\psi, E_j(\lambda')\psi) - (\psi, E_j(\lambda'')\psi)| \leq \frac{\varepsilon}{\|\psi\|^2}, \tag{8.11}$$

for

$$|\lambda' - \lambda''| \leq \delta. \tag{8.12}$$

In particular, for $\lambda' = \lambda + \delta$ e $\lambda'' = \lambda - \delta$

$$(\psi, E_j(\lambda + \delta)\psi) - (\psi, E_j(\lambda - \delta)\psi) \leq \frac{\varepsilon}{\|\psi\|^2}. \tag{8.13}$$

As a consequence, we have the estimate

$$\int_{-\infty}^{+\infty} d_\lambda(E_j(\lambda)\psi, \psi) \int_{\lambda-\delta}^{\lambda+\delta} d_\mu(\psi, E_j(\mu)\psi)$$
$$\leq \|\psi\|^2((\psi, E_j(\lambda + \delta)\psi) - (\psi E_j(\lambda - \delta)\psi))$$
$$\leq \|\psi\|^2 \frac{\varepsilon}{\|\psi\|^2} \leq \varepsilon. \tag{8.14}$$

$\square$

Let us note that for each $\varepsilon > 0$, there exists a set $D_\delta = \{(\lambda, \mu) \in R^2; |\lambda - \mu| < \delta\}$ which contains the line $\lambda = \mu$ and has measure less than ε in relation to the measure $d_\lambda(E_j(\lambda)\psi, \psi) \otimes d_\mu(E_j(\mu)\psi, \psi)$ as a result of Eq. (8.14). This shows our claim that $I = 0$ in our special case.

In the general case of $\tilde{\psi} \neq \psi$, we remark that solely the orthogonal component on the continuity sub-space $H_c(\mathcal{L})$ has a nonvanishing inner product with $\exp(it\mathcal{L})\psi$. By using now the polarization formulae, we reduce this case to the first analyzed result of $I = 0$.

At this point, we arrive at the complete RAGE theorem.[2]

RAGE Theorem. *Let K be a compact operator on $(H, (,.))$. We have thus the validity of the Ergodic limit*

$$\lim_{T \to \infty} \frac{1}{T} \int_0^T \|K \exp(it\mathcal{L})\psi\|_H^2 dt = 0, \tag{8.15}$$

with $\psi \in H_c(\mathcal{L})$.

We leave the details of the proof of this result for the reader, since any compact operator is the norm operator limit of finite-dimension operators and one only needs to show that

$$\lim_{T\to\infty} \frac{1}{T} \int_0^T \left\| \sum_{n=0}^N c_n(e^{it\mathcal{L}}\psi, e_n)g_n \right\|_H dt = 0, \tag{8.16}$$

for c_n constants and $\{e_n\}, \{g_n\}$ a finite set of vector of $(H, (,))$.

8.3. On the Boltzmann Ergodic Theorem in Classical Mechanics as a Result of the RAGE Theorem

One of the most important statement in physics is the famous zeroth law of thermodynamics: "any system approaches an equilibrium state." In the classical mechanics frameworks, one begins with the formal elements of the theory. Namely, the phase-space R^{6N} associated to a system of N-classical particles and the set of Hamilton equations

$$\dot{p}_i = -\frac{\partial H}{\partial q_i}; \quad \dot{q}_i = \frac{\partial H}{\partial p_i}, \tag{8.17}$$

where $H(q, p)$ is the energy function.

The above cited thermodynamical equilibrium principle becomes the mathematical statement that for each compact support continuous functions $C_c(R^{6N})$, the famous ergodic limit should hold true.[3]

$$\left\{ \lim_{T\to\infty} \frac{1}{T} \int_0^T f(q(t); p(t))dt \right\} = \eta(f), \tag{8.18}$$

where $\eta(f)$ is a linear functional on $C_c(R^{6N})$ given exactly by the Boltzmann statistical weight, and $\{q(t), p(t)\}$ denotes the (global) solutions of the Hamilton equation (8.12).

In this section, we point out a simple new mathematical argument of the fundamental equation (8.18) by means of Hilbert space techniques and the RAGEs theorem. Let us begin by introducing for each initial condition $(q(0), p(0))$, a function $\omega_{q_0, p_0}(t) \equiv (q(t), p(t))$, here $\langle q(t), p(t) \rangle$ is the assumed global unique solution of Eq. (8.17) with prescribed initial conditions. Let $U_t : L^2(R^{6N}) \to L^2(R^{6N})$ be the unitary operator defined by

$$(U_t f)(q, p) = f(\omega_{q_0 p_0}(t)). \tag{8.19}$$

We have the following theorem (the Liouville's theorem).[5]

Theorem 8.1. U_t *is a unitary one-parameter group whose infinitesimal generator is* $-i\bar{L}$, *where* $-iL$ *is the essential self-adjoint operator acting on* $C_0^\infty(R^{6N})$ *defined by the Poisson bracket.*

$$(Lf)(p,q) = \{f, H\}(q,p) = \sum_{i=1}^{3N} \left(\frac{\partial f}{\partial q_i} \frac{\partial H}{\partial p_i} - \frac{\partial f}{\partial p_i} \frac{\partial H}{\partial q_i} \right)(q,p). \qquad (8.20)$$

The basic result we are using to show the validity of the Ergodic limit equation (8.18) is the famous RAGEs theorem exposed in Sec. 8.1.

Theorem 8.2. *Let* $\phi \in H_c(-i\bar{L})(L^2(R^{6N})$, *here* $H_c(-i\bar{L})$ *is the continuity sub-space associated to self-adjoint operator* $-i\bar{L}$. *For every vector* $\beta \in L^2(R^{6N})$, *we have the result*

$$\lim_{T \to \infty} \frac{1}{T} \int_0^T |\langle \beta, U_t\psi \rangle|^2 dt = 0, \qquad (8.21)$$

or equivalently for every $\psi \in L^2(R^{6N})$

$$\lim_{T \to \infty} \frac{1}{T} \int_0^T \langle \beta, U_t\psi \rangle dt = \langle \beta, \mathbb{P}_{\ker(-i\bar{L})}\psi \rangle, \qquad (8.22)$$

where $\mathbb{P}_{\ker(-i\bar{L})}$ *is the projection operator on the (closed) sub-space* $\ker(-i\bar{L})$.

That Eq. (8.22) is equivalent to Eq. (8.21), is a simple consequence of the Schwartz inequality written below

$$\left| \int_0^T (\langle \beta, U_t\psi \rangle) dt \right| \leq \left(\int_0^T (\langle \beta, U_t\psi \rangle)^2 dt \right)^{\frac{1}{2}} \left(\int_0^T 1 dt \right)^{\frac{1}{2}}, \qquad (8.23)$$

or

$$\left| \frac{1}{T} \int_0^T \langle \beta, U_t\psi \rangle dt \right| \leq \left(\frac{1}{T} \int_0^T (\langle \beta, U_t\psi \rangle)^2 dt \right)^{\frac{1}{2}} \qquad (8.24)$$

As a consequence of Eq. (8.23), we can see that the linear functional $\eta(f)$ of the Ergodic theorem is exactly given by (just consider $\beta(p,q) \equiv 1$ on supp of $\mathbb{P}_{\ker(-i\bar{L})}(\psi)$)

$$\eta(f) = \mathbb{P}_{\ker(-i\bar{L})}(f)(q,p). \qquad (8.25)$$

By the Riesz's theorem applied to $\eta(f)$, we can rewrite (represent) Eq. (8.25) by means of a (kernel)-function $h_{\eta(H)}(q,p)$, namely

$$\eta(f) = \int_{R^{6N}} d^{3N}q \, d^{3N}p \, (f(p,q)) h_{\eta(H)}(p,q), \tag{8.26}$$

where the function $h_{\eta(H)}(q,p)$ satisfies the relationship

$$\{h_{\eta(H)}, L\} = \sum_{i=1}^{3N} \left(\frac{\partial h_\eta}{\partial q_i} \frac{\partial H}{\partial p_i} - \frac{\partial H}{\partial q_i} \frac{\partial h_\eta}{\partial p_i} \right) = 0, \tag{8.27}$$

or equivalently $h_{\eta(H)}(q,p)$ is a "smooth" function of the Hamiltonian function $H(q,p)$, by imposing the additive Boltzmann behavior for $h_{\eta(H)}(q,p)$ namely, $h_{\eta(H_1+H_2)} = h_{\eta(H_1)} \cdot h_{\eta(H_2)}$, one obtains the famous Boltzmann weight as the (unique) mathematical output associated to the Ergodic theorem on classical statistical mechanics in the presence of a thermal reservoir.[4]

$$h_{\eta(H)}(q,p) = \frac{\exp\{-\beta H(q,p)\}}{\int d^{3N}q \, d^{3N}p \exp\{-\beta H(q,p)\}}, \tag{8.28}$$

with β a (positive) constant which is identified with the inverse macroscopic temperature of the combined system after evaluating the system internal energy in the equilibrium state. Note that $\|h_{\eta(H)}\|_{L^2} = 1$ since $\|\eta(f)\| = 1$.

A last remark should be made related to Eq. (8.28). In order to obtain this result one should consider the nonzero value in ergodic limit

$$\lim_{T \to \infty} \frac{1}{T} \int_0^T dt (U_t \psi)(q,p) h_{\eta(H)}(q,p) = \langle h_{\eta(H)}, \mathbb{P}_{\ker(-i\bar{L})}(\psi) \rangle, \tag{8.29}$$

or by a point-wise argument (for every t)

$$(U_{-t} h_{\eta(H)}) \in \mathbb{P}_{\ker(-i\bar{L})}, \tag{8.30}$$

that is

$$h_{\eta(H)} \in \mathbb{P}_{\ker(-i\bar{L})} \Leftrightarrow Lh = 0. \tag{8.31}$$

8.4. On the Invariant Ergodic Functional Measure for Some Nonlinear Wave Equations

Let us start this section by considering the discreticized (N-particle) wave motion Hamiltonian associated to a nonlinear polynomial wave equation related to the vibration of a one-dimensional string of length $2a$

$$H(p_i, q_i) = \sum_{l=1}^{N} \left(\frac{2a}{N} \right) \left(\frac{p_i^2}{2} + \frac{1}{2}(q_{i+1} - q_i)^2 + g(q_i)^{2k} \right), \qquad (8.32)$$

with the initial one boundary Dirichlet conditions

$$q_i(0) = q_i,$$

$$p_i(0) = p_i,$$

$$q_i(-a) = q_i(a) = 0. \qquad (8.33)$$

Here k denotes a positive integer associated to the nonlinearity power and $g > 0$ the nonlinearity positive coupling constant.

Since one can argue that the above Hamiltonian dynamical system *possess unique global solutions on the time interval* $t \in [0, \infty)$ one can, thus, straightforwardly write the associated invariant (Ergodic) measure on the basis of Ergodic theorem exposed in Sec. 8.2 restricted to the configuration space after integrating out the term involving the canonical momenta

$$\lim_{T \to \infty} \frac{1}{T} \int_0^T dt F(q_1(t), \ldots, q_N(t)) = \int_{R^N} F(q_1, \ldots, q_N) d^{(inv)} \mu(q_1, \ldots, q_N) \qquad (8.34)$$

here the explicit expression for the Hamiltonian system invariant measure in configuration space is given by [with $\beta = 1$]

$$d_i^{(inv)} \mu(q_1, \ldots, q_N)$$

$$= \frac{1}{Z_n^{(0)}} \exp \left\{ -\frac{1}{2} \sum_{i=1}^{N} \left(\frac{2a}{N} \right) \left[\left(\frac{(q_{i+1} - q_1)^2}{a^2} \right) + \frac{g}{2k}(q_i)^{2k} \right] \right\}$$

$$\times (dq_1 \cdots dq_N), \qquad (8.35a)$$

$$Z_n^{(0)} = \int_{R^N} d^{(inv)} \mu(q_1, \ldots, q_N). \qquad (8.35b)$$

Now it is a nontrivial result on the theory of integration on infinite-dimensional spaces that the cylindrical measures (normalized to unity !) converges in a weak (star) topology sense to the well-defined nonlinear polynomial Wiener measure at the continuum limit $N \to \infty$, or $\sum_{l=1}^{N}() \to \int_{-L}^{L} dx()$ and $q_i(t) \to U(x,t)$ (see Appendix B for the relevant mathematical arguments supporting the mathematical existence of such limit).

As a consequence of Eqs. (8.35) at the continuum limit, we have the infinite-dimensional analogous of Ergodic result for the nonlinear polynomial wave equation (with the Boltzmann constant equal to unity — see Eq. (8.28) — Sec. 8.3 of this study)

$$\lim_{T \to \infty} \frac{1}{T} \int_0^T F(U(x,t))dt = \int_{C^\alpha(-a,a)} d^{\text{Wiener}}\mu[U(\sigma)] \times F(U(\sigma))$$

$$\times \exp\left\{-\frac{g}{2k} \int_{-a}^{a} dx(U(\sigma))^{2k}\right\}, \qquad (8.36)$$

where the wave field $U(x,t) \in L^\infty([0,\infty), L^{2k-1}((-a,a)))$ satisfies in a generalized weak sense the nonlinear wave equation below (see Appendix A):

$$\frac{\partial^2 U(x,t)}{\partial^2 t} = \frac{\partial^2 U(x,t)}{\partial^2 x} + g(U(x,t))^{2k-1}, \qquad (8.37)$$

$$U(x,0) = U^{(0)}(x) \in H^1((-a,a)), \qquad (8.38)$$

$$U_t(x,0) = 0, \qquad (8.39)$$

$$U(-a,t) = U(a,t) = 0, \qquad (8.40)$$

$d^{\text{Wiener}}\mu[U^{(0)}(x)]$ denotes the Wiener measure over the unidimensional Brownian field end-points trajectories $\{U^{(0)}(x)|$ with $U^{(0)}(-a) = U^{(0)}(a) = 0\}$, and $F(U^{(0)}(x))$ denotes any Wiener-integrable functional.

Note that $C^\alpha(-a,a)$ denotes the Banach space of the α-Holder continuous function (for $\alpha \leq \frac{1}{2}$) which is obviously the support of the Wiener measure $d^{\text{Wiener}}\mu[U^{(0)}(x)]$.

It is worth to remark the normalization factor Z of the nonlinear Wiener measure implicitly used on Eq. (8.36).

$$\frac{1}{Z} \int_{C^\alpha(-a,a)} d^{\text{Wiener}}\mu[U(\sigma)] \exp\left\{-\frac{g}{2k} \int_{-a}^{a} dx(U(\sigma))^{2k}\right\} = 1. \qquad (8.41)$$

Let us pass to the important case of existence of a dissipation term on the wave equation problem Eqs. (8.37)–(8.40) with ν the positive viscosity parameter and leading straightforwardly to solutions on the whole time propagation interval $[0, \infty)$:

$$\frac{\partial^2 U(x,t)}{\partial^2 t} = \frac{\partial^2 U(x,t)}{\partial^2 x} - \nu \frac{\partial U(x,t)}{\partial t} + g(U(x,t))^{2k-1}, \quad (8.42)$$

$$U(x,0) = U^{(0)}(x) \in H^1((-a,a)), \quad (8.43)$$

$$U_t(x,0) = U^{(1)}(x) \in L^2((-a,a)). \quad (8.44)$$

It is a simple observation that there is a bijective correspondence between the solution of Eq. (8.42) with the Klein–Gordon like wave equation for the rescaled wave field $\beta(x,t) = e^{\frac{\nu}{2}t}U(x,t)$, namely

$$\frac{\partial^2 \beta(x,t)}{\partial^2 t} - \frac{\partial^2 \beta(x,t)}{\partial^2 x} = \left(\frac{\nu^2}{4}\right)\beta(x,t) + ge^{-\frac{(2k-1)\nu t}{2}}(\beta(x,t))^{2k-1}, \quad (8.45)$$

$$\beta(x,0) = U^{(0)}(x) = H^1((-a,a)), \quad (8.46)$$

$$\beta_t(x,0) = U^{(1)}(x) - \frac{\nu}{2}U^{(0)}(x) = 0, \quad (8.47)$$

$$\beta(-a,t) = \beta(a,t) = 0. \quad (8.48)$$

As a result we have the analogous of the Ergodic theorem applied to Eq. (8.45) and a sort of Caldirola–Kanai action is obtained,[5] a new result of ours in this nonlinear dissipative case (with the path-integral identification $x \to \sigma$; $U^{(0)}(x) = X(\sigma)$)

$$\lim_{T \to \infty} \frac{1}{T} \int_0^T dt F(e^{+\frac{\nu}{2}t}U(x,t)) = \int_{C^{\frac{1}{2}}(-a,a)} d^{\text{Wiener}}\mu[X(\sigma)] F(X(\sigma))$$

$$\times \exp\left\{-\frac{g}{2k}\int_{-a}^a d\sigma e^{-\frac{(2k-1)\nu\sigma}{2}} \times (X(\sigma))^{2k}\right\}$$

$$\times \exp\left\{\left(\frac{\nu^2}{4}\right)\int_{-a}^a d\sigma(X(\sigma))^2\right\}. \quad (8.49)$$

Concerning the higher-dimensional case, let us consider A a strongly positive elliptic operator of order $2m$ associated to the free vibration of a domain Ω in a general space R^ν, and satisfying the Garding coerciviness condition

$$A = \sum_{\substack{|\alpha| \leq m \\ |\beta| \leq m}} (-1)^{|\alpha|} D_x^\alpha(a_{\alpha\beta}(x)) D_x^\beta, \quad (8.50)$$

with the Sobolev spaces operator's domain

$$D(A) = H^{2m}(\Omega) \cap H_0^m(\Omega). \tag{8.51}$$

Note there exists a constant $C_0(\Omega) > 0(C_0(\Omega) \to 0$ if vol $(\Omega) \to \infty)$ such that the Garding coerciviness condition holds true on $D(A)$

$$\text{Real}(AU, U)_{L^2(\Omega)} \geq C_0(\Omega)\|f\|_{H^{2m}(\Omega)}. \tag{8.52}$$

We can thus associate a Lipschitz nonlinear external vibration on the domain Ω as governed by the wave equation below

$$\frac{\partial^2 U(x,t)}{\partial^2 t} = (AU)(x,t) + G'(U(x,t)), \tag{8.53}$$

$$U(x,0) = U^{(0)}(x) \in H^{2m}(\Omega) \cap H_0^m(\Omega), \tag{8.54}$$

$$U_t(x,0) = g(x) \in L^2(\Omega). \tag{8.55}$$

Here $G(z)$ denotes a differentiable Lipschitizian function, like $G(z) = A\sin(Bz), e^{-\gamma z}$, etc. and thus, leading to the uniqueness and existence of global weak solutions on $C([0,\infty), H^{2m}(\Omega) \cap H_0^m(\Omega))$ by the fixed point theorem.[6]

Analogously to the discretization/N-particle technique of the $1+1$ case, one obtains as a mathematical candidate for the Ergodic invariant measure, the following rigorously measure on the Hilbert space of functions $H^{2m}(\Omega)$ (with $\beta = 1$):

$$d^{(\text{inv})}\mu(\phi(x)) = d_A\mu(\phi(x)) \times e^{-\frac{1}{2}\int_\Omega d^\nu x \, G(\phi(x))}. \tag{8.56}$$

Here $d_A\mu(\phi(x))$ denotes the Gaussian measure generated by the quadratic form associated to the operator A on its domain $D(A)$ for real initial conditions functions $\phi(x)$:

$$\int_{L^2(\Omega)} d_A\mu(\phi(x))(\phi(x_1)\phi(x_2)) = (A^{-1})(x_1, x_2). \tag{8.57}$$

Note that the Green function of the operator A belongs to the trace class operators on $H^{2m}(\Omega)$, which results on Eq. (8.57) through the application of the Minlo's theorem.[7]

At this point, we remark that $d_{\text{inv}}\mu(\phi(x))$ defines a mathematical rigorous Radon measure in $H^{2m}(\Omega)$ since the function (functional):

$$F(\phi(x)) = \exp\left\{-\frac{1}{2}\int_\Omega d^\nu x G(\phi(x))\right\}, \qquad (8.58)$$

belongs to $L^1(L^2(\Omega), d_A\mu(\phi(x)))$ due to the Lipschitz property of the nonlinearity $G(z)$, i.e. there exist constants c^+ and c^-, such that

$$c^-\phi(x) \le G(\phi(x)) \le c^+\phi(x). \qquad (8.59)$$

As a consequence, we have the obvious estimate below

$$e^{-c^+\int_\Omega d^n u x\phi(x)} \le e^{-\frac{1}{2}\int_\Omega d^\nu x G(\phi(x))} \le e^{-c^-\int_\Omega d^\nu x\phi(x)}, \qquad (8.60)$$

so by the Lebesgue comparison theorem, we get the functional integrability of the nonlinearity term of the Ergodic invariant measure Eq. (8.56):

$$\left|\int_{L^2(\Omega)} d_A\mu(\phi(x))e^{-\frac{1}{2}\int_\Omega d^\nu x G(\phi(x))}\right| \le e^{+\frac{1}{2}(c^-+c^+)^2 \text{Tr}_{H^{2m}(\Omega)}(A^{-1})} < \infty,$$
$$(8.61a)$$

and leading thus to the mathematical validity of Eq. (8.56).

At this point, it is worth to remark that one can easily generalize the type of our allowed Lipschitz nonlinearity for those of the following form

$$\left(\int_\Omega d^\nu x G(\phi(x))\right) \ge \left\{+\int_\Omega d^\nu x \frac{\gamma(x)}{2}(\phi)^2(x) + \int_\Omega d^\nu x C(x)\phi(x)\right\}, \qquad (8.61b)$$

with $\gamma(x)$ and $C(x)$ real positive $L^\infty(\Omega)$ functions.

This claim is a result of the straightforward Gaussian-infinite-dimensional integration

$$\left|\int_{H^{2m}(\Omega)} d_A\mu(\phi)\exp\left\{-\int_\Omega d^\nu x \frac{\gamma(x)}{2}(\phi\bar\phi)(x) + \int_\Omega d^\nu C(x)\bar\phi_{(x)}\right\}\right|$$

$$\le \left|\det^{-\frac{1}{2}}\left[1 + \int d^\nu y A^{-1}(x-y)\cdot\frac{\gamma(y)}{2}\right]\right.$$

$$\left.\times \exp\left\{+\frac{1}{2}\int_\Omega d^\nu x \int_\Omega d^\nu y C(x)\cdot A^{-1}(x,y)C(y)\right\}\right| < \infty. \quad (8.61c)$$

In the usual (Euclidean) Feynman path-integral formal notation, the Ergodic theorem takes the physicist's form after reintroducing the reservoir temperature parameter $\beta = 1/kT$

$$\lim_{T \to \infty} \frac{1}{T} \int_0^T F(U(x,t)) = \frac{1}{Z} \int_{L^2(\Omega)} D^F[\varphi(x)] e^{-\frac{1}{2}\beta \int_\Omega d^\nu x (\varphi(x) \cdot (A\varphi(x)))}$$
$$\times F(\varphi(x)) e^{-\beta \int_\Omega d^\nu x G(\varphi(x))}, \tag{8.62}$$

with the normalization factor

$$Z = \int_{H^{2m}(\Omega)} D^F[\varphi(x)] e^{\frac{1}{2}\beta \int_\Omega (\varphi(x)(A\varphi)(x)) d^\nu x} e^{-\beta \int_\Omega d^\nu x G(\varphi(x))}. \tag{8.63}$$

The functionals $F(\varphi(x))$ are objects belonging to the space $L^1(L^2(\Omega), d_A(\varphi(x))) \subset C(L^2(\Omega), R)$.

At this point, let us remark that by a direct application of the mean value theorem there is a sequence of growing times $\{t_n\}_{n \in \mathbb{Z}}$ (with $t_n \to \infty$) such that

$$\lim_{n \to \infty} F(\phi(x, t_n)) = \frac{1}{2} \int_{L^2(\Omega)} D^F[\varphi(x)] e^{-\frac{1}{2}\beta \int_\Omega d^\nu x (\varphi \cdot A\varphi)(x)}$$
$$\times F(\varphi(x)) e^{-\beta \int_\Omega d^\nu x G(\varphi(x))}. \tag{8.64}$$

It is worth to recall that the (positive) parameter β is not determined directly from the parameters of the underlying mechanical system and can be thought of as a remnant of the initial conditions "averaged out" by the Ergodic integral (see Sec. 8.3) and sometimes related to the presence of an intrinsic dissipation mechanism involved on the Ergodic-time average process responsible for the extensivity of the system's thermodynamics (see Eq. (8.28) — Sec. 8.3). For instance, at the limit of vanishing temperature (absolute Newtonian–Boltzmann–Maxwell zero temperature $\beta \to \infty$ and somewhat related to the so called Ergodic mixing microconomical ensemble), one can see the appearance of an "atractor" on the space $H^{2m}(\Omega) \cap H_0^m(\Omega)$ and exactly given by the stationary strong unique solution of the wave equation (8.53) since, we always have a soluble elliptic problem on $H^{2m}(\Omega) \cap H_0^m(\Omega)$ by means of the Lax–Milgram–Gelfand theorem.[6]

$$(A\varphi^*)(x) = G'(\varphi^*(x)) \tag{8.65}$$

$$\varphi^*(x)|_{\partial\Omega} = 0. \tag{8.66}$$

Note that the stationary strong solution $\varphi^*(x)$ is not a real atractor in the sense of the theory of dynamical systems, since we can only say that there is a specific sequence of growing $\{t_n\}_{n\in\mathbb{Z}}$, with $t_n \to \infty$, such that at the leading asymptotic limit $\beta \to \infty$ we have the time-infinite limit for every solution of Eqs. (8.65) and (8.66):

$$\lim_{t_n\to\infty} F(\varphi(x,t_n)) = F(\varphi^*(x)). \tag{8.67}$$

A (somewhat) formal path-integral calculation of the next term on the result (Eq. (8.41)) for $F(z)$ being smooth real-valued functions can be implemented by means of the usual path-integrals Saddle-point techniques applied to the (normalized) functional integral equation (8.62).[8]

$$\lim_{t_n\to\infty} F(\varphi(x,t_n))$$

$$= F(\varphi^*(x)) + \left\{ \frac{1}{\beta}\frac{\delta^2 F}{\delta^2\varphi}(\varphi^*(x)) \right.$$

$$\left. \times \frac{d}{d\alpha}\left\{ \lg\det\left[1 + A^{-1}\left(\frac{\delta^2 G}{\delta^2\varphi}(\varphi^*) + \alpha\right)\right]\right\}\Big|_{\alpha=0}\right\} + O(\beta^{-2}) \tag{8.68}$$

$$= F(\varphi^*(x)) + \frac{1}{\beta}\left\{ \frac{\delta^2 F}{\delta^2\varphi}(\varphi^*(x)) \right.$$

$$\left. \times \mathrm{Tr}_{H^{2m}(\Omega)}\left(\left[1 + A^{-1}\left(\frac{\delta^2 G}{\delta^2\varphi}(\varphi^*)\right)\right]^{-1}\right)\right\} + O(\beta^{-2})$$

$$= F(\varphi^*(x)) + \frac{1}{\beta}\left\{ \frac{\delta^2 F}{\delta^2\varphi}(\varphi^*(x)) \right.$$

$$\left. \times \sum_{n=0}^{\infty}\left[(-1)^n\mathrm{Tr}_{H^{2m}(\Omega)}\left(A^{-1}\left(\frac{\delta^2 G}{\delta^2\varphi}(\varphi^*)\right)\right)\right]^n\right\} + O(\beta^{-2}) \tag{8.69}$$

Another important remark to be made is related to the existence of time-independent simple constraints on the discreticized Hamiltonian equation (8.32), namely

$$H^{(a)}(q_1,\ldots,q_N) = 0; \quad a = 1,\ldots,n. \tag{8.70}$$

In this case, one should consider that the continuum version of Eq. (8.70) leads to (strongly) continuos functionals on $H^{2m}(\Omega) \cap H_0^m(\Omega)$:

$$H^{(a)}(\phi(x,t)) = 0; \quad a = 1,\ldots,n. \tag{8.71}$$

They have a direct consequence of restricting the wave motion to a closed nonlinear manifold of the original vector Hilbert space $H^{2m}(\Omega) \cap H_0^m(\Omega)$.

Now it is straightforward to apply the Ergodic theorem on the effective mechanical configuration space and obtain, thus, as the Ergodic invariant measure the constraint path-integral.

$$
\lim_{T \to \infty} \frac{1}{T} \int_0^T dt\, F(\phi(x,t))
$$

$$
= \frac{1}{Z} \int_{H^{2m}(\Omega)} D^F[\varphi(x)] e^{-\frac{1}{2}\beta \int_\Omega d^\nu x (\varphi(x) \cdot A\varphi(x))} e^{-\beta \int_\Omega d^\nu x G(\varphi(x))}
$$

$$
\times F(\varphi(x)) \left(\prod_{a=1}^n \delta^{(F)}(H^{(a)}(\varphi(x))) \right), \tag{8.72}
$$

where $\delta^{(F)}(\cdot)$ denotes the usual formal delta-functionals.

8.5. An Ergodic Theorem in Banach Spaces and Applications to Stochastic–Langevin Dynamical Systems

In this complementary Sec. 8.5, we intend to present our approach to study long-time/Ergodic behavior of infinite-dimensional dynamical systems by analyzing the somewhat formal diffusion equation with polynomial terms and driven by a white noise stirring.[4]

In order to implement such studies, let us present the author's generalization of the RAGE theorem for a contraction self-adjoint semi-group $T(t)$ on a Banach space X. We have, thus, the following theorem.

Theorem 8.3. *Let $f \in X$. We have the Ergodic generalized theorem*

$$
\lim_{T \to \infty} \frac{1}{T} \int_0^T dt (e^{-tA} f) = \mathbb{P}_{\ker(A)}(f), \tag{8.73}
$$

where A is the infinitesimal generator of $T(t)$, supposed to be self-adjoint.

Let us sketch the proof of the above claimed theorem of ours.

As a first step, one should consider Eq. (8.73) rewritten in terms of the "resolvent operator of A" by means of a Laplace transform (the

Hile–Yosida–Dunford spectral calculus)

$$\lim_{T \to \infty} \frac{1}{T} \int_0^T dt \left\{ \frac{1}{2\pi i} \int_{-i\infty}^{i\infty} dz e^{zt} ((z+A)^{-1} f) \right\}. \qquad (8.74)$$

Now it is straightforward to apply the Fubbini theorem to exchange the order of integrations (dt, dz) in Eq. (8.74) and get, thus, the result

$$\lim_{T \to \infty} \frac{1}{T} \int_0^T dt (e^{-tA} f) = \lim_{z \to 0^-} + ((z+A)^{-1} f). \qquad (8.75)$$

The $z \to 0^-$ limit of the integral of Eq. (8.75) (since Real $(z) \subset \rho(A) \subset (-\infty, 0)$) can be evaluated by means of saddle point techniques applied to Laplace's transforms. We have the following result

$$\lim_{z \to 0^-} ((z+A)^{-1} f) = \lim_{z \to 0^-} \int_0^\infty e^{-zt} (e^{-tA} f) \, dt$$

$$= \lim_{t \to \infty} (e^{-tA} f) = \mathbb{P}_{\ker(A)}(f). \qquad (8.76)$$

Let us apply the above theorem to the Lengevin equation. Let us consider the Fokker–Planck equation associated to the following Langevin equation:

$$\frac{dq^i}{dt}(t) = -\frac{\partial V}{\partial q_i}(q^j(t)) + \eta^i(t), \qquad (8.77)$$

where $\{\eta^i(t)\}$ denotes a white-noise stochastic time process representing the thermal coupling of our mechanical system with a thermal reservoir at temperature T. Its two-point function is given by the "fluctuation–dissipation" theorem

$$\langle \eta^i(t) \eta^j(t') \rangle = kT\delta(t - t'). \qquad (8.78)$$

The Fokker–Planck equation associated to Eq. (8.74) has the following explicit form

$$\frac{\partial P}{\partial t}(q^i, \bar{q}^i; t) = +(kT)\Delta_{q_i} P(q^i, \bar{q}^i; t) + \nabla_{q_i}(\nabla_{q_i} V \cdot P(q^i, \bar{q}^i; t)) \qquad (8.79)$$

$$\lim_{t \to 0^+} P(q^i, \bar{q}^i; t) = \delta^{(N)}(q^i - \bar{q}^i). \qquad (8.80)$$

By noting that we can associate a contractive semi-group to the initial value problem Eq. (8.80) in the Banach space $L^1(R^{3N})$, namely:

$$\int P(q^i, \bar{q}^i, t) f(\bar{q}^i) d\bar{q}^i = (e^{-tA} f). \qquad (8.81)$$

Here, the closed positive accretive operator A is given explicitly by

$$-A(\cdot) = kT\Delta_{q_i}(\cdot) + \nabla_{q_i}[(\nabla_{q_i}V)(\cdot)], \qquad (8.82)$$

and acts first on $X = C_\infty(R^{3N})$. It is instructive to point out that the perturbation accretive operator $B = \nabla_{qi} \cdot [(\nabla_{qi}V) \cdot (\cdot)]$ on $C_\infty(R^{3N})$ with $V(q^i) \in C_c^\infty(R^{3N})$ is such that it satisfies the estimate on $S(R^{3N})$: $\|Bf\|_{S(R^{3N})} \leq a\|\Delta f\|_{S(R^{3N})} + b\|f\|_{S(R^{3N})}$ with $a > 1$ and for some b. As a consequence $A(\cdot) = kT\Delta_{qi}(\cdot) + (B(\cdot))$ generates a contractive semi-group on $C_\infty(R^{3N})$ or by an extension argument on the whole $L^1(R^{3N})$ since the L^1-closure of $C_\infty(R^{3N})$ is the Banach space $L^1(R^{3N})$.

At this point, we may apply our Theorem 8.3 to obtain the Langevin–Brownian Ergodic theorem applied to our Fokker–Planck equation

$$\lim_{T\to\infty} \frac{1}{T} \int_0^T dt \left(\int_{R^{3N}} dq(e^{-tA}f)(q) \right) = \mathbb{P}_{\ker(A)}(f)(q)$$

$$= \int_{R^N} d^N \vec{q}^i f(\vec{q}^i)\mathbb{P}^{eq}(\vec{q}^i), \qquad (8.83)$$

where the equilibrium probability distribution is given explicitly by the unique normalizable element of the closed sub-space $\ker(A)$.

$$O = +(kT)\Delta_{q_i}\mathbb{P}^{eq}(\vec{q}^i) + \nabla_{\vec{q}^i}[(\nabla_{\vec{q}^i}V)\mathbb{P}(\vec{q}^i)] \qquad (8.84)$$

or exactly, we have the Boltzmann's weight for our equilibrium Langevin–Brownian probability distribution

$$\mathbb{P}^{eq}(\vec{q}^i) = \exp\left\{ -\frac{1}{kT}(V(\vec{q}^i)) \right\}. \qquad (8.85)$$

For the general Langevin equation in the complete phase space $\{q_i, p_i\}$ as in the bulk of this note, one should reobtain the complete Boltzmann statistical weight as the equilibrium ergodic probability distribution

$$\mathbb{P}^{eq}(\vec{q}^i, \vec{p}^i) = \frac{1}{Z} \exp\left\{ -\frac{1}{kT}\left[\sum_{l=1}^N \frac{1}{2}(\vec{p}^i)^2 + V(\vec{q}) \right] \right\}, \qquad (8.86)$$

with the normalization factor

$$Z = \int_{R^N} d\vec{p}^i \int_{R^N} d\vec{q}^i \mathbb{P}^{eq}(\vec{q}^i, \vec{p}^i). \qquad (8.87)$$

Let us apply the above exposed result for the following nonlinear (polynomial) stochastic diffusion equation on a one-dimensional domain $\Omega = (-a, a)$ with all positive coefficients $\{\lambda_j\}$ on the nonlinear polynomial term

$$\frac{\partial U(x,t)}{\partial t} = +\frac{1}{2}\frac{d^2 U(x,t)}{d^2 x} - \left(\sum_{\substack{j=0 \\ j=\text{odd}}}^{2k-1} \lambda_j U^j \right)(x,t) + \eta(x,t), \quad (8.88)$$

$$U(x,t)|_{\partial\Omega} = 0, \quad (8.89)$$

$$U(x,0) = f(x) \in L^2(\Omega). \quad (8.90)$$

Here, $\{\eta(x,t)\}$ are samplings of a white-noise stirring.

In this case, one can show by standard techniques that the global solution $U(x,t) \in C([0,\infty), L^2(-a,a))$ and by using the discreticized approach of Sec. 8.4, the invariant Ergodic measure associated to the nonlinear diffusion equation is given by the mathematically functional equilibrium Langevin–Brownian probability distribution

$$P^{\text{eq}}(U(x)) = \exp\left\{ -\int_{-a}^{a} dx \left(\sum_{\substack{j=0 \\ j\,\text{even}}}^{2k} \frac{\lambda_j}{j} U^j(x) \right) \right\} \times d_{-\frac{1}{2}\frac{d^2}{dx^2}} \mu(U(x)).$$

$$(8.91)$$

Here, $d_{-\frac{1}{2}\frac{d^2}{dx^2}} \mu(U(x))$ denotes the Gaussian functional measure (normalized to unity) on the Wiener space $C^\alpha(-a,a)$. In the Feynman path-integral notation (with $\beta = 1$)

$$\lim_{T\to\infty} \frac{1}{T}\int_0^T dt\, F(U(x,t)) = \int_{C^\alpha((-a,a))} D^F[U(x)] e^{-\frac{1}{2}\int_{-a}^{a}\left|\frac{dU}{dx}\right|^2(x)}$$

$$\times e^{-\left[\int_{-a}^{a} dx \left(\sum_{\substack{j=0 \\ j=\text{even}}}^{2k} \frac{\lambda_j}{j} U^j(x) \right) \right]} \times F[U(x)].$$

$$(8.92)$$

Note that Eq. (8.91) defines a rigorous mathematical object since it does not need to be renormalized from a calculational point of view on the Wiener space $C^\alpha(-a,a)$[10] as its nonlinear exponential term is bounded by the unity (Lebesgue theorem). It is worth to point out that for nonlinearities of Lipschitz type on general domains $\Omega C R^\nu$, one can see that the support of the Gaussian measure is readily $L^2(\Omega)$ and the equilibrium Langevin–Brownian probability distribution

$$P^{\mathrm{eq}}(U(x)) = \exp\left\{-\int_\Omega d^D x G(U(x))\right\} \times d_{(-\frac{1}{2}\Delta)}\mu, (U(x)), \qquad (8.93)$$

is a perfect well-defined random measure on $C^\alpha((-a,a))$ by the same Lebesgue convergence theorem.

8.6. The Existence and Uniqueness Results for Some Nonlinear Wave Motions in 2D

In this technical section, we give an argument for the global existence and uniqueness solution of the Hamiltonian motion equations associated first to Eq. (8.32) — Sec. 8.3 and by secondly to Eq. (8.53) — Sec. 8.3. Related to the two-dimensional case, let us equivalently show the weak existence and uniqueness of the associated continuum nonlinear polynomial wave equation in the domain $(-a,a) \times R^+$.

$$\frac{\partial^2 U(x,t)}{\partial^2 t} - \frac{\partial^2 U(x,t)}{\partial^2 x} + g(U(x,t))^{2k-1} = 0, \qquad (8.94)$$

$$U(-a,t) = U(a,t) = 0,$$
$$U(x,0) = U_0(x) \in H^1([-a,a]), \qquad (8.95)$$

$$U_t(x,0) = U_1(x) \in L^2([-a,a]). \qquad (8.96)$$

Let us consider the Galerkin approximants functions to Eqs. (8.95) and (8.96) as

$$\bar{U}_n(t) \equiv \sum_{\ell=1}^{n} U_\ell(t) \sin\left(\frac{\ell\pi}{a}x\right). \qquad (8.97)$$

Since there exists a $\gamma_0(a)$ positive such that

$$\left(-\frac{d^2}{d^2 x}\bar{U}_n, \bar{U}_n\right)_{L^2([-a,a])} \geq \gamma_0(a)(\bar{U}_n, \bar{U}_n)_{H^1([-a,a])}, \qquad (8.98)$$

we have the *a priori* estimate for any t

$$0 \leq \varphi(t) \leq \varphi(0), \qquad (8.99)$$

with

$$\varphi(t) = \frac{1}{2}\|\dot{\bar{U}}_n(t)\|_{L^2}^2 + \gamma_0(a)\|\bar{U}_n\|_{H^1}^2 + \frac{1}{2k}\|\bar{U}_n\|_{L^{2k}}^{2k}. \tag{8.100}$$

As a consequence of the bound equation (8.100), we get the bounds for any given T (with A_i constants)

$$\sup_{0\le t\le T} \mathrm{ess}\|\bar{U}_n\|_{H^1(-a,a)}^2 \le A_1, \tag{8.101}$$

$$\sup_{0\le t\le T} \mathrm{ess}\|\dot{\bar{U}}_n\|_{L^2(-a,a)}^2 \le A_2, \tag{8.102}$$

$$\sup_{0\le t\le T} \mathrm{ess}\|\bar{U}_n\|_{L^{2k}(-a,a)}^{2k} \le A_3. \tag{8.103}$$

By usual functional-analytical theorems on weak-compactness on Banach–Hilbert spaces, one obtains that there exists a sub-sequence $\bar{U}_n(t)$ such that for any finite T

$$\bar{U}_n(t)\overrightarrow{(\text{weak} - \text{star})}\bar{U}(t) \quad \text{in } L^\infty([0,T], H^1(-a,a)), \tag{8.104}$$

$$\dot{\bar{U}}_n(t)\overrightarrow{(\text{weak} - \text{star})}\bar{v}(t) \quad \text{in } L^\infty([0,T], L^2(-a,a)), \tag{8.105}$$

$$\bar{U}_n(t)\overrightarrow{(\text{weak} - \text{star})}\bar{p}(t) \quad \text{in } L^\infty([0,T], L^{2k}(-a,a)). \tag{8.106}$$

At this point, we observe that for any $p > 1$ (with $\tilde{A}_i$ constants) and $T < \infty$, we have the relationship

$$\int_0^T \|\bar{U}_n\|_{H^1(-a,a)}^p dt \le T(A_1)^{\frac{p}{2}} \Leftrightarrow \int_0^T \|\bar{U}_n\|_{L^2(-a,a)}^p \le \tilde{A}_1, \tag{8.107}$$

$$\int_0^T \|\dot{\bar{U}}_n\|_{L^2(-a,a)}^p dt \le T(A_2)^{\frac{p}{2}} \Leftrightarrow \int_0^T \|\dot{\bar{U}}\|_{L^{2k}(-a,a)}^p dt \le \tilde{A}_2, \tag{8.108}$$

since we have the continuous injection below

$$H^1(-a,a) \hookrightarrow L^2(-a,a), \tag{8.109}$$

$$L^{2k}(-a,a) \hookrightarrow L^2(-a,a). \tag{8.110}$$

As a consequence of the Aubin–Lion theorem,[6] one obtains straightforwardly the strong convergence on $L^P((0,T), L^2(-a,a))$ together with the almost everywhere point-wise equality among the solutions candidate

$$\bar{U}_n \to \bar{U}(t) = \bar{p}(t) = \int_0^t \bar{v}(s)ds. \tag{8.111}$$

By the Holder inequality applied to the pair (q,k)

$$\|U_n - \bar{U}\|_{L^q(-a,a)} \leq \|U_n - \bar{U}\|_{L^2(-a,a)}^{1-\theta} \|U_n - \bar{U}\|_{L^{2k}(-a,a)}^{\theta}, \tag{8.112}$$

with $0 \leq \theta \leq 1$

$$\frac{1}{q} = \frac{1-\theta}{2} + \frac{\theta}{2k}, \tag{8.113}$$

in particular with $q = 2k - 1$, one obtains the strong convergence of $U_n(t)$ in the general Banach space $L^\infty((0,T), L^{2k-1}(-a,a))$, with $\bar{\theta} = \frac{k}{1-k}(\frac{3-2k}{2k-1})$.

As a consequence of the above results, one can safely pass the weak limit on

$$C^\infty((0,T), L^2(-a,a)) \lim_{n \to \infty} \left\{ \frac{d^2}{d^2 t}(\bar{U}_n, v)_{L^2(-a,a)} \right.$$

$$+ \left(-\frac{d^2}{d^2 x}\bar{U}_n, v \right)_{L^2(-a,a)} + g(\bar{U}_n^{2k-1}, v)_{L^2(-a,a)} \right\}$$

$$= \frac{d^2}{d^2 t}(\bar{U}, v)_{L^2(-a,a)} + \left(-\frac{d^2}{d^2 x}\bar{U}, v \right)_{L^2(-a,a)} + g(\bar{U}^{2k-1}, v)_{L^2(-a,a)} = 0, \tag{8.114}$$

for any $v \in C^\infty((0,T), L^2(-a,a))$.

At this point, we shetch a rigorous argument to prove the problem's uniqueness.

Let us consider the hypothesis that the finite function

$$a(x,t) = \frac{((\bar{U})^{2k+1}(x,t) - (\bar{v})^{2k+1}(x,t))}{(\bar{U}(x,t) - \bar{v}(x,t))}, \tag{8.115}$$

is essentially bounded on the domain $[0,\infty) \times (-a,a)$ where $\bar{U}(x,t)$ and $\bar{v}(x,t)$ denote two hypothesized different solutions for the 2D-polynomial wave equations (8.94)–(8.96). It is straightforward to see that its difference

$W(x,t) = (\bar{U} - \bar{v})(x,t)$ satisfies the "Linear" wave equation problem

$$\frac{\partial^2 W}{\partial^2 t} - \frac{\partial^2 W}{\partial x} + (aW)(x,t) = 0, \tag{8.116}$$

$$W(0) = 0, \tag{8.117}$$

$$W_t(0) = 0. \tag{8.118}$$

At this point, we observe the estimate (where $H^1 \hookrightarrow L^2$!)

$$
\begin{aligned}
\frac{1}{2}\frac{d}{dt} &\left(\|\frac{d}{dt}W\|_{L^2(-a,a)}^2 + \|W\|_{H^1(-a,a)}^2 \right) \\
&\leq \|a\|_{L^\infty((0,T)(-a,a))} \left(\|W\|_{L^2(-a,a)} \times \left\|\frac{dW}{dt}\right\|_{L^2(-a,a)} \right) \\
&\leq M \left(\left\|\frac{dW}{dt}\right\|_{L^2(-a,a)}^2 + \|W\|_{L^2(-a,a)}^2 \right) \\
&\leq M \left(\left\|\frac{dW}{dt}\right\|_{L^2(-a,a)}^2 + \|W\|_{H^1(-a,a)}^2 \right),
\end{aligned} \tag{8.119}
$$

which after the application of the Gronwall's inequality give us that

$$
\begin{aligned}
\left(\left\|\frac{d}{dt}W\right\|_{L^2(-a,a)}^2 + \|W\|_{H^1(-a,a)}^2 \right)(t) \\
\leq \left(\left\|\frac{dW}{dt}\right\|_{L^2(-a,a)}^2 (0) + \|W\|_{H^1(-a,a)}^2 (0) \right) = 0,
\end{aligned} \tag{8.120}
$$

which proves the problem's uniqueness under the not proved yet hypothesis that in the two-dimensional case (at least for compact support infinite differentiable initial conditions)

$$\sup_{\substack{x \in (-a,a) \\ t \in [0,\infty)}} \|a(x,t)\| \leq M. \tag{8.121}$$

As a last comment on the $1 + 1$-polynomial nonlinear wave motion, we call the readers' attention that one can easily extend the technical results analyzed here to the complete polynomial nonlinearity

$$\frac{\partial^2 U}{\partial^2 t} - \frac{\partial^2 U}{\partial^2 x} = \sum_{\substack{j=1 \\ j,\text{odd}}}^{j<2k} c_j U^{2k-j} = C_1 U^{2k-1} + C_3 U^{2k-3} + \cdots, \tag{8.122}$$

with the set of couplings $\{C_j\}$ belonging to a positive set of real number. In this case, we have that the general solution belongs *a priori* to the space of functions $\bigcap_{\substack{j=1 \\ j<2k}} L^\infty((0,T), L^{2k-j}(-a,a))$.

References

1. W. Humzther, *Comm. Math. Phys.* **8**, 282–299 (1968).
2. V. Enss, *Comm. Math. Phys.* **61**, 285–291 (1978).
3. Ya. G. Sinai, *Topics in Ergodic Theory* (Princeton University Press, 1994).
4. A. B. Soussan and R. Teman, *J. Functional Anal.* **13**, 195 (1973); L. C. L. Botelho, *J. Math. Phys.* **42**, 1612 (2001); *Il Nuovo Cimento*, Vol. 117B(1), 15 (2002).
5. L. C. L. Botelho, *Phys. Rev.* E **58**, 1141 (1998); R. Teman, *Infinite-Dimensional Systems in Mechanics and Physics*, Vol. 68 (Springer Verlag, 1988).
6. O. A. Ladyzhenskaya, *The Boundary Value Problems of Mathematical Physics*, Applied Math Sciences, Vol. 49 (Springer Verlag, 1985).
7. Y. Yamasaki, *Measures on Infinite-Dimensional Spaces, Series in Pure Mathematics*, Vol. 5, (World Scientific, 1985).
8. B. Simon, *Functional Integration and Quantum Physics* (Academic Press, 1979); J. Glimm and A. Jaffe, *Quantum Physics — A Functional Integral Point of View* (Springer Verlag, 1981).
9. V. Rivasseau, *From Perturbative to Constructive Renormalization* (Princeton University, 1991).
10. C. L. Siegel and J. K. Moser, *Lectures on Celestial Mechanics* (Springer-Verlag, Berlin, Heidelberg, New York, 1971).
11. S. Sternberg, *Lectures on Differential Geometry* (Chelsea Publishing Company, New York, 1983).
12. V. Guillemin and A. Pollack, *Differential Topology* (Prentice-Hall, Inc., Englewood Cliffs, NJ, 1974).

Appendix A. On Sequences of Random Measures on Functional Spaces

Let us consider a completely regular topological space X and the Banach space of continuous function space

$$C(X) = \left\{ f : X \to R, f \text{ continuous on the } X\text{-topology with } \|f\|_\infty = \sup_{x \in X} |f(x)| < \infty \right\}. \tag{A.1}$$

Associated to the Banach space $C(X)$, there exists its dual $\mathcal{M}(X)$ which is the set of random measures on X with norm given by its total variation $\langle\langle\mu\rangle\rangle = |\mu|(X)$, where $\mu \in \mathcal{M}(X)$.

It is easy to see that if $\beta(X)$ is the Stone-chech compactification of X, we have the isometric immersion $\mathcal{M}(X) \hookrightarrow \mathcal{M}(\beta(X))$.

We have, thus, the important theorem of Prokhorov on the accumulation points of sets of random measures.[7]

Theorem (Prokhorov). *If a family of measures $\{\mu_\alpha\}$ on $\mathcal{M}(X)$ is such that it satisfies the following conditions*

(a) Bounded uniformly

$$\langle\langle\mu_\alpha\rangle\rangle = |\mu_\alpha|(X) \le M. \tag{A.2}$$

(b) Uniformly regular — For any given $\varepsilon > 0$, there exists a compact set $K_{(\varepsilon)} \subset X$ such that

$$\left| \int_X f(x) d\mu_\alpha(x) \right| \le \varepsilon \|f\|_\infty,$$

for any $f(x) \in C(X)$ satisfying the condition

$$(supp\ f(x)) \cap K_{(\varepsilon)} = \{\phi\}. \tag{A.3}$$

We have that the set of random measures is a relatively compact set on the weak-star topology on $\mathcal{M}(X)$. Namely, for any $f(x)C(X)$, we have that there exists a measure $\nu \in \mathcal{M}(X)$ such that for a given sequence $\{\alpha_k\}$

$$\limsup_{\{\alpha_K\}} \int_X f(x) d\mu_{\alpha_K}(x) = \int_X f(x) d\nu(x). \tag{A.4}$$

In order to apply such theorem to the set of random measures as given by Eq. (8.35), we introduce the completely regular space formed by the one-point compactification of the real line

$$X = \prod_{i\in\mathbb{Z}} (\dot{R})_i, \quad \left(\text{or } X = \prod_{i\in\mathbb{Z}} \beta(R) \right), \tag{A.5}$$

and note the strict positivity of the integrand (the path-integral weight) on Eq. (8.35) and its boundedness by the Gaussian free term since $\exp\{-\frac{1}{2}\frac{g(\frac{2a}{N})}{(2k)}\sum_{l=1}^{N}(q_i)^{2k}\} < 1$ for $\{q_i\} \in \prod_{l=1}^{N}(\dot{R})_i$.

We note either the chain property of the invariant measure set ensuring us the measure uniqueness limit on Eq. (B.4) when applied to Eq. (8.35) — Sec. 8.4, namely:

$$\int_{R^N} |f(q_1, \ldots, q_{N-1})| d^{(\mathrm{inv})} \mu(q_1, \ldots, q_N)$$

$$\geq \int_{R^{N-1}} |f(q_1, \ldots, q_{N-1})| d^{(\mathrm{inv})} \mu(q_1, \ldots, q_{N-1}). \qquad (A.6)$$

Finally, we remark that the well-defined nonlinear Wiener functional measure coincides with those finite-dimensional as given by Eqs. (8.35a) and (8.35b) when it restricts to the set of continuous polygonal paths and, thus, ensuring the convergence of the measure set Eqs. (8.35a) and (8.35b) to the full Wiener path measure Eq. (8.36), since its functional support is contained on the set of continuous functions on the interval $[-a, a]$.

Appendix B. On the Existence of Periodic Orbits in a Class of Mechanical Hamiltonian Systems — An Elementary Mathematical Analysis

B.1. Elementary May Be Deep — T. Kato

One of the most fascinating study in classical astronomical mechanics is that one related to the inquiry on how energetic-positions configurations in a given mechanical dynamical system can lead to periodic systems trajectories for any given period T, a question of basic importance on the problem of the existence of planetary systems around stars in dynamical astronomy.[10]

In this short note, we intend to give a mathematical characterization of a class of N-particle Hamiltonian systems possessing intrinsically the above-mentioned property of periodic trajectories for any given period T, a result which is obtained by an elementary and illustrative application of the famous Brower fixed point theorem.

As a consequence of this mathematical result, we conjecture that the solution of the existence of periodic orbits in classical mechanics of N-body systems moving through gravitation should be searched with the present status of our mathematical knowlegdment.

Let us, thus, start our note by considering a Hamiltonian function $H(x^i, p^i) \equiv H(z^i) \in C^\infty(R^{6N})$ of a system of N-particles set on motion in R^3. Let us suppose that the Hessian matrix determinant associated to the given Hamiltonian is a positive-definite matrix in the $(6N-1)$-dimensional energy-constant phase space $z^i = (x^i, p^i)$ of our given mechanical system

$$\left\| \frac{\partial^2 H}{\partial z^k \partial z^s} \right\|_{H(z^s)=E} > 0. \tag{B.1a}$$

In the simple case of N-particles with unity mass and a positive two-body interaction with a potential $V(|x_i - x_j|^2)$, such that the set in R^n, $\{x^i \,|\, V(x^i) < E\}$ is a bounded set for every $E > 0$. Our condition Eq. (B.1a) takes a form involving only the configurations variables ($\{x^p\}_{p=1,\dots,N}; x^p \in R^3$). Namely,

$$\|M_{ks}(\{x^p\})\| > 0, \tag{B.1b}$$

with

$$M_{ks}(\{x^p\}) = \sum_{i<j}^{N}[V''(|x_i - x_j|^2)4(x_i - x_j)^2(\delta_{is} - \delta_{js})(\delta_{ik} - \delta_{jk})$$

$$+ 2V'(|x_i - x_j|^2)(\delta_{ik} - \delta_{jk})(\delta_{is} - \delta_{js}))]. \tag{B.1c}$$

The class of mechanical systems given by Eqs. (B.1b) and (B.1c) is such that the energy-constant hypersurface $H(z^s) = E$ has always a positive Gaussian curvature and — as a straightforward consequence of the Hadamard theorem — such energy-constant hypersurface is a convex set of R^{6N}, besides obviously being a compact set. As a result of the above-made remarks, we can see that our mechanical phase space is a convex-compact set of R^{6N}.

The equations of motion of the particles of our given Hamiltonian system is given by

$$\frac{dx^i(t)}{dt} = \frac{\partial H}{\partial p^i}(x^i(t), p^i(t)),$$

$$\frac{dp^i(t)}{dt} = -\frac{\partial H}{\partial x^i}(x^i(t), p^i(t)), \tag{B.2}$$

with initial conditions belonging to our phase space of constant energy $\mathcal{H}_E = \{(x^i, p^i); \mathcal{H}(x^i, p^i) = E\}$, namely:

$$(x^i(t_0), p^i(t_0)) \in \mathcal{H}_E. \tag{B.3}$$

The existence of periodic solutions to the system of ordinary differential equations Eqs. (B.2) and (B.3); for any given period $T \in R^+$ is a consequence of the fact that the Poincaré recurrent application associated to the (global) solutions Eq. (B.2) of the given mechanical system. Note that $H(x^i(t), p^i(t)) = E$

$$P: \mathcal{H}_E \to \mathcal{H}_E,$$
$$(x^i(t_0), p^i(t_0)) \to (x^i(T), p^i(T)), \tag{B.4}$$

has always a fixed point as a consequence of the application of the Brower fixed point theorem.[12] This means that there is always initials conditions in the phase space Eq. (B.3) that produces a periodic trajectory of any period T in our considered class of mechanical N-particle systems.

As a physical consequence, we can see that the existence of planetary systems around stars may not be a rare physical astronomical event, but just a consequence of the mathematical description of the nature laws and the fundamental theorems of differential topology and geometry in R^N.

Chapter 9

Some Comments on Sampling of Ergodic Process: An Ergodic Theorem and Turbulent Pressure Fluctuations*

9.1. Introduction

A basic problem in the applications of stochastic processes is the estimation of a signal $x(t)$ in the presence of an additive interference $f(t)$ (noise). The available information (data) is the sum $S(t) = x(t) + f(t)$, and the problem is to establish the presence of $x(t)$ or to estimate its form. The solution of this problem depends on the state of our prior knowledge concerning the noise statistics. One of the main results on the subject is the idea of maximize the output signal-to-noise ratio,[1] the matched filter system. One of the most important results on the subject is that if one knows a priori the signal in the frequency domain $X(w)$ and, most importantly, the frequency domain expression for the noise correlation statistics function $S_{ff}(w)$, one has, at least in the theoretical grounds, the exact expression for the optimum filter transference function $H_{\text{opt}}(w) = k(X^*(w))(S_{ff}(w))^{-1} \cdot e^{-iw\bar{t}}$, here $\bar{t}$ denotes certain time on the observation interval process $[-A, A]$, supposed to be finite here.

As a consequence, it is important to have estimators and analytical expressions for the correlation function for the noise, specially in the physical situation of finite-time duration noise observation. Another very important point to be remarked is that in most of the cases of observed noise (as in the turbulence research[2]); one should consider the noise (at least in the context of a first approximation) as an Ergodic random process.

In Sec. 9.2 we aim in this note to present (of a more mathematical oriented nature) a rigorous functional analytic proof of an Ergodic theorem stating the equality of time-averages and Ensemble-averages for wide-sense mean continuous stationary random processes. In Sec. 9.3 somewhat

---

*Author's original results.

of electrical engineering oriented nature, we present a new approach for sampling analysis of noise, which leads to canonical and invariant analytical expressions for the Ergodic noise correlation function $S_{ff}(w)$ already taking into account the finite time-observation parameter which explicitly appears in the structure formulae, a new result on the subject, since it does not require the existence of Nyquist critical frequency on the sampling rate, besides removing, in principle, the aliasing problem in the computer-numerical sampling evaluations. In Sec. 9.4, we present a mathematical-theoretical application of the above theoretical results for a model of turbulent pressure fluctuations.

9.2. A Rigorous Mathematical Proof of the Ergodic Theorem for Wide-Sense Stationary Stochastic Process

Let us start our section by considering a wide-sense mean continuous stationary real-valued process $\{X(t), -\infty < t < \infty\}$ in a probability space $\{\Omega, d\mu(\lambda), \lambda \in \Omega\}$. Here, Ω is the event space and $d\mu(\lambda)$ is the underlying probability measure.

It is well-known[1] that one can always represent the above-mentioned wide-sense stationary process by means of a unitary group on the Hilbert space $\{L^2(\Omega), d\mu(\lambda)\}$. Namely, in the quadratic-mean sense in engineering jargon

$$X(t) = U(t)X(0) = \int_{-\infty}^{+\infty} e^{iwt} d(E(w)X(0)) = e^{iHt}(X(0)), \qquad (9.1)$$

here we have used the famous spectral Stone-theorem to rewrite the associated time-translation unitary group in terms of the spectral process $dE(w)X(0)$, where H denotes the infinitesimal unitary group operator $U(t)$. We also supposed that the σ-algebra generated by the $X(t)$-process is the whole measure space Ω, and $X(t)$ is a separable process.

Let us, thus, consider the following linear continuous functional on the Hilbert (complete) space $\{L^2(\Omega), d\mu(\lambda)\}$ — the space of the square integrable random variables on Ω

$$L(Y(\lambda)) = \lim_{T \to \infty} \frac{1}{2T} \int_{-T}^{T} dt E\{Y(\lambda)\overline{X(t, \lambda)}\}. \qquad (9.2)$$

By a straightforward application of the RAGE theorem,[3] namely:

$$L(Y(\lambda)) = \int_\Omega d\mu(\lambda)Y(\lambda) \left\{ \lim_{T\to\infty} \frac{1}{2T} \int_{-T}^{T} dt e^{-iwt}\overline{dE(w)X_0(\lambda)} \right\}$$

$$= \lim_{T\to\infty} \frac{1}{2T} \int_{-T}^{T} E\{Ye^{-iHt}\bar{X}\}dt$$

$$= \int_\Omega d\mu(\lambda)Y(\lambda)\overline{dE(0)X(0,\lambda)}$$

$$= E\{Y(\lambda)\overline{P_{\mathrm{Ker}(H)}(X(0,\lambda))}\}. \tag{9.3}$$

where $P_{\mathrm{Ker}(H)}$ is the (orthogonal projection) on the kernel of the unitary-group infinitesimal generator H (see Eq. (9.1)).

By a straightforward application of the Riesz-representation theorem for linear functionals on Hilbert spaces, one can see that $\overline{P_{\mathrm{Ker}(H)}(X(0))}d\mu(\lambda)$ is the searched time-independent Ergodic-invariant measure associated to the Ergodic theorem statement, i.e. for any square integrable time-independent random variable $Y(\lambda) \in L^2(\Omega, d\mu(\lambda))$, we have the Ergodic result $(X(0,\lambda) = X(0)$

$$\lim_{T\to\infty} \frac{1}{2T} \int_{-T}^{T} dt E\{X(t)\bar{Y}\} = E\{P_{\mathrm{Ker}(H)}(X(0))\bar{Y}\}. \tag{9.4}$$

In general grounds, for any real bounded borelian function it is expected that the result (not proved here)

$$\lim_{T\to\infty} \frac{1}{2T} \int_{-T}^{T} dt E\{f(X(t))\bar{Y}\} = E\{P_{\mathrm{Ker}(H)}f(X(0)) \cdot \bar{Y}\}. \tag{9.5}$$

For the auto-correlation process function, we still have the result for the translated time ζ fixed (the lag time) as a direct consequence of Eq. (9.1) or the process' stationarity property

$$\lim_{T\to\infty} \frac{1}{2T} \int_{-T}^{T} dt E\{X(t)X(t+\zeta)\} = E\{X(0)X(\zeta)\}. \tag{9.6}$$

It is important to remark that we still have the probability average inside the Ergodic time-averages (Eqs. (9.4)–(9.6)). Let us call the readers' attention that in order to have the usual Ergodic like theorem result without the probability average E on the left-hand side of the formulae, we proceed (as it is usually done in probability text-books[1]) by analyzing the probability convergence of the single sample stochastic-variables below (for instance)

$$\eta_T = \frac{1}{2T} \int_{-T}^{T} f(X(t)) dt, \tag{9.7}$$

$$R_T(\zeta) = \frac{1}{2T} \int_{-T}^{T} dt X(t) X(t + \zeta). \tag{9.8}$$

It is straightforward to show that if $E\{f(X(t))f(X(t+\zeta))\}$ is a bounded function of the time-lag, or, if the variance written below goes to zero at $T \to \infty$

$$\sigma_T^2 = \lim_{T\to\infty} \frac{1}{4T^2} \int_{-T}^{T} dt_1 \int_{-T}^{T} dt_2 [E\{X(t_1)X(t_1 + \zeta)X(t_2)X(t_2 + \zeta))$$

$$- E\{X(t_1)X(t_1 + \zeta)\}E\{X(t_1)X(t_1 + \zeta))\}] = 0, \tag{9.9}$$

one has that the random variables as given by Eqs. (9.7) and (9.8) converge at $T \to \infty$ to the left-hand side of Eqs. (9.4)–(9.6) and producing thus an Ergodic theorem on the equality of ensemble-probability average of the wide sense stationary process $\{X(t), -\infty < t < \infty\}$ and any of its single-sample $\{\bar{X}(t), -\infty < t < +\infty\}$ time average

$$\lim_{T\to\infty} \frac{1}{2T} \int_{-T}^{T} dt f(\bar{X}(t)) = E\{P_{\mathrm{Ker}(H)}(f(X_0))\} = E\{f(X(t))\},$$

$$\tag{9.10}$$

$$\lim_{T\to\infty} \frac{1}{2T} \int_{-T}^{T} dt \bar{X}(t)\bar{X}(t + \zeta) = E\{X(0)X(\zeta)\}$$

$$= E\{X(t)X(t + \zeta)\}$$

$$= R_{XX}(\zeta). \tag{9.11}$$

The above formulae will be analyzed in the next electrical engeneering oriented section.

9.3. A Sampling Theorem for Ergodic Process

Let us start this section by considering $F(w)$ as an entire complex-variable function such that

$$|F(w)| \le c e^{A|w|}, \quad \text{for } w = x + iy \in \mathbb{C} = R + iR, \tag{9.12}$$

and its restriction to real domain $w = x + i0$ satisfies the square integrable condition

$$\int_{-\infty}^{\infty} |F(x)|^2 dx < \infty. \tag{9.13}$$

By the famous Wienner theorem (Ref. 4) there exists a function $f(t) \in L^2(-A, A)$, vanishing outside the interval $[-A, A]$, such that

$$F(w) = \frac{1}{\sqrt{2\pi}} \int_{-A}^{A} dt f(t) e^{itw}. \tag{9.14}$$

The time-limited function $f(t)$ can be considered as a physically finite-time observed (time-series) of an Ergodic process on the sample space $\Omega = L^2(R)$ as considered in Sec. 9.1 (since $R_{XX}(0) < +\infty$). In a number of cases, one should use interpolation formulae in order to analyze several statistics aspects of such observed random process. One of the most useful result in this direction, however, based on the very deep theorem of Carleson on the pontual convergence of Fourier series of a square integrable function[1,5] is the well-known Shanon (in the frequency-domain) sampling theorem for time-limited signals, namely[1,6]

$$F(w) = \sum_{n=-\infty}^{+\infty} F\left(\frac{\pi n}{T}\right) \frac{\sin\left[T\left(w - \frac{\pi n}{T}\right)\right]}{T\left(w - \frac{\pi n}{T}\right)}. \tag{9.15}$$

Note in Eq. (9.15), however, the somewhat restrictive sampling condition hypothesis of the Nyquist interval condition $T \geq A$. Although quite useful, there are situations where its direct use may be cumbersome due to the Nyquist condition on the signal sampling besides the explicitly necessity of infinite time observations $\{\frac{\pi n}{T}\}_{n \in \mathbb{Z}}$.

At this point, let us propose the following more invariant sampling–interpolation result of ours. Let $f(t)$ be an observed time-finite random Ergodic signal (sample) as mathematically considered in Eqs. (9.12)–(9.14). It is well-known that we have the famous uniform convergent Fourier expansion for $t = A \sin\theta$, with $-\frac{\pi}{2} \leq \theta \leq +\frac{\pi}{2}$ for the Fourier kernel in terms of Bessel functions

$$e^{+iwt} = e^{+iw(A \sin\theta)} = \sum_{n=-\infty}^{+\infty} J_n(wA) e^{in\theta}. \tag{9.16}$$

As much as in the usual proof of the Shanon results Eq. (9.15), we introduce Eq. (9.16) into Eq. (9.14) and by using the Lebesgue convergence theorem, since $f(t) \in L^1(-A, A)$ either, we get the somewhat (canonical) interpolating formula without any Nyquist-like restrictive sampling frequency condition on the periodogran $F(w)$:

$$F(w) = \sum_{n=-\infty}^{+\infty} A J_n(wA) \left\{ \int_{-\pi/2}^{\pi/2} d\theta e^{in\theta} f(A\sin\theta)\cos\theta \right\}. \qquad (9.17)$$

It is worth to call the readers' attention that the sampling coefficients as given by Eq. (9.17) are exactly the values of the Fourier transform of the signal $g(\theta) = f(A\sin\theta)\cos\theta$ for $-\frac{\pi}{2} < \theta < \frac{\theta}{2}$, i.e.

$$g_n = \frac{1}{2\pi} \int_{-\pi/2}^{\pi/2} d\theta \, e^{in\theta} f(A\sin\theta)\cos\theta$$

$$= \frac{1}{2\pi} \int_{-A}^{A} dt \, e^{in\operatorname{arc} \sin\left(\frac{t}{A}\right)} f(t)$$

$$\simeq \sum_{i=1}^{N} \left(\frac{1}{2\pi} \exp\left(in\operatorname{arcsin}\left(\frac{t_i}{A}\right) \right) \cdot f(t_i) \right), \qquad (9.18)$$

which may be easily and straightforwardly evaluated by FFT algorithms, from arbitrary — finite on their number — chosen sampling values $\{f(A\sin\theta_i)\}_{i=1,\dots,N}$ with $\theta_i = \operatorname{arsin}(\frac{t_i}{A})$ (here t_i are the time observation process signal). Monte–Carlo integration techniques[7] can be used on approximated evaluations of Eq. (9.18) too. Note that $f(w)$ is completely determined by its samples Eq. (9.18) taken at arbitrary times $t_i \in [-A, A]$.

In the general case of a quadratic mean continuous wide-sense stationary process $\{X_t, 0 < t < \infty\}$,[1] we still have the quadratic mean result analogous to the above exposed result for those process with time-finite sampling[1]

$$X_t = \int_{-\infty}^{+\infty} e^{iwt} d\widehat{X}_w, \qquad (9.19)$$

with the spectral process possessing the canonical form (in the quadratic mean sense)

$$\widehat{X}(w) = \sum_{n=-\infty}^{+\infty} J_n(wA) \left\{ \int_{-\pi/2}^{\pi/2} e^{in\theta} d(X_{t=A\sin\theta}) \right\}. \qquad (9.20)$$

Let us expose the usefulness of the sampling result Eqs. (9.17) and (9.18) for the very important practical engineering problem of estimate the self-correlation function of a time-finite observed sampling of an Ergodic process already taking into consideration the time-limitation of the sampling observation in the estimate formulae. We have, thus, to evaluate the (formal) time-average with time-lag $\zeta > 0$

$$R_{ff}(\zeta) = \lim_{T \to \infty} \frac{1}{2T} \int_{-T}^{T} dt f(t) f(t + \zeta). \qquad (9.21)$$

Since we have observed the sampling-continuous function $\{f(t)\}$ only for the time-limited observation interval $([-A, A])$, it appears quite convenient at this point to rewrite Eq. (9.21) in the frequency domain (as a somewhat generalized process) (see Ref. 1) where $T \to \infty$ limit is already evaluated

$$R_{ff}(\zeta) = \frac{1}{2\pi} \int_{-\infty}^{+\infty} dw_1 dw_2 e^{iw_1\zeta} \left\{ \lim_{T \to \infty} \frac{1}{2T} \int_{-T}^{T} dt e^{it(w_1 + w_2)} F(w_1) F(w_2) \right\}$$

$$= \frac{1}{\sqrt{2\pi}} \int_{-\infty}^{+\infty} dw |F(w)|^2 e^{iw\zeta}. \qquad (9.22)$$

It is worth to point out that Eq. (9.22) has already "built" in its structure, the infinite time-Ergodic evaluation $\{\lim_{T\to\infty} \frac{1}{T} \int_{-T}^{T}(...)\}$, regardless of the time-limited finite duration nature of the observed sample $f(t)[\theta(t + A) - \theta(t - A)]$.

Now an expression for Eq. (9.22), after substituting Eq. (9.17) on Eq. (9.22) (without the use of somewhat artificial aliased A-periodic extensions of the observed sampling $f(t)$, as it is commonly used in the literature),[6] can straightforwardly be suggested possessing the important property of already taking into account (in explicit way) the presence of the observation time A in the result for the self-correlation function in the frequency domain

$$S_{ff}(w) = \frac{1}{\sqrt{2\pi}} \sum_{n=-\infty}^{+\infty} \sum_{m=-\infty}^{+\infty} \{ [A^2 (g_n \bar{g}_m)] \times J_n(wA) J_m(wA) \}, \qquad (9.23)$$

with

$$R_{ff}(\zeta) = \frac{1}{\sqrt{2\pi}} \int_{-\infty}^{+\infty} dw\, e^{iw\zeta} S_{ff}(w). \qquad (9.24)$$

The above equations are the main results of this section.

It is worth to call the readers' attention that after passing the random signal $f(t)$ by a causal linear system with transference function $H_{ff}(w)$,[6] one obtains the output self-correlation function statistics in the standard formulae

$$S_{yy}(w) = |H_{yf}(w)|^2 \times S_{ff}. \tag{9.25}$$

Finally, let us comment on the evaluation of Eq. (9.23) on the time-domain. This step can be implemented throughout the use of the well-known formula for Bessel functions (see Ref. 8, Eq. (6.626)), analytically continued in the relevant parameters formulae

$$\int_0^\infty dx e^{-ax} J_\mu(bx) J_\nu(cx) dx$$

$$= I_{\mu,\nu}(a) = \frac{b^\mu c^\nu}{\Gamma(\nu+1)} 2^{-\mu-\nu} (a)^{-1-\mu-\nu}$$

$$\times \left\{ \sum_{\ell=0}^\infty \frac{\Gamma(1+\mu+\nu+2\ell)}{\ell!\Gamma(\mu+\ell+1)} \, F\left(-\ell,-\mu-\ell,\nu+1,\frac{c}{b^2}\right) \left(-\frac{b^2}{4a^2}\right)^\ell \right\}. \tag{9.26}$$

We have, thus, the result:

$$\frac{1}{\sqrt{2\pi}} \int_{-\infty}^{+\infty} dw e^{iwt} J_n(Aw) \cdot J_m(wA) = \left[(-1)^{n+m} I_{n,m}\left(-\frac{t}{A}\right) + I_{n,m}\left(\frac{t}{A}\right)\right], \tag{9.27}$$

where

$$I_{n,m}(t) = \frac{1}{\Gamma(m+1)} 2^{-m-n} (-it)^{-1-m-n}$$

$$\times \left\{ \sum_{\ell=0}^\infty \frac{\Gamma(1+n+m+\ell)}{\ell!\Gamma(n+\ell+1)} F\left(-\ell,-n-\ell,m+1,1\right) \left(-\frac{1}{4t^2}\right)^\ell \right\}. \tag{9.28}$$

Another point to be called to attention and related to the integral evaluations of Fourier transform of Bessel functions are the recurrence set of Fourier integrals of Bessel functions

$$\int_{-\infty}^{+\infty} dt\, e^{iwt} J_n(w) = (1+(-1)^n)\left\{\int_0^\infty dt \cos(wt) J_n(w)\right\}$$

$$+ i(1+(-1)^{n+1})\left\{\int_0^\infty dt \sin(wt) J_n(w)\right\}, \tag{9.29}$$

with, here, the explicitly expressions $T_n(t) = 2^{-n}[(t + \sqrt{t^2 - 1})^n + (t - \sqrt{t^2 - 1})^n]$.

$$\int_0^\infty \cos(wt) J_n(w) dw$$

$$= \begin{cases} (-1)^k \dfrac{1}{\sqrt{1-t^2}} T_n(t), & \text{for } 0 < t < 1 \text{ and } n = 2k, \\ 0, & \text{for } 0 < 1 < t, \\ \displaystyle\int_0^\infty \cos(wt) \dfrac{4k}{w} J_{n-1}(w) dw, & \\ -\displaystyle\int_0^\infty \cos(wt) J_{n-2}(w) dw, & \text{for } n = 2k+1, \end{cases} \tag{9.30}$$

$$\int_0^\infty \sin(wt) J_n(w) dw$$

$$= \begin{cases} (-1)^k \dfrac{1}{\sqrt{1-t^2}} T_n(t), & \text{for } 0 < t < 1 \quad \text{and} \quad n = 2k+1, \\ 0, & \text{for } 0 < 1 < t, \\ 2(2k+1)\displaystyle\int_0^\infty \dfrac{\sin(wt)}{w} J_{n-1}(w) & \\ -\displaystyle\int_{0'}^\infty \sin(wt) J_{n-2}(w) dw, & \text{for } n = 2k+2. \end{cases}$$

$$\tag{9.31}$$

9.4. A Model for the Turbulent Pressure Fluctuations (Random Vibrations Transmission)

One of the most important studies of pressure turbulent fluctuations (random vibrations) is to estimate the turbulent pressure component transmission inside fluids.[9,10] In this section, we intend to propose a simple analysis of a linear model of such random pressure vibrations.

Let us consider an infinite beam backed on the lower side by a space of depth d which is filled with a fluid of density ρ_2 and sound speed v_2. On the upper side of the beam there exists a supersonic boundary-layer turbulent pressure $P(x,t)$. The fluid on the upper side of the beam which is on the turbulence steadily regime is supposed to have a free stream velocity U_∞, density ρ_1, and sound speed v_1.

The effective equation governing the "outside" pressure in our model is thus given by ($d \leq z < \infty$):

$$\left(\frac{\partial}{\partial t} + U_\infty \frac{\partial}{\partial x}\right)^2 p_1(x,z,t) - v_1^2 \left(\frac{\partial^2}{\partial^2 x} + \frac{\partial^2}{\partial^2 z}\right) p_1(x,z,t) = 0 \qquad (9.32)$$

with the boundary condition (Newton's second law)

$$-\rho_1 \left(\frac{\partial}{\partial t} + U_\infty \frac{\partial}{\partial x}\right)^2 W(x,t) = \left.\frac{\partial p_1(x,z,t)}{\partial z}\right|_{z=d}. \qquad (9.33)$$

Here, the beam's deflection $W(x,t)$ is assumed to be given by the beam small linear deflection equation

$$B\frac{\partial^4 W(x,t)}{\partial^4 x} + m\frac{\partial^2 W(x,t)}{\partial^2 t} = P(x,t) + (p_1 - p_2)(x,d,t), \qquad (9.34)$$

with B denoting the bending rigidity, m the mass per unit length of the beam, and $p(x,z,t)$ the (supersonic) boundary-layer pressure fluctuation of the exterior medium.

The searched induced pressure $p_2(x,z,t)$ in the fluid interior medium ($0 \leq z \leq d$) is governed by

$$\frac{\partial^2 p_2}{\partial^2 t}(x,z,t) - (v_2)^2 \left(\frac{\partial^2}{\partial^2 x} + \frac{\partial^2}{\partial^2 z}\right) p_2(x,z,t) = 0, \qquad (9.35)$$

$$\frac{\partial p_2(x,d,t)}{\partial z} = -\rho_2 \frac{\partial^2}{\partial^2 t} W(x,t). \qquad (9.36)$$

The solution of Eqs. (9.32) and (9.33) is straightforward obtained in the Fourier domain

$$\tilde{p}_1(k,z,w) = -\rho_1 \left\{\frac{v_1(w + U_\infty k)^2}{\sqrt{v_1^2 k^2 - (w + U_\infty k)^2}}\right\} \tilde{w}(k,w)$$

$$\times \exp\left(-\frac{1}{v_1} \times \sqrt{v_1^2 k^2 - (w + U_\infty k)^2}\right)(z - d), \qquad (9.37)$$

where the deflection beam $\tilde{W}(k,w)$ is explicitly given by

$$\tilde{W}(k,w) = (\tilde{P}(k,w) - \tilde{p}_2(k,w,d)$$

$$\times \left\{Bk^4 - mw^2 + \frac{\rho_1(w + U_\infty k)^2 v_1}{\sqrt{v_1^2 k^2 - (w + U_\infty \cdot k)^2}}\right\}^{-1}. \qquad (9.38)$$

At this point, we solve our problem of determining the pressure $\tilde{p}_2(k, w, z)$ in the interior domain Eqs. (9.35) and (9.36) if one knows the pressure $\tilde{p}_2(k, w, d)$ on the boundary $z = d$. Let us, thus, consider the Taylor's series in the z-variable $(0 \leq z \leq d)$ around $z = d$, namely

$$\tilde{p}_2(k, w, z) = \tilde{p}_2(k, w, d) + \left.\frac{\partial \tilde{p}_2(k, w, z)}{\partial z}\right|_{z=d} (z - d)$$

$$+ \cdots + \frac{1}{k!} \frac{\partial^k \tilde{p}_2(k, w, z)}{\partial^k z} (z - d)^k + \cdots . \qquad (9.39)$$

From the boundary condition Eq. (9.36), we have the explicitly expression for the second-derivative on the depth z

$$\left.\frac{\partial \tilde{p}_2(k, w, z)}{\partial z}\right|_{z=d} = +\rho_2 w^2 \tilde{W}(k, w) = +\rho_2 w^2 [\tilde{P}(k, w) - \tilde{p}_2(k, w, d)]$$

$$\times \left\{ Bk^4 - mw^2 + \frac{\rho_1 (w + U_\infty k)^2 v_1}{\sqrt{v_1 k^2 - (w + U_\infty k)^2}} \right\}^{-1}. \qquad (9.40)$$

The second z-derivative of the interior pressure (and the higher ones!) is easily obtained recursively from the wave equation (9.35) $(k \geq 0, k \in \mathbb{Z}^+)$ and Eq. (9.40)

$$\left.\frac{\partial^{2+k}}{\partial^{2+k} z} \tilde{p}_2(k, z, w)\right|_{z=d}$$

$$= \left(\frac{1}{(v_2)^2} \times (-w^2 + (v_2)^2 k^2) \right) \left(\left.\frac{\partial^k}{\partial^k z} \tilde{p}_2(k, z, w) \right) \right|_{z=d}. \qquad (9.41)$$

Let us finally make the connection of this random transmission vibration model with Sec. 9.3 by calling the readers' attention that the general turbulent pressure is always assumed to be expressible in an integral form

$$P(x, t) = \int_{-\infty}^{+\infty} dw \int_{-\infty}^{+\infty} dk \, e^{ikx} e^{i(w - ku)t} \tilde{G}(k) \cdot F(w), \qquad (9.42)$$

where $F(w)$ is the Fourier transform of a time-limited finite duration sample function of an Ergodic process[1] simulating the stochastic-turbulent nature of the pressure field acting on the fluid with the exactly interpolating formulae Eqs. (9.17) and (9.18).

References

1. E. Wong and B. Hajek, *Stochastic Processes in Engineering System* (Springer-Verlag, 1985); E. Wong, *Introduction to Random Processes* (Springer-Verlag, 1983); H. Cramér and M. R. Leadbetter, *Stationary and Related Stochastic Processes* (Wiley, USA).

2. G. L. Eyink, *Turbulence Noise J. Stat. Phys.* **83**, 955 (1996); V. Yevjevich, *Stochastic Processes in Hydrology* (Water Resources Publications, Fort Collins, Colorado, USA, 1972); H. Tennekes and J. L. Lunley, *A First Course in Turbulence* (The MIT Press, 1985); L. C. L. Botelho, *Int. J. Mod. Phys. B* **13**(11), 363–370 (1999); S. Fauve *et al.*, Pressure fluctuations in swirling turbulent flows, *J. Phys.* **II-3**, 271–278 (1993).

3. L. C. L. Botelho, *Mod. Phys. Lett. B* **17**(13–14), 733–741 (2003).

4. W. Ruldin, *Real and Complex Analysis* (Tata McGraw-Hill Publishing, New Delhi, 1974).

5. Y. Katznelson, *An Introduction to Harmonic Analysis* (Dover Publishing, 1976).

6. A. Papoulis, A note on the predictability of band limited processes, *Proceedings of the IEEE*, Vol. 73 (1985); E. Marry, Poisson sampling and spectral estimation of continuous-time processes, *IEEE (Trans. Information Theory)*, Vol. II-24 (1978); J. Morlet, Sampling theory and wave propagation, *Proc. 51st. Annu. Meet. Soc. Explan. Geophys.* Los Angeles (1981).

7. M. Kalos and P. Whitlock, *Monte Carlo Methods*, Vol. 1 (Wiley, 1986); W. Press, S. Teukolsky, W. Vetterling and B. Flannery, *Numerical Recipes in C*, 2nd ed. (Cambridge Univ. Press., 1996).

8. I. S. Gadshteyn and I. M. Ryzhik, *Table of Integrals, Series, and Products* (Academic Press, Inc., 1980).

9. D. J. Mead, Free wave propagation in periodically supported infinite beams *J. Sound Vib*, **11**, 181 (1970); R. Vaicaitis and Y. K. Lin, Response of finite periodic beam to turbulence boundary-layer pressure excitation, *AIAA J* **10**, 1020 (1972).

10. L. Maestrello, J. H. Monteith, J. C. Manning and D. L. Smith, Measured response of a complex structure to supersonic turbulent boundary layers, *AIAA 14th Aerospace Science Meeting*, Washington, DC (1976).

11. W. Rudin, *Real and Complex Analysis* (Tata Mc Graw-Hill, 1974).

12. W. Rudin, *Functional Analysis* (Tata Mc Graw-Hill, 1974).

13. L. Gillman and M. Lewison, *Rings of Continuous Functions* (D. Van Nostrand Company, Inc., 1960).

14. S. M. Flatte, *Sound Transmission Through a Fluctuation Ocean* (Cambridge University Press, Cambridge, 1979); N. S. Van Kampen, *Stochastic Process in Physics and Chemistry* (North-Holland, Amsterdam, 1981); R. Dashen, *J. Math. Phys.* **20**, 892 (1979).

15. P. Sheng (ed), *Scattering and Localization of Classical Waves in Random Media, World Scientific Series on Directions in Condensed Matter Physics*, Vol. 8, (World Scientific, Singapore, 1990); L. C. L. Botelho *et al.*, *Mod. Phys.*

266   *Lecture Notes in Applied Differential Equations of Mathematical Physics*

Lett. B **12**, 301 (1998); L. C. L. Botelho *et al.*, *Mod. Phys. Lett. B* **13**, 317 (1999), J. M. F. Bassalo, *Nuovo Cimento B* **110** (1994).

16. L. C. L. Botelho, *Phys. Rev. E* **49R**, 1003 (1994).
17. L. C. L. Botelho, *Mod. Phys. Lett. B* **13**, 363 (1999).
18. B. Simon, *Functional Integration and Quantum Physics* (Academic, New York, 1979).
19. B. Simon and M. Reed, *Methods of Modern Mathematical Physics*, Vol. 2 (Academic, New York, 1979,).
20. B. Simon and M. Reed, *Methods of Modern Mathematical Physics*, Vol. 3 (Academic, New York, 1979).
21. T. Kato, *Analysis Workshop* (Univ. of California at Berkeley, USA, 1976).
22. Any standard classical book on the subject as for instance
    E. C. Titchmarsh, *The Theory of Functions*, 2nd ed. (Oxford University Press, 1968); H. M. Edwards, *Riemann's Zeta Function, Pure and Applied Mathematics* (Academic Press, New York, 1974); A. Ivic, *The Riemann Zeta-Function* (Wiley, New York, 1986); A. A. Karatsuba and S. M. Doramin, *The Riemann Zeta-Function* (Gruyter Expositions in Math, Berlin, 1992).

Appendix A. Chapters 1 and 9 — On the Uniform Convergence of Orthogonal Series — Some Comments

A.1. Fourier Series

It is well-known that point-wise convergence of the partial sums of the Fourier series of a function in a given class (continuous, square integrable, etc.) constitutes a cumbersome and difficult mathematical problem.

A useful elementary result in this direction is the well-known Tauberian theorem of G. Hardy.

Hardy Theorem. *Let* $f \in L^1([0, 2\pi])$ *and its complex Fourier coefficients satisfying an estimate as*

$$|b_n| \leq O\left(\frac{1}{n}\right). \tag{A.1}$$

Then, the usual partial sums $S_n(f)(x)$ *and* $\sigma_n(f)(x)$ *(Fejer partial sum of order n) converge to the sum limit.*

We have now the following sorts of variants of the above enunciated Hardy theorem.

Theorem A.1. *Let* $f(x) \in L^1([0, 2\pi])$ *and thus usual Fourier coefficients obeying the estimates*

$$|a_n| \le \frac{\tilde{a}_n}{n}, \qquad (A.2)$$

$$|b_n| \le \frac{\tilde{b}_n}{n}, \qquad (A.3)$$

where

$$\sum_{k=0}^{n} (\tilde{a}_k)^2 \le M_a(n)^{2-\varepsilon}, \qquad (A.4)$$

$$\sum_{k=0}^{n} (\tilde{b}_k)^2 \le M_b(n)^{2-\varepsilon}, \qquad (A.5)$$

with (M_a, M_b) *denoting (positive) real n-independent constants and* $\varepsilon \in (0, 2]$.

Then, there is a subsequence $S_{n_k}(f)(x)$ which converges almost everywhere to the given function $f(x)$.

Proof. It is straightforward that the following estimate holds true

$$\frac{1}{\pi} \|S_n(f)(x) - \sigma_n(f)(x)\|_{L^2}^2 = \sum_{n=1}^{n} \frac{k^2}{n^2} (a_n^2 + b_n^2) \le (M_A + M_B) n^{-\varepsilon}. \quad (A.6)$$

We have as well as

$$\|S_n(f)(x) - f(x)\|_{L^1} \le \frac{\sqrt{2\pi}}{\pi} (M_A + M_B) n^{-\varepsilon}. \qquad (A.7)$$

As a consequence $S_n(f) \xrightarrow{L^1} f(x)$, so it exists a basic result that there exists a subsequence $S_{n_h}(f)(x)$ converges almost everywhere to $f(x)$ in $[0, 2\pi]$. $\qquad \square$

Theorem A.2. *Let* $f(x) \in L^2([0, 2\pi])$ *and suppose that the estimate below holds true*

$$\|f - S_n(f)(x)\|_{L^2}^2 = \left(\sum_{\ell \ge n}^{\infty} |b_\ell|^2 \right) \le M \cdot n^{\varepsilon - 2}, \qquad (A.8)$$

for M *and* ε *denoting positive real numbers* $(0 \le \varepsilon < 2)$. *Then, there exists a subsequence* $S_{n_k}(f)(x)$ *which converges uniformly to* $f(x)$ *almost everywhere.*

Proof. Let

$$a_n = ||f - S_n(f)||_{L^1} \leq \sqrt{2\pi} \left(\sum_{\ell \geq n}^{\infty} |b_\ell|^2 \right)^{\frac{1}{2}} \tag{A.9}$$

and

$$g_n(x) = S_n(f)(x) + a_n.$$

Let us apply the Markov–Tchebbicheff inequalite with m being the Lebesgue measure

$$m\{x \in [0, 2\pi] : |(S_n(f) + a_n) - f|(x) \geq n^{-\varepsilon}\}$$

$$\times \frac{1}{n^\varepsilon} \left(\int_{-\pi}^{\pi} dx |S_n(f) + a_n - f|^2(x) \right)^{\frac{1}{2}} \sqrt{2\pi}$$

$$\leq \frac{\sqrt{2\pi}}{n^\varepsilon} \sqrt{2} \left\{ \left(\int_{-\pi}^{\pi} dx |S_n(f) - f|^2 + 2\pi |a_n|^2 \right)^{\frac{1}{2}} \right\}$$

$$\leq \frac{\bar{C}}{n^\varepsilon} \left\{ \sum_{\ell \geq n}^{\infty} |b_\ell|^2 \right\} \leq \frac{\bar{C}}{n^2}, \tag{A.10}$$

where $\bar{C}$ is an over-all positive constant.

We have thus that Eq. (A.10) means that the Markov–Luzin condition is satisfied, since

$$\lim_{n \to \infty} \frac{1}{n^\varepsilon} = 0 \quad \text{and} \quad \sum_{n=1}^{\infty} \frac{\bar{C}}{n^2} = \bar{C} \left(\frac{\pi^2}{6} \right) < \infty. \tag{A.11}$$

By the Luzin theorem and the obvious result that $a_n \to 0$ $(S_n(f) \xrightarrow{L^1} f)$, we have that there is a subsequence $S_{n_k}(f)(x)$ almost everywhere converging to $f(x)$ uniformly. □

Finally, we have the basic theorem of uniform convergence of the Fourier series, very useful on linear partial differential equations.

Theorem A.3. *Let $f(x)$ be an absolutely continuous function such that $f'(x) \in L^2([0, 2\pi])$ $(f(x) \in \mathcal{H}^1([0, 2\pi]))$, with exception of first-order discontinuities. Then, the Fourier series converges uniformly to $f(x)$ in any interval not containing the discontinuity points.*

Proof. Without loss of generality, let us consider a unique interior point x_0 as a point of discontinuity of our given function $f(x)$. Namely

$$\lim_{\substack{h \to 0 \\ h > 0}} (f(x_0 + h) - f(x_0 - h)) = [f(x_0)]^- < \infty. \tag{A.12}$$

We now have the estimates:

$$|a_n| = \frac{1}{\pi} \left| \int_{-\pi}^{\pi} f(x) \cos(nx) dx \right| \le \frac{1}{\pi} \left| \frac{\sin nx_0}{n} [f(x_0)]^- \right| + \frac{|b_n'|}{n},$$

$$|b_n| = \frac{1}{\pi} \left| \int_{-\pi}^{\pi} f(x) \sin(nx) \right| \le \frac{1}{\pi} \left| \frac{\cos nx_0}{n} [f(x_0)]^- \right| + \frac{|a_n'|}{n}. \tag{A.13}$$

Here a_n' and b_n' are the associated Fourier coefficients of the derivative of the function $f(x)$.

Now it a straightforward estimate to see that the Weierstrass uniform convergence criterion holds true in our case, since

$$\sum_{n=1}^{\infty} \frac{\cos(nx_0)}{n} = -\ell n \left| 2 \sin \left(\frac{1}{2} x_0 \right) \right|, \tag{A.14}$$

$$\sum_{n=1}^{\infty} \frac{\sin(nx_0)}{n} = \frac{x_0}{2} - \pi 2. \tag{A.15}$$

As a consequence, we have the estimate

$$\left| \sum_{h=1}^{n} a_k \cos(kx) + b_k \sin(kx) \right|$$

$$= |S_n(f)(x)| \le \sum_{h=1}^{n} |a_k| + \sum_{h=1}^{n} |b_k| \le \sum_{h=1}^{n} \frac{|b_k'|}{k} + \sum_{k=1}^{n} \frac{|a_k'|}{k}$$

$$+ \left| \frac{f(x_0)]^-}{\pi} \left(\sum_{h=1}^{n} \frac{\sin kx_0 + \cos kx_0}{k} \right) \right|, \tag{A.16}$$

where $\sum_{n=1}^{\infty} M_n < \infty$ (see Eqs. (A.12) and (A.13)).

On basis of the above result one can see that there exists a unique solution of the "brick" linear string wave equation in $C^2([0, \ell[\times (0, \infty))$ for initial conditions satisfying the above constraints:

(A) $U(x, 0) = f(x)$ with $f^{(3)}(x) \in L^2([0, \ell])$ and $f(x)$ and $f^{(1)}(x)$ absolutely continuous functions and $f^{(3)}$ possessing first-order discontinuities (piecewise continuous) $(f(\ell) = f(-\ell) = f^{(1)}(\ell) = f^{(1)}(-\ell) = f^{(2)}(\ell) = f^{(2)}(-\ell) = f^{(3)}(\ell) = f^{(3)}(-\ell) = 0)$.

(B) $U_t(x,0) = g(x)$ with $g^{(2)}(x) \in L^2([0,\ell])$ and $g(x)$ being an absolutely continuous with $g^{(2)}(x)$ piecewise continuous in $[0,\ell]$ ($g(\ell) = g(-\ell) = g^{(1)}(\ell) = g^{(1)}(-\ell) = g^{(2)}(-\ell) = g^{(2)}(\ell) = 0$).

$\square$

Finally, let us point out the following L. Schwartz distributional Fourier series theorem.

Theorem A.4. *Let $f(x) \in L^1_{\text{loc}}(R)$ and $f(x)$ periodic of period T. Let its formal Fourier complex coefficients c_n satisfy the polynomial bound*

$$|c_n| = \left| \frac{1}{T} \int_{-T/2}^{T/2} f(x) e^{-\frac{2\pi i n}{T} x} dx \right| \leq M(n)^L,$$

for a fixed positive integer L and $M \in R^+$. Then, the associated partial sums $S_n(f)(x)$ converges in the distributional sense in $D'_{\text{periodic}}(R)$.

On the light of this distributional sense, one can see that it is always possible to write a solution of the string linear wave equation in the space $C(D'_{\text{periodic}}(R), (0, \infty))$. (Exercise for our diligent readers.) For any given datum $f(x)$ and $g(x)$ in $D'_{\text{periodic}}(R)$.

A.2. Regular Sturm–Liouville Problem

In the case of an orthonormal set associated to a regular self-adjoint Sturm–Liouville problem in the open interval (a, b)

$$-\frac{d}{dx}\left(p(x) \frac{dy_n}{dx} \right) + g(x) y_n(x) = \lambda_n y_n(x),$$

$$a_{11} y_n'(a) + a_{12} y_n(a) = 0, \tag{A.17}$$

$$a_{21} y_n'(b) + a_{22} y_n(b) = 0,$$

we have the following result on the uniform convergence of the eigenfunctions $y_n(x)$.

Theorem A.5. *Let $f(x) \in \text{Dom}(L)$, where L is the second-order differential self-adjoint operator associated to the Sturm–Liouville problem (Eq. (A.17)). Then, its series orthogonal expansion in terms of the eigenfunctions $y_n(x)$ is a uniformly convergent series.*

Proof. Since $f \in \mathrm{Dom}(f)$, we have that

$$\langle f, \mathsf{L}f \rangle_{L^2} = \int_a^b dx f(x) \overline{(\mathsf{L}f)}(x) = \sum_{n=0}^{\infty} \lambda_n |c_n|^2 < \infty. \qquad (A.18)$$

Here

$$c_n = \int_a^b dx f(x) y_n(x).$$

Now straightforward to see that Cauchy partial sum

$$\left(\sum_{n=M}^{N} c_n y_n(x) \right)^2 = \sum_{n=M}^{N} (\sqrt{\lambda_n} c_n) \left(\frac{y_n(x)}{\sqrt{\lambda_n}} \right)$$

$$\leq \left(\sum_{n=M}^{N} \lambda_n |c_n|^2 \right) \left(\sum_{n=M}^{N} \frac{y_n(x) \bar{y}_n(x)}{\lambda_n} \right)$$

$$\leq \left(\sup_{x \in (a,b)} |y_n(x)|^2 \right) \langle f, \mathsf{L}f \rangle_{L^2} \left(\sum_{n=m}^{N} \frac{1}{\lambda_n} \right). \qquad (A.19)$$

By the usual Sturm comparison theorems, we have the well-known asymptotic limits for large M, N

$$\left| \sum_{n=M}^{N} \frac{1}{\lambda_n} \right| \leq \sum_{n=M}^{N} \frac{(b-a)^2}{n^2}, \qquad (A.20)$$

which leads to our result on the uniform convergence as a consequence of the Weierstrass M-text. In the case of an infinite interval $[a, b]$, we need that the Green function $\mathcal{L}^{-1}$ should be a compact-trace class operator with all eigenfunctions satisfying a n-uniform supremum bound as in Eq. (A.17). $\qquad \square$

Our reader can apply our result to the following Sturm–Liouville problem in $(-1, 1)$

$$\frac{d}{dx} \left[(1 - x^2) \frac{dy_n}{dx} \right] - \frac{m}{1 - x^2} y_n(x) = -\lambda_n y_n(x). \qquad (A.21)$$

In this case, we have that the problem Green function has a kernel bounded

$$\left| \sum_{n=0}^{\infty} \frac{y_n(x) y_n(\bar{x})}{\lambda_n} \right| \leq \frac{1}{2m}. \qquad (A.22)$$

As a consequence of Theorem A.5, we have the uniform convergence of the orthogonal series expansion in terms of the eigenfunctions $y_n(x)$ if

$$\int_{-1}^{1} dx \left(\left| \frac{df}{dx} \right|^2 + \frac{m^2 f^2}{1 - x^2} \right) < \infty.$$

Appendix B. On the Müntz–Szasz Theorem on Commutative Banach Algebras

B.1. Introduction

The classical theory of Müntz–Szasz states that the set of functions $\{1, t^{\alpha_1}, t^{\lambda_2}, \ldots, t^{\lambda_n}, \ldots\}$ span $C([0,1], R)$ if $\sum \frac{1}{\lambda_n} = +\infty$. In this short note, we intend to point out that the usual elementary proof of such a theorem works out with minor changes to produce an abstract result in certain commutative self-adjoint Banach algebras.

Let us start our note by considering A be a semi-simple (self-adjoint) commutative Banach algebra possessing a simple generator f such that its Gelfand transform $\widehat{f} \in C(\Delta, R)$ is a (continuous) function with the property that the polynomial algebra generated by f satisfies those conditions of the Stone–Weirstrass theorem on $C(\Delta, R)$.

We now state our generalization of the Müntz–Szasz theorem for the abstract case.

Theorem B.1. *Let $\{\lambda_n\}$ be a sequence of positive real numbers such that $\sum \frac{1}{\lambda_n} = +\infty$. Then, the span of the set $\{1, \widehat{f}^{\lambda_1}, \widehat{f}^{\lambda_2}, \ldots, \widehat{f}^{\lambda_n}, \ldots\}$ is the whole $C(\Delta, R)$.*

Proof. Through the Hahn–Banach theorem and the Riesz representation theorem applied to the topological dual of $C(\Delta)$, our result will be a straightforward consequence of the following proposition (Ref. 11, Theorem 15.26):

If $\sum \frac{1}{\lambda_n} = +\infty$, and u is a complex Borel measure on the compact maximal ideal space Δ such that

$$\int_{\Delta} (\widehat{f})^{\lambda_n}, d\mu = 0, \quad (n = 1, 2, 3, \ldots), \tag{B.1}$$

then also

$$\int_{\Delta} (\widehat{f})^{n} d\mu = 0, \quad (n = 1, 2, 3, \ldots), \tag{B.2}$$

a contradiction with our hypothesis about the denseness of polynomial algebra generated by $\widehat{f}$ on $C(\Delta, R)$.

In order to show this result, let us consider without loss of generality our function $\widehat{f}$ such that $\|f\|_{C(\Delta)} \leq 1$. Now it is straightforward to see that the function

$$h(z) = \int_{\Delta} f^z d\mu, \tag{B.3}$$

is a bounded holomorphic function in the right half-plane $H^+ = \{z = x + iy, x > 0\}$ by an application of Morera theorem. Note that $g(z) = f(\frac{1+z}{1-z}) \in H^\alpha(U)$. Here $U = \{z \mid |z| < 1\}$ is the unit disc and $g(\alpha_n) = 0$, with $\alpha_n = \frac{\lambda_n - 1}{\lambda_n + 1}$. Note that either $g(z) \equiv 0$ if $\Sigma(1 - |\alpha_n|) = \infty$ or equivalently: $\sum \frac{1}{\lambda_n} = +\infty$. This means that $h(z) \equiv 0$. $\qquad\square$

As an interesting consequence of Theorem B.1, we have the following generalization of the famous Wiener theorem on inversion of absolute summable trigono/metric polynomials on the torus $T = \{(e^{i\theta}), 0 \leq \theta < 2\pi\}$, useful in linear system theory.

Theorem B.2. *Suppose that*

$$\ell(e^{i\theta}) = \sum_{n=-\infty}^{+\infty} c_n e^{in\theta}; \quad \sum_{-\infty}^{+\infty} |c_n| < \infty \tag{B.4}$$

and $\ell(e^{i\theta}) \neq 0$ *for* $e^{i\theta} \in T$. *Then, for any Müntz–Szasz sequence* $|\lambda_n|$, *we have the result*

$$\frac{1}{\ell(e^{i\theta})} = \sum_{-\infty}^{+\infty} \gamma_n e^{i\lambda_n \theta} \quad \text{with} \quad \sum_{n=-\infty}^{+\infty} |\gamma_n| < \infty. \tag{B.5}$$

Proof. The element $\widehat{f}(\theta) = e^{i\theta}$ satisfies the conditions of Theorem B.1.

In the case of the Müntz–Szasz sequence $\{\lambda_n\}$ be the prime numbers sequence or that one made up by their logarithms, the result may have interesting applications in the theory of Zeta functions.[13]

In the case of the existence of a polynomial subalgebra generated by a finite number of generators $\{\widehat{f}_1, \ldots, \widehat{f}_p\}$ we can use the Hahn–Banach theorem for complex homomorphism of our Banach algebra through the Gleason, Kahane, Zelazuo theorem (Ref. 12, Theorem 10.9).

Finally in the Banach algebra of the bounded continuous functions vanishing at the infinity in a locally compact space X, it is straightforward to see that if $f(x)$ is an arbitrary continuous function satisfying the condition $f(x) \geq 1$ and such that the polynomial algebra generated by the bounded function $h = f\exp(-f)$ is dense on $C_0(X)$, then the set $\{1, h^{\lambda_1}, h^{\lambda_2}, \ldots, h^{\lambda_n}, \ldots\}$ remains dense in $C_0(X)$ the only difference with the standard proof presented in Theorem B.1 is related to the estimate on the boundedness of the function $h(z)$ on H^+, which now takes the following form:

$$\sup_{z\in H^+} |h(z)| \leq \sup_{z\in H^+} \left\{ |fe^{-f}|^z(m)d\mu(m) \right\}$$

$$\leq \mu(X) \times \left\{ \sup_{\substack{0<x<\infty \\ f(m)\in[1,\infty)}} |f|^x \cdot e^{-x|f|}(m) \right\}$$

$$\leq \mu(X) \times \left\{ \sup_{\substack{0<x<\infty \\ f(m)\in[1,\infty)}} |e^{x(\ell g f - f)}|(m) \right\}$$

$$\leq \mu(X) \times \left\{ \sup_{0<x<\infty} |e^{-x}| \right\} = \mu(X) < \infty, \qquad \Box$$

Appendix C. Feynman Path-Integral Representations for the Classical Harmonic Oscillator with Stochastic Frequency

C.1. Introduction

The problem of the (random) motion of a harmonic oscillator in the presence of a stochastic time-dependent perturbation on its frequency is of great theoretical and practical importance.[14,15] In this appendix, we propose a formal path-integral solution for the above-mentioned problem by closely following our previous studies.[16,17] In Sec. C.2, we write a Feynman path-integral representation for the external forcing problem. In Sec. C.3, we consider a similar problem for the initial-condition case.

C.2. The Green Function for External Forcing

Let us start our analysis by considering the classical motion equation of a harmonic oscillator subject to an external forcing

$$\left\{ \frac{d^2}{dt^2} + w_0^2(1+g(t)) \right\} x(t) = F(t). \qquad (C.1)$$

Here, $w_0^2(1 + g(t))$ is the time-dependent frequency with stochastic part given by the random function $g(t)$ obeying the Gaussian statistics

$$x(t, [g]) = \int_0^t G(t, t', [g]) F(t') dt', \qquad (C.3)$$

where $G(t, t', [g])$ denotes the Green problem functionally depending on the random frequency $g(t)$ and the notation emphasizes that the objects under study are functionals of the random part $g(t)$ of the harmonic oscillator frequency.

In order to write a path-integral representation for the Green function equation (C.3), we follow our previous study[16] by using a "proper-time" technique by introducing related Schrödinger wave equations with an initial point source and $-\infty \le t \le +\infty$:

$$i \frac{\partial \bar{G}(x; (t, t'))}{\partial s} = - \left[\frac{d^2}{dt^2} + w_0^2(1 + g(t)) \right] \bar{G}(s; (t, t')), \qquad (C.4)$$

$$\lim_{s \to 0} \bar{G}(s; (t, t')) = \delta(t - t'), \qquad (C.5)$$

$$\lim_{s \to x} \bar{G}(s; (t, t')) = 0. \qquad (C.6)$$

At this point, we note the following identity between the Schrödinger equations (C.4)–(C.6) and the searched harmonic oscillator Green function:

$$G[(t, t', [g])] = -i \int_0^\infty ds \bar{G}(s; (t, t')). \qquad (C.7)$$

Let us, thus, write a path-integral for the associated Schrödinger equations (C.4)–(C.7) by considering $\bar{G}(s; (t, t'))$ in the operator form (the Feynman–Dirac propagator):

$$\bar{G}(s; (t, t')) = \langle t | \exp(isH) | t' \rangle, \qquad (C.8)$$

where H is the differential operator

$$H = - \left\{ \frac{d^2}{dt^2} + w_0^2(1 + g(t)) \right\}. \qquad (C.9)$$

As in quantum mechanics, we write Eq. (C.8) as an infinite product of short-time s propagations (see Chap. 3)

$$\left\langle t | \exp(isH) | t' \right\rangle = \lim_{N \to x} \prod_{i=1}^N \int_{-\infty}^{+\infty} dt_i \langle t_i | \exp i \left(\frac{s}{N} H \right) t_{i-1} \right\rangle. \qquad (C.10)$$

The standard short-time expansion in the s-parameter for Eq. (C.10) is given by

$$\lim_{s \to 0^+} \langle t_i | e^{isH} | t_{i-1} \rangle d = \lim_{s \to 0^+} \int dw_i \exp\{is[(w_i)^2 + (w_0)^2(1 + g^2(t_{i-1}))]\}$$
$$\times \exp[iw_i(t_i - t_{i-1})]. \tag{C.11}$$

If we substitute Eq. (C.11) into Eq. (C.10) and take the Feynman limit of $N \to \infty$, we will obtain the following path-integral representation after evaluating the w_i-Gaussian integrals of the representation equation (C.8):

$$\bar{G}(s;t,t')) = \int \left(\prod_{\substack{0 < \sigma s \\ t(0)=t':t(s)=t}} dt(\sigma) \right) \exp\left\{ \frac{i}{2} \int_0^2 d\sigma \left[\left(\frac{dt(\sigma)}{d\sigma} \right)^2 \right] \right\}$$
$$\times \exp\left\{ i \int_0^s [w_0^2(1 + g(t(\sigma)))]d\sigma \right\}. \tag{C.12}$$

The averaged out Eq. (C.7) is thus given straightforwardly by the following Feynman polaron-like path-integral:

$$\langle G(t,t',[g]) \rangle_g = -i \int_0^\infty ds(e^{i(w_0)^2 s}) \int_{t(0)=t':t(s)=t} D^F[t(\sigma)]$$
$$\times \exp\left\{ \frac{i}{2} \int_0^s d\sigma \left[\left(\frac{dt}{d\sigma} \right)^2 \right] \right\}$$
$$\times \exp\left\{ -w_0^4 \int_0^s d\sigma \int_0^s d\sigma' K(t(\sigma); t(\sigma')) \right\}. \tag{C.13}$$

The two-point correlation function is still given by a two-full similar path-integral, namely

$$\langle G(t_1, t'_1, [g]) G(t_2, t'_2, [g]) \rangle_g$$
$$= \int_0^\infty ds_1 ds_2 e^{i(w_0)^2(s_1+s_2)}$$
$$\times \int_{t_1(0)=t'_1:t_2(s_1)=t_1:t_2(0)=t'_2:t_2(s_2)=t_2} D^F[t_1(\sigma), t_2(\sigma)]$$
$$\times \exp\left\{ \frac{i}{2} \left(\int_0^{s_1} d\sigma(t_1(\sigma))^2 + \int_0^{s_2} d\sigma(i_2(\sigma))^2 \right) \right\}$$

$$\times \exp \left\{ -w_0^4 \left[\int_0^{s_1} d\sigma \int_0^{s_1} d\sigma' K(t_1(\sigma) \cdot t_1(\sigma')) \right. \right.$$

$$+ \int_0^{s_1} d\sigma \int_0^{s_2} d\sigma' (K(t_1(\sigma) \cdot t_2(\sigma')) + K(t_2(\sigma) \cdot t_1(\sigma')))$$

$$\left. \left. + \int_0^{s_1} d\sigma \int_0^{s_2} d\sigma' K(t_2(\sigma) \cdot t_2(\sigma')) \right] \right\}. \tag{C.14}$$

Similar N-iterated path-integral expressions hold true for the N-point correlation function $\langle x(t_1, [g]) \cdots x(t_N, [g]) \rangle_g$. Explicit and approximate evaluations of the path-integral equations (C.13) and (C.14) follow procedures similar to those used in the usual contexts of physics statistics, quantum mechanics, and random wave propagation (last reference of Ref. 14).

Let us show such exact integral representation for Eq. (C.13) in the case of the practical case of a slowly varying (even function) kernel of the form:

$$K(t) \sim K(0) - \frac{\ell_0}{2} |t|^2 \quad |t| \ll \left(+\frac{K(0)}{\ell_0} \right)^{1/2} = L, \tag{C.15}$$

$$K(t) \sim 0 \quad |t| \gg L.$$

In this case, we have the following exact result for the path-integral in Eq. (C.13):

$$\langle G(t, t', [g]) \rangle_g$$

$$= \int_0^\infty ds e^{-s(-iw_0^2)} e^{-(u_0^4 K(0))s^2} \left\{ (2\pi i s)^{1/2} \left[\frac{w_0^2 (-\frac{\ell}{2})^{1/2} S^{3/2}}{\sin[s^{3/2} w_0^2 (-\frac{\ell_0}{2})^{1/2}]} \right] \right.$$

$$\left. \times \exp \left(\left[\frac{i w_0^2}{2} s^{1/2} \left(-\frac{\ell_0}{2} \right)^{1/2} \cot \left(w_0^2 \left(-\frac{\ell_0}{2} \right)^{1/2} s^{3/2} \right) \right] (t - t')^2 \right) \right\}. \tag{C.16}$$

Another useful formula is that related to the "mean-field" averaged path-integral when the kernel $K(t, t')$ has a Fourier transform of the general form:

$$K(t, t') = \frac{1}{\sqrt{2\pi}} \int_{-\infty}^{+\infty} dp \cdot e^{ip(t-t')} \tilde{K}(p). \tag{C.17}$$

The envisaged integral representation for Eq. (C.13) is, thus, given by

$$\langle G(t, t', [g]) \rangle_g = -i \int_0^\infty ds e^{isu_0^2} \exp \left\{ -\frac{w_0^4}{\sqrt{2\pi}} \int_{-\infty}^{+\infty} \tilde{K}(p) \times M(p, s, t, t') \right\}, \tag{C.18}$$

where

$$M(p, s, t, k') = \int_0^s d\sigma \int_0^s d\sigma' \left\{ \int_{\substack{t(0)=t' \\ t(s)=t}} D^F[t(\sigma)] \exp \left\{ \frac{i}{2} \int_0^s \left[\left(\frac{dt}{d\sigma} \right) \right]^2 \right\} \right.$$

$$\left. \times \exp[ip(t(\sigma) - t(\sigma'))] \right\}. \tag{C.19}$$

C.3. The Homogeneous Problem

Let us start this section by considering the problem of determining two linearly independent solutions of the homogeneous harmonic oscillator problem

$$\left\{ \frac{d^2}{dt^2} + w_0^2(1 + g(t)) \right\} x(t) = 0, \tag{C.20}$$

with the initial conditions

$$x(0) = x_0 \quad x'(0) = v_0. \tag{C.21}$$

It is straightforward to see that two linearly independent solutions are given by the following expressions:

$$x_1(t, [g]) = \exp \left\{ \int_0^t y^2(\sigma, [g]) d\sigma \right\}, \tag{C.22}$$

$$x_2(t, [g]) = x(t, [g]) \int_0^t (x_1(\sigma, [g]))^{-2} d\sigma, \tag{C.23}$$

where $y(t, [g])$ satisfies the first-order nonlinear ordinary differential equation

$$\frac{dy}{dt}(t) + (y(t)^2 = -w_0^2(1 + g(t)). \tag{C.24}$$

In order to obtain a path-integral representation for Eq. (C.24), we remark that the whole averaging (stochastic) information is contained in the characteristic functional

$$Z[j(t)] = \left\langle \exp \left\{ i \int_0^\infty dt\, y(t, [g]) j(t) \right\} \right\rangle_g. \tag{C.25}$$

In order to write a path-integral representation for the characteristic functional equation (C.25) we rewrite (C.25) as a Gaussian functional integral in $g(t)$:

$$Z[j(t)] = \int D^F[g(t)] \exp\left(-\frac{1}{2}\int_0^\infty dt\, dt'\, g(t)K^{-1}(t,t')g(t')\right)$$

$$\times \exp\left\{i\int_0^\infty dt\, y(t,[g])\right\}. \tag{C.26}$$

At this point, we observe the validity of the following functional integral representation for the characteristic functional equation (C.26) after considering the functional change $g(t) \to y(t)$ defined by Eq. (C.24), namely

$$Z[j(t)] = \int D^F[y(t)]$$

$$\times \exp\left(-\frac{1}{2(w_0)^4}\int_0^\infty dt\, dt' \left[\left(\frac{dy}{dt} + y^2\right)(t) + w_0^2\right] K(t,t')\right.$$

$$\times \left[\left(\frac{dy}{dt'} + y^2\right)(t') + w_0^2\right] \exp\left\{i\int_0^\infty dt\, j(t)y(t)\right\}, \tag{C.27}$$

where we have used the fact that the Jacobian associated with the functional change $g(t) \to y(t)$ is unity:

$$\det_F\left[\frac{d}{dt} + 2y\right] = \frac{\delta g(t)}{\delta y(t)} = 1. \tag{C.28}$$

At this point, it is instructive to remark that in the important case of a white-noise frequency process with strength γ

$$K(t,t') = \gamma\delta(t - t), \tag{C.29}$$

the path-integral representation for the characteristic functional equation (C.27) takes the more amenable form

$$Z[j(t)] = \int D^F[y(t)] \exp\left\{-\frac{\gamma}{2(w_0)^4}\int_0^\infty dt\left[\frac{dy}{dt} + y^2(t) + w_0^2\right]^2\right\}$$

$$\times \exp\left\{i\int_0^\infty dt\, y(t)j(t)\right\}, \tag{C.30}$$

we obtain, thus, the standard $\lambda\varphi^4$ zero-dimensional path-integral as a functional integral representation for the characteristic functional equation (C.25) in the white-noise case

$$Z[j(t)] = \int D^F[\bar{y}(t)] \exp\left\{ -\frac{\gamma}{2(w_0)^4} \int_0^\infty dt \left[\frac{d\bar{y}}{dt} + 2w_0^2\bar{y}^2 + \bar{y}^4 \right](t) \right\}$$

$$\times \exp\left\{ i \int_0^\infty dt\, \bar{y}(t)j(t) \right\}. \tag{C.31}$$

In this paragraph, we discuss some mathematical points related to the final condition on the quantum propagator equations (C.4)–(C.6) (its vanishing in the limit $s \to \infty$). Let us first remark that by imposing the trace class condition on the correlation equation (C.2) $k(t,t')(\int_0^\infty k(t,t)dt < \infty)$ one has the result that all realizations (sampling) $g(t)$ of the associated stochastic process are square integrable functions by a direct application of the Minlos theorem on the domain of functional integrals.[18] At this point, we note that if one restricts $g(t)$ further to be a $\frac{d^2}{dt^2}$-perturbation, namely with $g(t) \in L^2(0,\infty)$:

$$\|w_0^2(1 + g(t))h\|_{L^2} \le a \left\| \frac{d^2}{dt^2}h \right\|_{L^2} + b\|h\|_{L^2} \tag{C.32}$$

and $0 < a < 1$ and b arbitrary, one can apply the Kato–Rellich theorem[19] to be sure that the domain of the differential operator $-\frac{d^2}{dt^2} + w_0^2 + w_0^2 g(t)$ is (at least) contained on the domain of $-\frac{d^2}{dt^2}$ which in turn has a purely continuous spectrum on $L^2(R)$. As a consequence of the above-exposed remarks one does not have bound states on the Schrödinger operator spectrum of Eq. (C.4) for each realization of $g(t)$ on the above-cited functional class. As a consequence, we have that the evolution operator

$$\exp\left(is \left[-\frac{d^2}{dt^2} + w_0^2(1 + g(t)) \right] \right), \tag{C.33}$$

is a unitary operator on $\mathrm{Dom}(-\frac{d^2}{dt^2})$ and $\bar{G}(s,t,t')$ in turn is expected to have the same behavior of $\exp(is[-\frac{d^2}{dt^2}])$ (asymptotic completeness[20]) at $s \to \infty$, which in turn vanishes at $s \to \infty$, making our condition Eq. (C.6) highly reasonable from a mathematical physicist's point of view.

At this point, we remark that the *different problem* of *determining* the *quantum propagator* of a particle under the influence of a harmonic potential $V(x) = \frac{1}{2}w_0^2 x^2$ and a stochastic potential $g(x)$ may be possible to handle in our framework. Note that the "únperturbed" Hamiltonian $-\frac{h^2 d^2}{2m dx^2} + \frac{1}{2} m w_0^2 x^2$ has a pure point spectrum (bound states) and its perturbation by a $g(x) \in L^2(R)$ potential does not alter the spectrum behavior. However, one can proceed in a mathematical physicist's (formal) way as exposed in our letter and see that Eq. (C.3) adapted to this case is still (formally) correct. Namely

$$\langle G(t, (x, x')) \rangle_g = \int \left(\prod_{\substack{0 < \sigma < t \\ x(0) = x' : x(t) = x}} dx(\sigma) \right) \exp\left\{ \frac{i}{2} m \int_0^t d\sigma \left[\left(\frac{dx(\sigma)}{d\sigma} \right)^2 \right] \right\}$$

$$\times \exp\left\{ \frac{i}{2} m w_0^2 \int_0^t d\sigma [x(\sigma)]^2 \right\}$$

$$\times \exp\left\{ -w_0^4 \int_0^t d\sigma \int_0^t d\sigma' K(x(\sigma); x(\sigma)) \right\}. \tag{C.34}$$

Finally, we want to point out that in the general case where $k(t, t')$ is not defined on an operator of the trace class, the realizations $g(t)$ will be distributional objects. Unfortunately in this case there is no rigorous spectral perturbation mathematics for differential operators acting on nuclear spaces $(s'(R), D'(R))$, etc, which is the natural mathematical setting to understand Eq. (C.1) of this letter.

In this complementary paragraph, we complete our study by considering the problem of a harmonic oscillator in the presence of a damping term $(0 < t < \infty)$

$$\left\{ \frac{d^2}{dt^2} + v \frac{d}{dt} + w_0^2 (1 + g(t)) \right\} x(t) = F(t). \tag{C.35}$$

In order to map the above differential equation in the analysis presented in Sec. C.2, we implement in Eq. (C.35) the following time-variable change:[16]

$$\zeta = \frac{m}{v} (1 - e^{-t \frac{v}{m} u}),$$

$$y(\zeta) = x(t). \tag{C.36}$$

We obtain, thus, the following pure harmonic oscillator differential equation without the damping term in place of the original equation (C.35), namely

$$\left\{ \frac{d^2}{d\zeta^2} + w_0^2 [1 + \tilde{g}(\zeta)] \right\} y(\zeta) = \tilde{F}(\zeta). \tag{C.37}$$

Here

$$\tilde{g}(\zeta) = g\left(-\left(\frac{m}{v}\right)\lg\left(1 - \left(\frac{v}{m}\right)\zeta\right)\right), \tag{C.38}$$

$$\tilde{F}(\zeta) = F\frac{\left(-\left(\frac{m}{v}\right)\lg\left(1 - \left(\frac{v}{m}\right)(\zeta)\right)\right)}{\left(1 - \left(\frac{v}{m}\right)\zeta\right)^2}, \tag{C.39}$$

and the correlator stochastic frequency

$$K(\zeta, \zeta') = K((\zeta, \zeta')) = K\left(\frac{m}{v}\left|\lg\left[\frac{(1 - (\frac{v}{m})\zeta)}{(1 - (\frac{v}{m})\zeta')}\right]\right|\right). \tag{C.40}$$

From here on the analysis goes as in the bulk of this appendix under the condition that the range associated with this new time variable is the finite interval $[0, \frac{m}{v}]$.

Appendix D. An Elementary Comment on the Zeros of the Zeta Function (on the Riemann's Conjecture)

D.1. Introduction — "Elementary May Be Deep"

The Riemann's series $\zeta(x) = \sum_{n=1}^{\infty} n^{-x}$ converges uniformly for all real numbers x greater than or equal to a given (fixed) abscissa $\bar{x}$: $x > \bar{x} > 1$. It is well-known that the complex-valued (meromorphic) continuation to complex values ($z = x + iy$) throughout the complex plane $z \in \mathbb{C}$ is obtained from standard analytic (finite-part) complex variables methods applied to the integral representation.[21]

$$
\begin{aligned}
\zeta(z) &= \frac{i\Gamma(1-z)}{2\pi}\left(\int_C \left(\frac{(-w)^{z-1}}{e^w - 1}\right) dw\right) \\
&= \frac{1}{\Gamma(z)}\left(\int_0^\infty w^{z-1} \times \left[\frac{1}{2^w - 1} - \frac{1}{w}\right] dw\right) \\
&= \frac{i\Gamma(1-z)}{2\pi}\left\{\int_C \left[(-w)^{z-2} - \frac{1}{2}(-w)^{z-1}\right.\right. \\
&\qquad\qquad \left.\left. + \sum_{n=1}^{\infty}\frac{(-1)^{n+z-2}B_n/w^{z+2n-2}}{(2n)!}\right]\right\},
\end{aligned}
\tag{D.1}
$$

here B_n are the Bernoulli's numbers and C is any contour in the complex plane, coming from positive infinity and encircling the origin once in the positive direction.

An important relationship resulting from Eq. (D.1) is the so-called functional equation satisfied by the zeta function, holds true for any $z \in \mathbb{C}$

$$\frac{\zeta(z)}{\zeta(1-z)} = 2^z \cdot \pi^{z-1} \cdot \sin\left(\frac{\pi z}{2}\right) \cdot \Gamma(1-z). \tag{D.2}$$

In applications to number theory, where this special function plays a special role, it is a famous conjecture proposed by B. Riemann (1856) that the only nontrivial zeros of the zeta function lie in the so-called critical line Real $(z) = x = \frac{1}{2}$.

In the next section, we intend to propose an equivalent conjecture, hoped to be more suitable for handling the Riemann's problem by the standard methods of classical complex analysis,[22] besides of proving a historical clue for the reason that led B. Riemann to propose his conjecture.[22]

D.2. On the Equivalent Conjecture D.1

Let us state our conjecture

Conjecture D.1. *In each horizontal line of the complex plane of the form Imaginary* $(z) = y = b = constant$, *the zeta function* $\zeta(z)$ *possess at most a unique zero.*

We show now that the above conjecture leads straightforwardly to a proof of the Riemann's conjecture.

Theorem D.1 (The Riemann's Conjecture). *All the nontrivial zeros of the Riemann zeta function lie on the critical line Real* $(z) = x = \frac{1}{2}$.

Proof. Let us consider a given nontrivial zero $\bar{z} = \bar{x} + i\bar{y}$ on the open strip $0 < x < 1$, $-\infty < y < +\infty$. It is a direct consequence of the Schwartz's reflection principle since $\zeta(x)$ is a real function in $0 < x < 1$ that $(\bar{z})^* = \bar{x} - i\bar{y}$ is another nontrivial zero of the Riemann function on the above pointed out open strip. The basic point of our proof is to show that $1 - (\bar{z})^* = (1 - \bar{x}) + i\bar{y}$ is another zero of $\zeta(z)$ in the same horizontal line $\text{Im}(z) = \bar{y}$. This result turns out that $(1 - (\bar{z})^*) = \bar{z}$ on the basis of the validity of our conjecture. As a consequence, we obtain straightforwardly that $1 - \bar{x} = \bar{x}$. In others words, Real $(\bar{z}) = \frac{1}{2}$.

At this point, we call the readers' attention, on the result that if $\bar{z}$ is a zero of the Riemann zeta function, then $1 - \bar{z}$ must be another nontrivial zero is a direct consequence of the functional Eq. (D.2), since $\sin(\frac{\pi z}{2})$ and $\Gamma(1 - z)$ never vanish both on the open strip $0 < \text{Real}(z) < 1$.

At this point of our note, we want to state clearly the significance of replacing the Riemann's original conjecture by our complex oriented conjecture rests on the possibility of progress in producing sound results for its proof, which is not claimed in our elementary note. However, we intend to point out directions (arguments) in its favor.

First, it is worth to call the readers' attention on the validity of the elementary expansion below on the open interval $0 < x < 1$

$$\zeta(x + ib) = \frac{1}{\Gamma(x)} \left\{ \exp\left(ib \frac{d}{dx} \right) (\zeta(x)\Gamma(x)) \right\}. \tag{D.3}$$

As a consequence, one can easily see that if $\zeta'(x + ib)$ (the zeta function derivative on the horizontal line $\text{Im}(z) = b = $ constant) does not possesses zeros between two consecutive zeros of the zeta function $\zeta(x + ib)$ in $0 < x < 1$, then the Rieman's conjecture is proved. $\square$

At this point, we wish to remark the following weaker Conjecture D.2 is equivalent to our strong Conjecture D.1.

Conjecture D.2. *If there is a zero of the Riemann's zeta function $t \in (0, 1)$ such that the set of real number $\{\frac{d^n}{d^n x}\zeta(t)\}_{n=0,1,2,...}$ is contained entirely on the positive or negative real exist (these numbers all have the same signal), then the Conjecture D.1 holds true.*

On basis of the above-stated Conjecture D.2, we can easily show an argument of our conjecture based on the power series expansion around the fixed zero t $(\bar{x} > t)$

$$\zeta(\bar{x}) = 0 = \sum_{n=1}^{\infty} \frac{\zeta^{(n)}(t)}{n!} (\bar{x} - t)^n > 0, \tag{D.4}$$

here $\bar{x}$ is the supposed different zero of $\zeta(x)$ on the open interval $(0, 1)$.

Finally, let us point out the following formula of ours, related to the zeros of the Riemann zeta function on the complex plane $\mathbb{C}$.

Lemma D.1. *Let* $\{z_n\}_{n=0,1,2,\ldots}$ *denote the zeros of the Riemann's zeta function* $\zeta(z)$. *We have thus the following result, for any integer* $p > 0$.

$$-\left\{\sum_{\substack{n=1\\n\neq k}}^{\infty}\frac{1}{(z_n-z_k)^{p+1}}\right\}$$

$$= \lim_{z\to z_k}\left\{\frac{1}{2\pi}\int_0^{2\pi}d\theta\left(\frac{1}{z-1}+\frac{\zeta'(z)}{\zeta(z)}\right)(z_k+e^{i\theta})\cdot(z_k-z)^{p+1}e^{-ip\theta}\right\}.$$

$$(D.5)$$

Proof. This result can be seen from the fact that $h(z) = (z-1)\zeta(z)$ is an integral function and from any integral function of order k, with zeros $\bar{z}_1, \bar{z}_2, \ldots, (h(0) \neq 0)$

$$\lim_{z\to z_k}\left[(z-\bar{z}_k)^{p+1}\left(\frac{d}{dz}\right)^p\left\{\frac{h'(z)}{h(z)}\right\}\right] = -p!\sum_{\substack{n=1\\n\neq k}}^{\infty}\frac{1}{(\bar{z}_n-\bar{z}_k)^{p+1}}.\qquad(D.6)$$

By rewriting the derivative term in the left-hand side of Eq. (D.6) by means of the Cauchy theorem

$$\left\{\frac{h'(z)}{h(z)}\right\}^{(p)} = \frac{p!}{2ni}\left\{\int_{\partial D(z,1)}dw\left[\frac{h'(w)}{h(w)}\right][(w-z)^{-(p+1)}]\right\},\qquad(D.7)$$

where $D(z,1) = \{w \mid |w-z| \leq 1\}$, and arranging terms of the form $(z-\bar{z}_k)^{p+1}/(w-z)^{(p+1)}$ inside Eqs. (D.6) and (D.7), we obtain Eq. (D.5). $\qquad\square$

Lemma D.2. *Let* $h(z) = \zeta(z)(z-1)$ *be an (analytic) function on the strip* $0 < Real(z) < 1$. *Let us consider its expression on the unit disk conformally equivalent to the above considered strip, including the boundaries correspondence.*

$$h\left(\frac{1}{\pi i}lg\left(\frac{i(i-w)}{i+w}\right)\right) = h(g(w)) = \bar{h}(w).\qquad(D.8)$$

We have that, the generalized Jensen's formula

$$\int_0^1\frac{\eta(x)}{x}dx = \frac{1}{2\pi}\int_0^{2\pi}d\theta\left(lg|\bar{h}(e^{i\theta})|\right)\left(\frac{g'}{g}\right)(e^{i\theta})d\theta,\qquad(D.9)$$

where $\eta(x)$ is the number of zeros of the zeta function on the region $g^{-1}(|w| < x)$, contained on the strip $0 < \text{Real}(z) < 1$.

Proof. Since $h(z)$ is an analytic function in the strip $0 < \text{Real}(z) < 1$, we are going to prove the general result for a region Ω conformally equivalent to a disc of radius 1, including its boundary correspondence, namely

$$g: \Omega \to D(0,1),$$
$$g(\partial\Omega) = \partial D(0,1),$$

(D.10)

with $w = g(z)$ a conformal mapping applying $\partial\Omega$ in $\partial D(0,1)$ diffeomorphically.

Let $\eta(x)$ denotes the number of zeros of $f(g^{-1}(w)) = h(w)$ on $D(0,x)$. By the usual Jensen's formula for $r < 1$, with the hypothesis of $h(w)$ has no zero in $\partial D(0,r)$.

$$\int_0^r \frac{\eta(x)}{x} dx = \frac{1}{2\pi i} \int_{\partial D(0,r)} \frac{h'(w)}{h(w)} dw.$$

(D.11)

Since $\eta(x)$ coincides with the number of zeros of the function $f(z)$ in the region $g^{-1}(D(0,1))$, we obtain the Jensen's formula in Ω in the general case

$$\int_0^r \frac{\eta(x)}{x} dx = \left(\frac{1}{2\pi i} \int_{\partial\Omega} lg\left\{ |l(z)| \frac{g'(z)}{g(z)} \right\} \cdot dz \right) - lg|f(0)|.$$

(D.12)

Equation (D.8) comes from the fact that the strip $0 < \text{Real}(z) < 1$ is conformally mapped — with the boundary correspondence — on the unit circle $\{w|j \mid |w| \le 1\}$ by the function $z = \frac{1}{\pi i} lg(\frac{i(i-w)}{i+h})$. $\square$

Chapter 10

Some Studies on Functional Integrals Representations for Fluid Motion with Random Conditions[*]

10.1. Introduction

The main task of the statistical approach to random fluid dynamics[1] is to solve the set of infinite hierarchy equations for the random fluid velocity correlation functions. One important scheme to solve these equations consists in considering directly for the appropriate flux equation the random conditions generating the flux stochasticity in the hope that the fluid turbulence is appropriately described in this statistical approach at least as an effective analytical theory.[2]

In this chapter, our aim is to present (formal) functional integral representations for the Navier–Stokes equation in the following random conditions:

(1) A pure white-noise initial fluid velocity condition in Sec. 10.2.
(2) A soluble Beltrami flux with appropriate Gaussian random stirrings in Sec. 10.3.
(3) The Burger–Beltrami one-dimensional equation with a general Gaussian random stirring in Sec. 10.4.

Finally, in Appendix A we show via path-integral techniques the appearance of vortex phase factors as an important object in the advection physics of scalars on fluid fluxes relevant to the studies presented in Sec. 10.3.

10.2. The Functional Integral for Initial Fluid Velocity Random Conditions

Let us start this section by writing the Navier–Stokes equation for the velocity field of an incompressible fluid in the presence of a non-random external force $F_i(x, \tau)$ with a Gaussian (ultra-local) random initial

---

[*] Author's original results.

condition

$$\frac{\partial}{\partial \tau} v_i - \nu \Delta_x v_i + \left(v_k \frac{\partial}{\partial x_k} v^i \right)^{\mathrm{Tr}} = F_i, \tag{10.1a}$$

$$v_i(x, 0) = \varphi_i(x), \tag{10.1b}$$

$$\langle \varphi_{i1}(x_1) \varphi_{i2}(x_2) \rangle = \lambda \delta^{(3)}(x_1 - x_2) \delta_{i_1 i_2}. \tag{10.1c}$$

Let us remark that we have eliminated the pressure term $-\frac{1}{\rho} \vec{\Delta} \cdot \vec{p}$ by using the incompressibility condition $(\vec{\Delta v}) \equiv 0$ which, in turn, leads us to consider only the transverse part of the force and non-inertial field terms in Navier–Stokes equation. The transverse part of a generical vector field $\vec{W}(x, \tau)$ is defined by the relations

$$\vec{W}^{\mathrm{Tr}}(x, \xi) = \vec{W}(x, \xi) + \frac{1}{4\pi} \vec{\Delta}_x^{-1} \left(\int_{R^3} dy \frac{(\vec{\Delta}_y \vec{W})(y, \xi)}{|x - y|} \right),$$

$$\vec{W}(x, \xi) = \{ \vec{W}_i(x, \xi); \ i = 1, 2, 3 \}. \tag{10.2}$$

Our task, now, is to compute the φ-averge of the N-point fluid velocity field equation (10.1a) for an arbitrary space-time points, by means of a functional integral representation for the characteristic functional of the random fluid velocity fields $Z[J_i(x, \xi)]$; namely[3]

$$\langle V_{i1}(x_1, \xi_1, [\varphi]) \cdots V_{iN}(x_N, \xi_N, [\varphi]) \rangle_{\varphi}$$

$$= (-1)^N \frac{\delta^{(N)}}{\delta J_{i1}(x_1, \xi_1) \cdots \delta J_{iN}(x_N, \xi_N)} Z[J_i(x, \xi)]\big|_{J_i(x,\xi)=0}, \tag{10.3a}$$

where

$$Z[J_i(x, \xi)] = \int_M d\mu[V_i] \exp\left(-\int_{R^3} dx \int_0^\infty d\xi (V_i \cdot J_i)(x, \xi) \right). \tag{10.3b}$$

The functional measure $d\mu[V_i]$ in Eq. (10.3b) is defined over the functional space M of all possible realizations of the random fluid motion defined by Eq. (10.1). An explicit (formal) expression for the above functional measure should be given by the product of the usual Feynman measure weighted by a certain functional $S[V_i]$ to be determined,

$$d\mu[V_i] = D^F[V^i] \exp(S[V^i]), \tag{10.4a}$$

$$D^F[V^i] = \prod_{\substack{x \in R^3 \\ 0 < \xi < \infty \\ i=1,2,3}} (dV_i(x, \xi)). \tag{10.4b}$$

In order to determine the Weight Functional $S[V_i]$, we first rewrite the Navier–Stokes equation as a pure integral equation which has an explicit term taking into account the initial condition[4]

$$A_i[\tilde{v}] = B_i[\varphi],\tag{10.5}$$

with

$$A_i[v] = V_i(x,\xi) - \int_0^\infty ds \int_{R^3} dy \mathcal{O}_{ijk}(x-y,\xi-s)(V_j V_k)(y,s)^-$$
$$- \int_0^\infty ds \int_{R^3} dy H_{(1)}(x-y),(\xi-s)F_i(y,s)\tag{10.6a}$$

$$B_i[\varphi] = \int_{R^3} dy H_{(0)}(x-y,\xi)\varphi_i(y).\tag{10.6b}$$

Here, the Kernels $\mathcal{O}_{ijk}, H_{(1)}, H_{(0)}$ are given, respectively, by

$$\mathcal{O}_{ik\ell}(z,\xi) = -\frac{1}{2}\left(\frac{\partial}{\partial z_\ell}\bar{\mathcal{O}}_{ik} + \frac{\partial\bar{\mathcal{O}}_{i\ell}}{\partial z_k}\right)(z,\xi),\tag{10.7a}$$

$$\bar{\mathcal{O}}_{pq}(z,\xi) = \delta_{pq}\theta(\xi)H_{(0)}(|z|,\xi) + \frac{\partial^2}{\partial z_p \partial z_q}\left(\frac{2\nu\xi}{|z|}\int_0^{|z|} H_0(|z'|,\xi)dz'\right),$$
$$\tag{10.7b}$$

$$H_{(0)}(|z|,\xi) = \frac{1}{(4\pi\nu\xi)^{3/2}}\exp\left(-\frac{|z|^2}{4\pi\nu\xi}\right),\tag{10.7c}$$

$$H_{(1)}(|z|,\xi) = \theta(\xi)H_{(0)}(|z|,\xi).\tag{10.7d}$$

Let us now introduce the following functional representation for the generating functional $Z[J_i(x,\xi)]$[4,5]

$$Z[J_i(x,\xi)] = \int D^F[V^i]\langle\delta^{(F)}(V^i - \tilde{V}^i[\varphi])\rangle_\varphi$$

$$\times \exp\left(-\int_{R^3}dx\int_0^\infty d\xi(V^i\cdot J^i)(x,\xi)\right),\tag{10.8}$$

where $\delta^{(F)}(\cdot)$ denotes the delta-functional integral representation defined by the rule

$$\int_M D^F[V_i]\delta^{(F)}(V_i - A_i)\Sigma(V_i) = \Sigma(A_i),\tag{10.9}$$

with $\Sigma(V_i)$ being an arbitrary functional defined on M.

By writing the φ-average in Eq. (10.9) by means of a Gaussian functional integral in $\varphi(x)$, we obtain the following functional integral representation for the weight $S[V^i]$:

$$\exp(-S[V^i]) = \int D^F[\varphi^i] \exp\left[-\frac{1}{2\lambda}\int_{R^3} dx(\varphi^i \cdot \varphi^i)(x)\right]$$

$$\times \left\{\int D^F[K^i] \exp i \int_{R^3} dx \int_0^\infty d\xi K_i \cdot (A^i[v] - B^i|\varphi|)\right\},$$

$$(10.10)$$

where we have used the Fourier functional integral representation for the delta-functional in Eq. (10.8)

$$\delta^{(F)}(V^i - V^i[\varphi]) = \delta^{(F)}(A_i[v] - B_i[\varphi])$$

$$= DET_F\left(\frac{\delta}{\delta V_i}A_k[v]\right)\int D^F[K^i] \exp\left(i\int_{R^3} dx\right.$$

$$\left.\times \int_0^\infty d\xi K_i(A^i[v] - B^i[\varphi])(x,\xi)\right). \qquad (10.11)$$

It is worth to remark that the functional determinant in Eq. (10.11) is unity as a straightforward consequence of the fact that the Green function of the operator $\partial/\partial\xi$ is the step function (see Appendix B).

We, then, face the problem of evaluating φ and K functional integrals in Eq. (10.11).

The φ-functional integral is of Gaussian type and easily evaluated

$$\int D^F[\varphi^i] \exp\left(-\frac{1}{2\lambda}\int_{R^3} dx(\varphi_i \cdot \varphi_i)(x)\right)$$

$$\times \exp\left(-\left(i\int_{R^3} dx \int_0^\infty d\varepsilon(K_i \cdot B^i[\varphi])(x,\varepsilon)\right)\right)$$

$$= \exp\left\{-\frac{\lambda}{2}\int_{R^3} dx_1 \int_{R^3} dx_2 \int_0^\infty d\xi_1 \int_0^\infty d\xi_2 K^i(x_1,\xi_1)\right.$$

$$\left.\times \delta^{(3)}(x_1 - x_2)C(x_1,\xi_1;x_2,\xi_2)K^i(x_2,\xi_2)\right\}, \qquad (10.12)$$

where the kernel $C(x_1,\xi_1;x_2,\xi_2)$ is given by

$$C(x_1,\xi_1;x_2,\xi_2) = \int_{R^3} dz H_{(0)}(x_1 - z,\xi_1)H_{(0)}(z - x_2, -\xi_2) \qquad (10.13)$$

and is the (formal) Green function of the self-adjoint extension of the square $B^{i*}B_i$ diffusion operator

$$(B^i)^*B_i = \left(-\frac{\partial}{\partial\xi_1} - \nu\Delta_{x_1}\right)\left(\frac{\partial}{\partial\xi_1} - \nu\Delta_{x_1}\right) \qquad (10.14)$$

with the (well-posed) initial and boundary conditions

$$\lim_{\xi_1 \to 0^+} C(x_1, \xi_1; x_2, \xi_2) = \delta^{(3)}(x_1 - x_2). \tag{10.15}$$

Its explicit expression in K-momentum space is given by (see Ref. 6)

$$\tilde{C}(k; \xi_1, \xi_2) = -\frac{1}{\nu k^2} [e^{-\nu k^2 |\xi_1 - \xi_2|} - e^{-\nu k^2 (\xi_1 + \xi_2)}]. \tag{10.16}$$

As a consequence of Eq. (10.12), we have represented the weight $S[v^i]$ by a Gaussian functional integral in the $K_i(x, \xi)$ field

$$\exp(-S[v^i]) = \int D^F[K^i] \exp\left[-\frac{\lambda}{2} \int_{R^3} dx_1 dx_2 \int_0^\infty d\xi_1 d\xi_2\right.$$

$$\left. \times \left(K^i(x_1, \xi_1) C(x_1, \xi_1; x_2, \xi_2) \delta^{(3)}(x_1 - x_2) K_i(x_2, \xi_2)\right)\right]$$

$$\times \exp\left(i \int_{R^3} dx \int_{R^3} dx \int_0^\infty d\xi (K^i A_i[v])(x, \xi)\right) \tag{10.17}$$

By evaluating Eq. (10.17) we, thus, obtain the result

$$\exp(-S[v^i]) = \exp\left(-\frac{1}{2\lambda} \int_{R^3} dx_1 dx_2 \int_0^\infty d\xi_1 d\xi_2 A^i[v](x_1, \xi_1) C^{-1}\right.$$

$$\left. \times (x_1, \xi_1; x_2, \xi_2) \delta^{(3)}(x_1 - x_2) A_i[v](x_2, \xi_2)\right) \tag{10.18}$$

By noting that (see Ref. 4)

$$B_i^{-1}[A[v]] = \left(\frac{\partial}{\partial \xi} - \nu \Delta_k\right) A^i[v]$$

$$= \left(\frac{\partial}{\partial \xi} - \nu \Delta_x\right) v^i + \left(v_k \frac{\partial}{\partial x_k} v^i\right)^{\text{Tr}} - F_i^{\text{Tr}}. \tag{10.19}$$

We finally obtain the expression for the weight $S[v^i]$

$$S[v^i] = \frac{1}{2\lambda} \int_{R^3} dx \int_0^\infty d\xi d\xi' \left[\left(\frac{\partial}{\partial \xi} - \nu \Delta_x\right) A^i[v]\right]^* (x, \xi)$$

$$\times \left[\left(\frac{\partial}{\partial \xi'} - \nu \Delta_x\right) A_i[v]\right] (x, \xi')$$

$$= \frac{1}{2\lambda} \int_{R^3} dx \int_0^\infty d\xi \int_0^\infty d\xi'$$

$$\times \left[\left(\frac{\partial}{\partial \xi} - \nu \Delta_x \right) v^i + \left(v_k \frac{\partial}{\partial x_k} v^i \right)^{\mathrm{Tr}} + F_i^{\mathrm{Tr}} \right] (x, \xi)$$

$$\times \left[\left(\frac{\partial}{\partial \xi'} - \nu \Delta_x \right) v^i + \left(v_k \frac{\partial}{\partial x_k} v^i \right)^{\mathrm{Tr}} + F_i^{\mathrm{Tr}} \right] (x, \xi'). \quad (10.20)$$

We obtain, thus, our proposed functional integral representation for Eqs. (10.1a)–(10.1c)

$$Z[J_i(x, \xi)] = \int D^F[v^i] \exp(-S[v^i]) \exp \left(- \int_{R^3} dx \int_0^\infty d\xi (J_i \cdot V_i)(x, \xi) \right). \quad (10.21)$$

The above written functional integral is the main result of this section.

A perturbative analysis for Eq. (10.21) may be implemented by using the free propagator Eq. (10.16) in the context of a background field decomposition $V_i = \bar{V}_i + \beta V_i^q$ where $\bar{V}_i$ satisfies the non-random Navier–Stokes equation

$$\frac{\partial}{\partial \xi} \bar{V}_i = \nu \Delta_x \bar{V}^i - \left(\bar{V}_k \frac{\partial}{\partial x_k} \bar{V}^i \right)^{\mathrm{Tr}} + F^{\mathrm{Tr}}, \quad (10.22a)$$

with

$$\bar{V}^i(x, 0) \equiv 0, \quad (10.22b)$$

with β being a coupling constant ($\beta \ll 1$). It is worth remarking that the cross term

$$\int_{R^3} dx \int_0^\infty d\xi (\partial_i v^i \Delta_x v_i)(x, \xi) \quad (10.23a)$$

vanishes in $S[v^i]$ as a result of the boundary condition

$$v_i(x, 0) = v_i(x, \infty) = 0 \quad (10.23b)$$

for the pure random diffusion free propagator (Eq. (10.16)).

Finally, we point out that our proposed functional integral equation (10.21) differs from that proposed in Ref. 7.

10.3. An Exactly Soluble Path-Integral Model for Stochastic Beltrami Fluxes and its String Properties

In this section, our aim is to present, in our framework, an exactly soluble path-integral model for stochastic hydrodynamic motions defined here to be random regime of the *physical Navier–Stokes* equation (10.1a) in the incompressible case dominated by generalized Beltrami fluxes defined by the condition rot $\mathbf{v} = \lambda \mathbf{v}$ with λ a positive parameter.

Let us, thus, start with the usual Navier–Stokes equation (10.1a)

$$\frac{\partial \mathbf{v}}{\partial t} + \left(\frac{1}{2}\text{grad}(v^2) - (v \times \text{rot } v) \right) = -\frac{\text{grad}P}{\rho} + \nu \Delta v + \mathbf{F}^{\text{ext}}, \quad (10.24)$$

where the random stirring force is such that its satisfies the following spatially non-local Gaussian statistics in our reduced model for turbulence, i.e.

$$\langle (F^{\text{ext}})_i(\mathbf{r}, t)(F^{\text{ext}})_j(\mathbf{r}', t') \rangle = \lambda^2 \delta_{ij}((\Delta_r^{-1})\delta(\mathbf{r} - \mathbf{r}')\delta(t - t') \quad (10.25)$$

where Δ_r^{-1} denotes the Laplacean Green function.

At this point we take curl of Eq. (10.24) and consider the already mentioned Beltrami flux condition and its direct consequence, namely:

$$\lambda^2 v = \text{rot}(\text{rot } v) = \text{grad}(div \; v) - \Delta v = -\Delta v, \quad (10.26a)$$

$$v \times \text{rot } v = v \times (\lambda v) = 0, \quad (10.26b)$$

in order to replace the Navier–Stokes equation, Eq. (10.24) by the exact soluble Langevin equation for the fluid flux stirred by the external force $\mathbf{\Omega}^{\text{ext}} = \text{rot}(\mathbf{F}^{\text{ext}})$ in our proposed model of Navier–Stokes turbulence dominated by the generalized Beltrami fluxes

$$\frac{\partial v(r, t)}{\partial t} = -\nu \lambda^2 v(r, t) + \frac{1}{\lambda}\mathbf{\Omega}^{\text{ext}}(r, t) . \quad (10.27)$$

The new external stirring $\mathbf{\Omega}^{\text{ext}} = \text{rot}(\mathbf{F}^{\text{ext}})(\mathbf{r}, t)$ satisfies a Gaussian process with the following two-point correlation function:

$$\langle \Omega_\ell^{\text{ext}}(r, t)\Omega_{\ell'}^{\text{ext}}(r', t') \rangle = (\varepsilon^{\ell j k} \partial_j^{(r)})(\varepsilon^{\ell' j' k'} \partial_{j'}^{(r')}) \langle F_k^{\text{ext}}(r, t)F_{k'}^{\text{ext}}(r', t') \rangle$$

$$= \lambda^2 (\delta^{\ell\ell'}\delta^{ljj'} - \delta^{\ell j'}\delta^{\ell' j})\partial_j^{(r)}\partial_{j'}(\Delta_r^{-1}\delta^{(3)}(r - r') \times \delta(t - t'))$$

$$= \lambda^2 \delta^{\ell\ell'}\delta^{(3)}(r - r')\delta(t - t') - \lambda^2 \partial_\ell^{(r)}\partial_{\ell'}^{(r')}(\Delta_r^{-1}\delta(r - r'))\delta(t - t'). \quad (10.28)$$

It is obvious that Eq. (10.28) satisfies the incompressibility condition necessary for the incompressibility consistency of our Brownian–Langevin fluid equation (10.27) and its stochastic version below.

It is important to remark that the formal wave vectors of the Beltrami hydrodynamical motions have eddies of a fixed scale $|\mathbf{k}| = \lambda$ in our reduced model. As a consequence of this fact, we assume implicitly the same wave vector constraint in our random strings Eqs. (10.25) and (10.28).

Proceeding as in Sec. 10.2, it leads to the exact generating path integral for our Brownian reduced model, where we have used the incompressibility constraint $\partial_i^{(r)} v^i(\boldsymbol{r}, t) = 0$ to see that the spatially non-local piece of Eq. (10.28) does not contribute to the final path-integral weight equation (10.29)

$$Z[\boldsymbol{j}(\boldsymbol{r},t)] = \int D[\boldsymbol{v}(\boldsymbol{r},t)] \exp\left(i \int_{-\infty}^{+\infty} d^3\boldsymbol{r} \int_0^\infty dt (\boldsymbol{j} \cdot \boldsymbol{v})(\boldsymbol{r},t)\right)$$

$$\times \det\left[\frac{\partial}{\partial t} - \nu\lambda^2\right] \delta^{(F)}(div\ \boldsymbol{v}) \exp\left\{-\frac{1}{2} \int_{-\infty}^{+\infty} d^3\boldsymbol{r} d^3\boldsymbol{r}' \int_0^{+\infty} dt dt'\right.$$

$$\times \left(\frac{\partial v_i}{\partial t} + \nu\lambda^2 v_i\right)(\boldsymbol{r}-\boldsymbol{r}')\delta^{ii'}\delta^{(3)}(\boldsymbol{r}-\boldsymbol{r}')\delta(t-t')$$

$$\left. - \partial_i^{(r)}\partial_{i'}^{(r)}(\Delta_r^{-1}\delta(\boldsymbol{r}-\boldsymbol{r}'))\delta(t-t')\right]\left(\frac{\partial v_{i'}}{\partial t} + \nu\lambda^2 v_{i'}\right)(\boldsymbol{r}-t')\right\}$$

$$= \int D[(\boldsymbol{v}(\boldsymbol{r},t)] \exp\left(i \int_{-\infty}^{+\infty} d^3\boldsymbol{r} \int_0^\infty dt (\boldsymbol{j} \cdot \boldsymbol{v})(\boldsymbol{r},t)\right)$$

$$\times \exp\left\{-\frac{1}{2} \int_{-\infty}^{+\infty} d^3\mathbf{r} \int_0^{+\infty} dt \left(\frac{\partial \mathbf{v}}{\partial t} + \nu\lambda^2\mathbf{v}\right)^2(\mathbf{r},t)\right\}. \quad (10.29)$$

At this point it is worth to compare the exact soluble path-integral written above (note the fixed wave vector $|\boldsymbol{k}| = \gamma$ imposed implicitly in Eq. (10.29)) with that the one associated with the complete Navier–Stokes equation for ultra-local random external stirring with strength D, namely: $\langle F_i(\boldsymbol{r},t)F_j(\boldsymbol{r}',t')\rangle = D\delta^{(3)}(\boldsymbol{r}-\boldsymbol{r}')\delta(t-t')\delta_{ij}$ and full range scale $0 \leq |\mathbf{k}| < \infty$ (see Sec. 10.2)

$$Z[\mathbf{j}(\mathbf{r},t)] = \int D[\boldsymbol{v}(\boldsymbol{r},t)]\det\left[\left(\frac{\partial}{\partial t} - \nu\Delta\right)\delta_{\ell k} + \sqrt{D}\frac{\delta}{\delta v_\ell}((\boldsymbol{v}\cdot\boldsymbol{\Delta})v)_k\right]$$

$$\times \left\{-\frac{1}{2}\int_{-\infty}^{+\infty} d^3r \int_0^{+\infty} dt \left(\frac{\partial}{\partial t} - \nu\Delta\boldsymbol{v} + \sqrt{D}(\boldsymbol{v}\cdot\boldsymbol{\nabla})\boldsymbol{v}\right.\right.$$

$$\left.\left. + \frac{\mathrm{grad}P}{\rho}\right)^2(\boldsymbol{r},t)\right\}\exp\left\{i\sqrt{D}\int_{-\infty}^{+\infty} d^3r \int_0^{+\infty}(\boldsymbol{j}\cdot\boldsymbol{v})(\boldsymbol{v},t)\right\}.$$

$$(10.30a)$$

Let us remark that it is possible to eliminate the pressure term $-(1/\rho)\mathrm{grad}\,P$ in this path-integral framework by using the incompressibility condition $div(\boldsymbol{v}) = 0$, which, by its turn leads one to consider only the transverse part of the external force and of the nonlinear term in the effective action in Eq. (10.30a) (see Sec. 10.2)

$$Z[\mathbf{j}(\mathbf{r},t)] = \int D^F[\mathbf{v}(\mathbf{r},t)] \exp\left\{ -\frac{1}{2}\int_{-\infty}^{+\infty} d^3r \right.$$

$$\left. \times \int_0^{+\infty} dt \left(\frac{\partial}{\partial t}\boldsymbol{v} - \nu\Delta\boldsymbol{v} + \sqrt{D}((\boldsymbol{v}\cdot\boldsymbol{\nabla}\boldsymbol{v})^{\mathrm{tr}})^2 \right) \right\}. \quad (10.30\mathrm{b})$$

Here the transverse part of a generic vector field is defined by the expression (see Eq. (10.2))

$$(\boldsymbol{W}(r,t))^{\mathrm{tr}} = \boldsymbol{W}(r,t) - \frac{1}{4\pi}\mathrm{grad}_r(\Delta^{-1}(div\,\boldsymbol{W})). \quad (10.31)$$

Note that now one should postulate the local two-point correlation function in order to get Eq. (10.30b) $\langle F_{\mathrm{i}}^{\mathrm{tr}}(\boldsymbol{r},t)F_{\mathrm{j}}^{\mathrm{tr}}((\boldsymbol{r}',t'))\rangle = \mathcal{D}\delta^{(3)}(\boldsymbol{r}-\boldsymbol{r}')\delta(t-t')\delta_{ij}$.

It is worth to remark that Eqs. (10.3a) and (10.3b) applied to the Burger equation lead to a different path-integral than that proposed in Ref. 8, since in the path-integral framework the viscosity is not a perturbative parameter which, in our case, is $\sqrt{D}$. Besides, the propagator in the free case for *the time parameter in the range* $0 \leq t \leq \infty$ is given by (see Eq. (10.16))

$$\left(\left(\frac{\partial}{\partial t} - \nu\Delta \right)^{-1} \cdot \left(-\frac{\partial}{\partial t} - \nu\Delta \right)^{-1} \right)(k,t,t')$$

$$= -\frac{1}{\nu k^2}\left[e^{-\nu k^2|t-t'|} - e^{-\nu k^2|t+t'|} \right] \quad (10.32)$$

and differing from the Dominicis–Martin propagator suitable for the range $-\infty \leq t \leq \infty$ (Ref. 8)

$$\left(\left(\frac{\partial}{\partial t} - \nu\Delta \right)^{-1} \cdot \left(-\frac{\partial}{\partial t} - \nu\Delta \right)^{-1} \right)(k,t,t')$$

$$= \int_{-\infty}^{+\infty} dw(e^{iw(t-t')})\frac{1}{w^2 + \nu^2|\boldsymbol{k}|^4}. \quad (10.33)$$

Let us now evaluate the vortex phase factor defined by a fixed-time spatial loop $\ell = \{\ell(\sigma), a \leq \sigma \leq b\}$ in our exactly soluble model equation (10.27) in order to see the connection with strings (random surfaces) (see Appendix A for the relevance of theses non-local objects for advection phenomema)

$$\left\langle \exp\left(i \oint \mathbf{v}(\ell(\sigma),t)d\ell(\sigma) \right) \right\rangle_v$$

$$\equiv \int D^F[\mathbf{v}(\mathbf{r},t)] \exp\left\{ -\frac{1}{2} \int_{-\infty}^{+\infty} d^3\mathbf{r} \int_0^{+\infty} dt \left(\frac{\partial \mathbf{v}}{\partial t} + \nu\lambda^2\mathbf{v} \right)^2 (\mathbf{r},t) \right\}$$

$$\times \exp\left(i \oint \mathbf{v}(\ell(\sigma),t)d\ell(\sigma) \right). \tag{10.34}$$

Since the flux is of a Beltrami type in our soluble model equation (10.29), we propose to rewrite the circulation phase factor as a sum over all surfaces bounding the fixed loop ℓ by making use of Stokes theorem and by taking into account again the Beltrami condition, i.e.

$$\left\langle e^{i \oint_c \mathbf{v}\cdot d\ell} \right\rangle = \int \mathcal{D}^F[\mathbf{v}(\mathbf{r},t)] \exp\left\{ -\frac{1}{2} \int_{-\infty}^{+\infty} d^3r \int_0^{+\infty} dt \left[\mathbf{v} \left(-\frac{\partial}{\partial t^2} + \nu\lambda^2 \right) \mathbf{v} \right] \right.$$

$$\left. \times \left(\sum_S \exp\left(i\lambda \int\int_S \mathbf{v}(x,t) \cdot d\mathbf{A}(x) \right) \right) \right\}. \tag{10.35}$$

By observing now that the two-point correlation of our Brownian–Beltrami turbulent flux is exactly given by

$$\langle v_i(\mathbf{r},t)v_j(\mathbf{r}',t') \rangle_v = \int_{|\mathbf{k}|=\lambda} e^{-i\mathbf{k}\cdot(\mathbf{r}-\mathbf{r})} \frac{e^{-\nu\lambda^2|t-t'|}}{\nu\lambda^2} \theta(t-t')\delta_{ij}, \tag{10.36}$$

with $t,t' \in [0,\infty]$ and $\theta(0) = 1/2$ in this initial value problem, we can easily evaluate the average equation (10.35) and producing a strongly coupled area dependent functional for the spatial vortex phase factor in our proposed turbulent flux regime

$$W[\ell,\boldsymbol{v}] \equiv \left\langle e^{i \oint_\ell \boldsymbol{v}(\ell,t)\cdot d\ell} \right\rangle$$

$$= \sum_{\{S\}} \exp\left\{ -\frac{\lambda}{\nu} \int\int_S d\boldsymbol{A}(y) \frac{\sin(\lambda|x-y|)}{|x-y|} d\boldsymbol{A}(y) \right\}. \tag{10.37}$$

Just for completeness of our study and in order to generalize the Beltrami flux turbulence analysis represented in the main text, for the physical case of the complete wave vector range $0 \leq |\boldsymbol{k}| < \infty$ in our turbulent path-integral soluble model studies, we propose to consider a kind of generalized Beltrami condition to overcome this possible drawback of our turbulence modeling, namely:

$$\operatorname{rot} \boldsymbol{v}(\boldsymbol{r}, t) = \lambda(\boldsymbol{r}) \boldsymbol{v}(\boldsymbol{r}, t), \tag{10.38}$$

where $\lambda(\boldsymbol{r})$ is a positive function varying in the space and to be determined from a phenomenological point of view. Note that the Fourier transformed (wave vector) condition now takes the general form

$$|\boldsymbol{k}| \cdot |\tilde{\boldsymbol{v}}(\boldsymbol{k}, t)| = \int_{R^3} d^3 \boldsymbol{p} |\tilde{\lambda}(\boldsymbol{p} - \boldsymbol{q})| \cdot |\tilde{\boldsymbol{v}}(\boldsymbol{p}, t)| \tag{10.39}$$

which, by its turn, leads to the full range scale $0 < |\boldsymbol{k}| < \infty$ for the eddies hydrodynamical motions under study. By supposing that the "vortical" stirring equation (10.28) is a pure white noise process with strength D,

$$\langle \Omega_\ell^{\text{ext}}(\boldsymbol{r}, t) \Omega_{\ell'}^{\text{ext}}(\boldsymbol{r}, t') \rangle = D \cdot \delta^{\ell\ell'} \delta^3(\boldsymbol{r} - \boldsymbol{r}') \delta(t - t'). \tag{10.40}$$

It is a straightforward deduction by following our procedures as exposed in the text to arrive at an analogous Gaussian path-integral for the generalized Beltrami random hydrodynamical defined by Eq. (10.38). The generalized effective motion equation is given, in this new situation, by

$$\left[\left(\left(\frac{\partial}{\partial t} - \nu \left(\frac{\Delta\lambda}{\lambda} \right) \right) (\boldsymbol{r}) - \frac{\nu}{\lambda(\boldsymbol{r})} \frac{\partial\lambda(\boldsymbol{r})}{\partial x_e} \frac{\partial}{\partial x_e} + \nu\lambda^2(\boldsymbol{r}) \right) \delta^{ik} \right.$$

$$\left. + \nu \left(\varepsilon^{ijk} \frac{\partial\lambda(\boldsymbol{r})}{\partial x_j} \right) \right] v_k(\boldsymbol{r}, t) = \Omega_i^{\text{ext}}(\boldsymbol{r}, t). \tag{10.41}$$

The Gaussian path-integral, thus, is exactly written as

$$Z[j_i(\boldsymbol{r}, t)] = \int \prod_{i=1}^3 D^F[v_i(\boldsymbol{r}, t)] \exp\left(i \int_\infty^{+\infty} d^3 \boldsymbol{r} \int_0^\infty dt (j^i v_i)(\boldsymbol{r}, t) \right)$$

$$\times \exp\left[-\frac{1}{2D} \int_{-\infty}^{+\infty} d^3 \boldsymbol{r} \int_0^\infty dt v_k(\boldsymbol{r}, t) (M^{*ki} \cdot M^{is}) v_s(\boldsymbol{r}, t) \right]. \tag{10.42}$$

Here, the differential operators entering in the kinetic term of the turbulent path-integral are

$$M^{*ki} = \left(-\frac{\partial}{\partial t} + \frac{\nu}{\lambda(\boldsymbol{r})} \frac{\partial\lambda(\boldsymbol{r})}{\partial x_e} \frac{\partial}{\partial x^e} + \frac{\nu}{\lambda(\boldsymbol{r})} \cdot \Delta\lambda \boldsymbol{r} \right.$$

$$\left. - \nu \frac{\Delta\lambda(\boldsymbol{r})}{\lambda(\boldsymbol{r})} + \nu\lambda^2(\boldsymbol{r}) \right) \delta^{ki} + \nu\varepsilon^{kji} \frac{\partial\lambda(\boldsymbol{r})}{\partial x_j} \tag{10.43}$$

and

$$M^{is} = \left(+\frac{\partial}{\partial t} - \frac{\nu}{\lambda(r)} \frac{\partial \lambda(r)}{\partial x_e} \frac{\partial}{\partial x^e} + \nu \lambda^2(r) - \nu \frac{\Delta \lambda(r)}{\lambda(r)} \right) \delta^{is} + \nu \varepsilon^{ijs} \frac{\partial \lambda(r)}{\partial x_j}.$$
(10.44)

It is worth pointing out that the exact evaluation of the variance in Eq. (10.42) depends on the *exact* form of our rotation $\lambda(r)$ defining the Beltrami condition (10.38).

The vortex phase factor equation (10.34), takes now a form closely related to the pure self-avoiding string theory in the case of a slowly varying function $|\text{grad}\,\lambda(r)| \ll \lambda(r)$ and $\lambda(r) \sim 1$ (a very slowly r-varying function: for instance as $\lambda(r) = \lambda_0 \exp(-10^{-5}|r|^2)$)

$$\langle e^{i \oint v(\ell,t)d\ell} \rangle = \sum_{\{S\}} \exp \left\{ -\frac{1}{\nu} \int_S \int_S dA(x) \cdot \delta^{(3)}(x-y) \cdot dA(y) \right\}$$

$$\sim \exp \left(-\frac{1}{\nu} \text{area}(S) \right).$$
(10.45)

Now, if we follow Ref. 9 it is an easy task to deduce that the above written *time-fixed* vortex phase factor satisfies the famous loop wave equation for Abelian Q.C.D. at very low energy and a large number of colors. It may be written in the geometrical (infinitely differentiable loops $\ell(\sigma)$) as the following:

$$\partial_\mu^x \frac{\delta}{\delta \sigma_{\mu\nu}(x)} \left(\langle e^{i \oint v(\ell,t)d\ell} \rangle \right) = \frac{1}{\nu} \oint dy \delta^{(3)}(x-y) \langle e^{i \oint v(\ell,t)d\ell} \rangle.$$
(10.46)

The above obtained results rise hopes again that a string theory may be relevant to understand turbulence modeled as an amalgamation of "rough" roll up of random stirred fluid motions.

10.4. A Complex Trajectory Path-Integral Representation for the Burger–Beltrami Fluid Flux

The Hopf wave equation for turbulence is a master functional compressing the infinite hierarchy fluid velocity correlation functions in a single functional differential equation.[3]

In this section, our aim is to a certain extent complete the previous path-integral studies by presenting a complex *trajectory* path-integral representation for a reduced model simulating "Burger turbulence" by considering directly the "experimental observable" N-point grid velocity observable as a fundamental object of the proposed dynamically reduced Burger–Beltrami turbulent flux model below defined.

Let us start with the dynamical equation defining our one-dimensional Brownian-like fluid flux

$$\frac{\partial v(x,t)}{\partial t} + \left(v\, \frac{\partial v}{\partial x} \right)(x,t) = -(\nu\lambda^2)v(x,t) + f(x,t),$$
$$v(x,0) = g(x), \tag{10.47}$$

where we have replaced the usual fluid viscosity term $\nu d^2 v(x,t)/dx^2$ by the pure damping term $-\nu\lambda^2 v(x,t)$ (the reader should compare our proposed Brownian-like flux with that of the Navier–Stokes–Beltrami studied in Sec. 10.3).

One of the most important observable objects in fluid turbulence is the fixed velocities measurements at the grid points (x_ℓ) and *at a common observation time* t

$$\left\langle \prod_{\ell=1}^{N} \delta(v(x_\ell;t) - v_\ell) \right\rangle \tag{10.48}$$

where the average $\langle\ \rangle$ is defined by the random stirring satisfying the gaussian statistics[10]

$$\langle f(x,t)f(x',t') \rangle = k(x - x')\delta(t - t'). \tag{10.49}$$

In momentum space, the observable equation (10.48) is given by the following (grid dependent) characteristic functional (the Hopf wave functional restricted on the N-point grid)

$$\psi((x_1,\ldots,x_N);(p_1,\ldots,p_N);t) = \left\langle \exp\left(i\sum_{\ell=1}^{N} p_\ell v(x_\ell,t) \right) \right\rangle. \tag{10.50}$$

The Hopf wave equation associated with our model equation (10.48) is given in a closed form by applying straightforwardly the methods of Ref. 10

$$-i\frac{\partial}{\partial t}\psi((x_1,\ldots,x_N);(p_1,\ldots,p_N);t)$$

$$= \left\{ \sum_{\ell=1}^{N} \left[p_\ell\, \frac{\partial}{\partial p_\ell} \left(\frac{1}{p_\ell}\, \frac{\partial}{\partial x_\ell} \right) - (\nu\lambda^2)p_\ell\, \frac{\partial}{\partial p_\ell} \right] + \sum_{\ell=1,\ell'=1}^{N} (k(x_\ell - x_{\ell'})p_\ell p_{\ell'} \right\}$$

$$\times \psi((x_1,\ldots,x_N);(p_1,\ldots,p_N);t) \tag{10.51}$$

Note that we must add the deterministic initial date condition to Eq. (10.51)

$$\psi(x_1, \ldots, x_N); (p_1, \ldots, p_N); t \to 0^+) = \exp\left(i \sum_{\ell=1}^{N} p_\ell g(x_\ell)\right) \qquad (10.52)$$

Let us remark that in the physical grid on $R^3 = \{x_k^{(a)}; a = 1, 2, 3; k = 1, \ldots, N\}$, Eq. (10.51) naturally reads

$$-i\frac{\partial}{\partial t}((x_1^{(a)}, \ldots, x_N^{(a)}); (p_1^{(a)}, \ldots, p_N^{(a)}); t)$$

$$\times \sum_{\ell=1}^{N} \sum_{a=1}^{3} \left[p_\ell^{(a)} \frac{\partial}{\partial p_\ell^{(a)}} \left(\frac{1}{p_\ell^{(a)}} \frac{\partial}{\partial x_\ell^{(a)}} \right) - \nu\lambda^2 p_\ell^{(a)} \frac{\partial}{\partial p_\ell^{(a)}} \right]$$

$$+ \psi((x_1, \ldots, x_N^{(a)}); (p_1^{(a)}, \ldots, p_N^{(a)}); t)$$

$$+ \left[\sum_{\ell=1, \ell'=1}^{N} K_{ab}(x_\ell^{(a)} - x_{\ell'}^{(a)}) p_\ell^{(a)} p_\ell^{(b)} \right]$$

$$\times \psi((x_1^{(a)}, \ldots, x_N^{(a)}); (p_1^{(a)}, \ldots p_N^{(a)}); t) \qquad (10.53)$$

Hereafter as said earlier we will present our study of Eq. (10.53) for the one-dimensional case equation (10.51). By introducing the mixed coordinates defined by the transformation law.

$$p_j + x_j = u_j; \quad p_j - x_j = v_j. \qquad (10.54)$$

The turbulent wave equation (10. 51) takes the more invariant form similar to a many-particle Schrödinger equation in quantum mechanics.

$$-i\frac{\partial}{\partial t}(\psi(u_1, \ldots, u_N); (v_1, \ldots v_N); t)$$

$$= \sum_{\ell=1}^{N} \frac{1}{4} \left[\frac{\partial^2}{\partial^2 u_\ell} - \frac{\partial^2}{\partial^2 v_\ell} - \left(\frac{2}{u_\ell + v_\ell} \right) \left(\frac{\partial}{\partial u_\ell} - \frac{\partial}{\partial v_\ell} \right) \right.$$

$$\left. - \frac{\nu\lambda^2}{2}(u_\ell + v_\ell) \left(\frac{\partial}{\partial u_\ell} + \frac{\partial}{\partial v_\ell} \right) \right] \psi((u_1, \ldots, u_N); (v_1, \ldots, v_N); t)$$

$$+ \frac{1}{4} \left[\sum_{\ell=1; \ell^1=1}^{N} (u_\ell + v_\ell)(u_{\ell'} + v_{\ell'}) K\left(\frac{u_\ell - u_{\ell'} + (v_{\ell'} - v_\ell)}{2} \right) \right]$$

$$\times \psi((u_1, \ldots, u_N); (p_1, \ldots, p_N); t) \qquad (10.55)$$

and

$$\psi((u_1,\ldots,u_N);(v_1,\ldots v_N);0) = \exp\left(i\sum_{\ell=1}^{N}\left(\frac{u_\ell+v_\ell}{2}\right)g\left(\frac{u_\ell-v_\ell}{2}\right)\right)$$

(10.56)

The above written closed partial differential equation is the basic result of this section. At this point we can implement perturbative calculations for our turbulent wave equation by considering a physical slowly varying (even function) correlation function of the form

$$K(x) \approx K(0) - \frac{\kappa_0}{2}x^2; \quad |x| \ll \left(\frac{K(0)}{\kappa_0}\right)^{1/2} \equiv L,$$
$$0; \quad |x| \gg L,$$

(10.57)

which by its turn leads to the harmonic and quartic anharmonic potential which is written as

$$\sum_{\ell=1;\ell'=1}^{N} \left\{ K(0)(u_\ell u_{\ell'} + v_\ell v_{\ell'} + u_\ell v_{\ell'} + v_\ell u_{\ell'}) \right.$$
$$\left. -\frac{1}{8}\kappa_0(u_\ell + v_\ell)(u_{\ell'} + v_{\ell'}[u_\ell^2 + u_{\ell'}^2 + v_{\ell'}^2 + v_\ell^2 - 2u_\ell u_{\ell'} - 2v_\ell v_{\ell'}]) \right\}.$$

(10.58)

In the important case of the single fluid velocity average, our turbulent wave equation takes the following form, after making an analytic continuation $v \to iv$; namely

$$i\frac{\partial}{\partial t}\psi(u,v;t) = (\mathcal{L}_0 + \mathcal{L}_1)\psi(u,v;t),$$

(10.59)

with the initial condition

$$\psi(u,v;t \to 0^+) = \exp\left[i\left(\frac{u+iv}{2}\right)g\left(\frac{u-iv}{2}\right)\right].$$

(10.60)

Here the kinetic and perturbation terms are

$$\mathcal{L}_0 = -\frac{1}{4}\left(\frac{\partial^2}{\partial u^2} + \frac{\partial^2}{\partial v^2}\right) + \frac{k(0)}{4}(u^2 - v^2),$$
$$\mathcal{L}_1 = \frac{2}{u+iv}\left(\frac{\partial}{\partial u} - \frac{1}{i}\frac{\partial}{\partial v}\right) - \nu\lambda^2(u+iv)\left(\frac{\partial}{\partial u} + \frac{1}{i}\frac{\partial}{\partial v}\right).$$

(10.61)

The harmonic oscillator propagator of the kinetic term equation (10.61) is determined in a straightforward way and a Feynman diagramatic analysis may be easily implemented for $\nu \ll 1$ by the same perturbative procedure used in quantum mechanical problems. Similar remarks hold true in the general case equation (10.55).

It is worth to point out that analogous results are easily obtained in the physical case of turbulent Beltrami flux in the three-dimensional case.

Let us comment the case of general turbulent flux. In this case, although being impossible to write a closed partial differential equation as we did in Sec. 10.4, we can develop approximate schemes to solve the full functional Hopf equation by approximating the fluid shear stress tensor by finite differences, namely:

$$\left\{ \nu \frac{d^2 v(x_j; t)}{dx_j^2} \approx \frac{\nu}{\Delta}(-2v(x_j, t) + v(x_{j+1}; t) + v(x_{j-1}, t)) \right\} \quad (10.62)$$

With the uniform grid spacing $\Delta = |x_{j+1} - x_j|$.

Let us now write a trajectory functional integral representation for the initial-value problem equation (10.55) after taking into account the analytic continuation $v_\ell \to i v_\ell$ there.

As a first step to achieving our goal, we write the associated Green functional of Eq. (10.55) in an operator form (the Feynman–Dirac propagation) for the free case $k \equiv 0$

$$\overline{G}[(u_\ell, v_\ell); (u'_\ell, v'_\ell; t)]$$

$$= \left\langle (u_\ell, v_\ell) \middle| \exp\left(it \left[\sum_{\ell=1}^{N} \frac{1}{4} \Delta_{(u_\ell, v_\ell)} - \left(\frac{2}{u_\ell + i v_\ell} \right) \left(\frac{\partial}{\partial v_\ell} - \frac{1}{i} \frac{\partial}{\partial v_\ell} \right) \right. \right. \right.$$

$$\left. \left. \left. - \frac{\nu \lambda^2}{2}(u_\ell + i v_\ell) \left(\frac{\partial}{\partial u_\ell} + \frac{1}{i} \frac{\partial}{\partial v_\ell} \right) \right] \right) \middle| (u'_\ell, v'_\ell) \right\rangle. \quad (10.63)$$

As in the usual Feynman analysis we write Eq. (10.63) as an infinite product of short-time t-propagation and consider the standard short–time expansion

$$\lim_{s \to 0^+} \langle (u_\ell^{(I)}, v_\ell^{(I)}) | \exp(isH) | (v_\ell^{(I-1)}, u_\ell^{(I-1)}) \rangle$$

$$= \lim_{s \to 0^+} \int d^N p_I d^N q_I \exp \left\{ is \left[\frac{p_I^2 + q_I^2}{4} - \left(\frac{2}{u_\ell^{(I)} + i v_\ell^{(I)}} \right) \left(i p_I^{(\ell)} - q_I^{(\ell)} \right) \right] \right\}$$

$$-\frac{\nu\lambda^2}{2}(u_\ell^{(I)} + iv_\ell^{(I)})\left(ip_I^{(\ell)} + q_I^{(\ell)}\right) - \exp\left\{\sum_{\ell=1}^{N}\left[ip_I^{(\ell)}(u_I^{(\ell)} - u_{I-1}^{(\ell)})\right]\right\}$$

$$\times \exp\left\{\sum_{\ell=1}^{N}\left[iq_I^{(\ell)}(v_I^\ell - v_{I-1}^{\ell-1})\right]\right\}, \tag{10.64}$$

where H denotes the second-order differential operator inside the brackets of Eq. (10.63).

If we substitute Eq. (10.64) into the short-time product expansion of Eq. (10.63), namely

$$\langle u_\ell, v_\ell | e^{itH} | u'_{\ell'}, v'_\ell \rangle$$

$$= \prod_{I=1}^{M}\int_{-\infty}^{+\infty} du_I dv_I \left\langle (u_\ell^I, v_\ell^I) \left| \exp\left(i\left(\frac{t}{M}\right)H\right) \right| (u_\ell^{I-1}, v_\ell^{I-1}) \right\rangle \tag{10.65}$$

and evaluate the (p_I, q_I)-momenta functional integrals (see Ref. 11 for a detailed exposition), we get our searched trajectory path-integral representation for the Green-function of Eqs. (10.56)–(10.65) in the free case $k \equiv 0$

$$\bar{G}\left[(u_\ell, v_\ell); ((u_{\ell'}, v_{\ell'}); t]\right.$$

$$= \int_{\substack{\bar{U}_\ell(0)=u_{\ell'} \\ \bar{U}_\ell(t)=u_\ell}} \int_{\substack{\bar{v}_\ell(0)=v_{\ell'} \\ \bar{v}_\ell(t)=v_\ell}} \exp\left\{\frac{i}{4}\int_0^t d\sigma \left[\left(\frac{d}{d\sigma}\bar{U}(\sigma)\right)^2\right.\right.$$

$$-2\sum_{\ell=1}^{N}\left(\frac{d}{d\sigma}\bar{U}^{(\ell)}(\sigma)\right)\left(-\frac{2}{\bar{U}_\ell(\sigma) + i\bar{V}_\ell(\sigma)} - (\nu\lambda^2)(\bar{U}_\ell(\sigma) + i\bar{V}_\ell(\sigma))\right)$$

$$+\sum_{\ell=1}^{N}\left(-\frac{2}{\bar{U}_\ell(\sigma) + i\bar{V}_\ell(\sigma)} - (\nu\lambda^2)(\bar{U}_\ell(\sigma) + i\bar{V}_\ell(\sigma))\right)^2\right]\right\}$$

$$\times \exp\left\{\frac{i}{4}\int_0^t d\sigma \left[\left(\frac{\partial}{\partial\sigma}\bar{V}_\ell(\sigma)\right)^2 - 2\sum_{\ell=1}^{N}\left(\frac{\partial}{\partial\sigma}\bar{V}_\ell(\sigma)\right)\right.\right.$$

$$\times \left(-\frac{2 \cdot i}{\bar{U}_\ell(\sigma) + i\bar{V}_\ell(\sigma)} + i(\nu\lambda^2)(\bar{U}_\ell(\sigma) + i\bar{V}_\ell(\sigma))\right)^2$$

$$\times \sum_{\ell=1}^{N}\left(-\frac{2i}{\bar{U}_\ell(\sigma) + i\bar{V}_\ell(\sigma)} + i(\nu\lambda^2)(\bar{U}_\ell(\sigma) + iV_I^{(\ell)}(\sigma))\right)\right\}. \tag{10.66}$$

Note that the discrete index $I = 1, \ldots, M$ has become the continuous time parameter σ ranging in $[0, t]$.

The general $k \not\equiv 0$ case is straightforwardly obtained from Eq. (10.66) by only considering the additional weight

$$\exp\left\{i\left[\sum_{\ell=1,\ell'=1}^{N}(\bar{U}^{\ell}(\sigma)+i\bar{V}^{\ell}(\sigma))(\bar{U}^{\ell'}(\sigma)+i\bar{V}^{\ell'}(\sigma))\right.\right.$$
$$\left.\left.\times K\left(\frac{(\bar{U}^{\ell}(\sigma)-\bar{U}^{\ell'}(\sigma))+i(\bar{V}^{\ell'}(\sigma)-\bar{V}^{\ell}(\sigma))}{2}\right)\right]\right\}. \quad (10.67)$$

It is obvious from the above written N-body (complex valued!) trajectory path-integrals representations that any analytical analysis will be somewhat cumbersome. However, its numerical (Monte-Carlo and F.F.T algorithms) studies may be useful to implement approximate evaluations on applied problems.

References

1. R. Kraichnan and D. Montgomery, *Rep. Prog. Phys.* **43** (1980).
2. A. A. Migdal, *Mod. Phys. Lett. A* **6**(11), 1023 (1991).
3. A. S. Monin and A. M. Yaglom, *Statistical Fluid Mechanics* (MIT. Press, Cambridge, MA, 1971).
4. G. Rosen, *J. Math. Phys.* **12**, 812 (1971).
5. L. C. L. Botelho, *J. Phys. A: Math. Gen.* **23**, 1829 (1990).
6. W. G. Faris and G. J. Lasinio, *J. Phys. A: Math. Gen.* **15**, 3025 (1982).
7. P. C. Martin, E. O. Siggia and H. A. Rose, *Phys. Rev.* **A8**, 423 (1973); E. Medina, T. Hwa and M. Kardar, *Phys. Rev.* **39A** (1989); V. Gurarie and A. Migdal, Phys. Rev. E54, 4908 (1996); A. A. Migdal, *Int. J. Mod. Phys.* **A9**, 1197 (1994).
8. L. C. L. Botelho, *J Math. Phys.* **42**, 1689 (2001).
9. L. C. L. Botelho, *J. Math. Phys.* **30**, 2160 (1989).
10. A. M. Polyakov, *Phys. Rev. E* **52**, 4908 (1996); L. C. L. Botelho, *Int. J. Mod. Phys. B* **13**, 1663, (1999).
11. L. C. L. Botelho and R. Vilhena, *Phys. Rev. E* **49**, 1003 (1994); L. C. L. Botelho, *J. Phys. A: Math. Gen.* **34**, L131–L137 (2001).

Appendix A. A Perturbative Solution of the Burgers Equation Through the Banach Fixed Point Theorem

A very important result in the theory of metric spaces is the famous Banach–Picard fixed point theorem.

Theorem of Banach–Picard. *Let* $T: (X, d) \rightarrow (X, d)$ *be a contractive application in a given complete metric space* $(X, d) \equiv X$. *This means that for any pair* $(x, y) \in X \times X$, *we have the uniform bound with* $L < 1$

$$d(Tx, Ty) \leq L d(x, y). \tag{A.1}$$

As a consequence, we have the following results:

(a) *There is a unique solution of the functional abstract equation*

$$T\hat{x} = \hat{x}. \tag{A.2}$$

(b) *For any* $\bar{x} \in X$, *the sequence* (x_n) *defined by the recurrrence relationship*

$$x_{n+1} = T^n \bar{x} \tag{A.3}$$

converges to the (unique) solution of Eq. (A.2).

(c) *For each* n, *one has the mathematical control of the error from the approximate sequence equation (A.3) to the real solution* $\hat{x}$ *Eq. (A.2)*

$$d(x_n, \hat{x}) \leq \frac{L^{n-1}}{1 - L} d(\bar{x}, T\bar{x}). \tag{A.4}$$

We left its (easy) detailed proof to our readers.

Another important result is the following theorem related to the continuous dependence of the solution equation (A.2) on parameters.

Theorem A.1. *Let* X *be a complete metric space and* A *be a general topological space. Let* $T: X \times A \rightarrow X$ *be a family of locally contractive applications. This means that for any* $\lambda \in A$, *there is a topological neighbourhood of* λ, *namely* $V_\lambda \in A$, *and a constant* $L_\lambda < 1$, *such that* $d(T_\mu x, T_\mu y) \leq L_\lambda d(x, y)$, *for* $\mu \in V_\lambda$. *Then, we have that all fixed points of* T_λ *are continuous functions of the parameter* λ *in* A.

A curious interplay among compacity and "almost" contractive applications is the following lemma.

Lemma A.1. *Let* X *be a compact metric space and* $T: X \rightarrow X$ *an "almost" contractive application* $d(Tx, Ty) < d(x, y)$ *for* $x \neq y$. *(Example:* $T: [0, 1] \rightarrow$

$[0, 1]$; $T(x) = x - x^2$). *Let us consider the following real function on X*

$$\varphi \colon X \to R$$
$$x \to d(x, Tx). \tag{A.5}$$

Now $\varphi(x)$ has a minimum value of δ in X due to its compacity. On the other side, if $\delta > 0$, one can consider the restriction of φ to a new compact domain set $W = \{x \in X \mid d(x, Tx) \le \frac{\delta}{2}\}$. One finds in $W \subset X$, another different minimum of φ in X. So $\delta = 0$, which means that there is a $\hat{x} \in X$, such that

$$d(\hat{x}, T\hat{x}) = 0 \Leftrightarrow T\hat{x} = \hat{x}. \tag{A.6}$$

Let us examplify the Banach–Picard theorem by considering the following initial value problem for differential equations in Banach Space E, where all initial conditions are contained in the Ball of radius $R(||U_0|| \le R)$ and the nonlinearity is supposed to be Lipschitz in any bounded set in E

$$\frac{dU}{dt} = F(t, U(t)),$$
$$U(t_0) = U_0. \tag{A.7}$$

We introduce the equivalent integral form of Eq. (A.7) in our proposed complete metric space $E_{(T,R)}$ of continuous function taking values in E

$$E_{(T,R)} = \Big\{ u \in C([0, T], E); \sup_{0 \le t \le T} ||U(t)||_E \le K(R);$$
$$d(u, v) = \sup_{0 \le t \le T} \{||u(t) - v(t)||_E\} \Big\}. \tag{A.8}$$

Namely,

$$U(t) = U_0 + \int_0^t F(s, U(s))ds = \hat{F}(U)(t). \tag{A.9}$$

Let us now choose the parameter T, R in the definition of Eq. (A.8) in such a way that $\hat{F}$ maps $E_{(R,T,\alpha)}$ into $E_{(R,T,\alpha)}$ and $\hat{F}$ be a contractive application there.

Now, we can see that

$$d((\hat{F}U)(t), 0) \le ||U_0||_E + \int_0^t ||F(s, U(s))||_E ds$$

$$\le ||U_0||_E + \int_0^t ||F(U) - F(0) + F(0)||_E(s)ds$$

$$\le R + (||F(0)||_E + 2L(K_R) \cdot K_R)t. \tag{A.10}$$

Here, for $(x, y) \in E_{(T,R)}$ we have the Lipschitz property

$$\|F(x,t) - F(y,t)\|_E \leq \overbrace{\sup_{0 \leq t \leq T} \{\ell(\bar{R}, t)\}}^{L(K_R)} \cdot \|x - y\|_E$$

$$\leq 2L(K_R) \cdot K(R). \tag{A.11}$$

Let us now choose the global time evolution T in such a way that

$$R + T(\|F(0)\|_E + 2L(K_R) \cdot K_R) \leq K(R) \tag{A.12a}$$

or

$$T \leq \frac{K(R) - R}{\|F(0)\|_E + 2K(R)L(K_R)}. \tag{A.12b}$$

We have the following additional estimate:

$$d(\hat{F}U, \hat{F}V) = \sup_{0 \leq t \leq T} \left\{ \int_0^t ds \|F(s, U(s)) - F(s, V(s))\|_E \right.$$

$$\left. \leq T \cdot L(K_R) \cdot d(U, V) \right\}. \tag{A.13}$$

If one chooses the global time satisfying again the bound for any $\delta > 1$.

$$T \leq \frac{1}{L(K_R)(1 + \delta)} \tag{A.14}$$

we have a unique solution of Eq. (A.7) in $E_{(T,R)}$. For instance, if $F(0) = 0$ and $K(R) = R + b$, the compatibility of Eqs. (A.12b) and (A.14) means the existence of $b > 0$ such that $b < 2R/(\delta - 1)$.

Related to the material exposed in this chapter, let us apply the Banach–Picard fix point theorem to a Burger equation with a viscosity time-dependent parameter and zero initial conditions in $R \times R^+$

$$U(x,t) = \frac{1}{\nu} \left\{ \int_0^t d\tau \int_{-\infty}^{+\infty} dy' \right.$$

$$\left. \times \left[-\frac{\partial}{\partial y'} \left(\exp\left(-\frac{(x - y')^2}{4\tau} \right) \right) U(y', \tau)^2 \cdot \tau^{(-\frac{1}{2} + \beta)} \right] \right\} \stackrel{\text{def}}{=} F(u) \tag{A.15}$$

Let us introduce the complete metric space defined below

$$E_{(T,R)} = \{U(\cdot, t) \in C([0, T], R); \sup_{0 \leq t \leq T} (\|U(\cdot, t)\|_{L^\infty(R)}) \leq R\}. \tag{A.16}$$

Proceeding as in the previous example, we will choose T in such a way that $F(E_{(T,R)}) \subset E_{(T,R)}$. This means that (for $\beta > -\frac{1}{2}$)

$$\sup_{0 \leq t \leq T} \|(FU)(x,t)\|_{L^\infty(R)}$$

$$\leq \frac{1}{\nu} \sup_{0 \leq t \leq T} \left| \int_0^t d\zeta \left(\int_{-\infty}^{+\infty} dy |e^{-(x-y)^2/4\zeta} \frac{2}{4} |x-y| \right. \right.$$

$$\left. \left. \times \left(\frac{\zeta^\beta}{\zeta^{1/2} \cdot \zeta} \right) \right) \|U^2(y,\zeta)\|_{L^\infty} \right|$$

$$\leq \frac{R^2}{\nu} \left\{ \sup_{0 \leq t \leq T} \left| \int_0^t d\zeta \zeta^{(\beta - \frac{3}{2})} \left(\int_0^\infty dz \cdot e^{-(z)^2/4\zeta} \cdot z \right) \right| \right\}$$

$$\leq \frac{R^2}{\nu} \left\{ \sup_{0 \leq t \leq T} \left| \int_0^t d\zeta \zeta^{(\beta - \frac{1}{2})} \right| \right\} \overbrace{\left(\int_0^\infty d\bar{z} e^{-\bar{z}^2/4} \bar{z} \right)}^{c}$$

$$\leq \frac{R^2 c}{\nu(\beta + \frac{1}{2})} T^{\beta + \frac{1}{2}} < R \Leftrightarrow T^{\beta + \frac{1}{2}} < \frac{\nu(2\beta + 1)}{2cR} \tag{A.17}$$

Let us further make a choice of the global time T in such a way that $\hat{F}$ is a contractive application in $E_{(T,R)}$. In order to show such a result, let us consider the estimate below

$$\sup_{0 \leq t \leq T} \left\{ \|(TU)(x,t) - (TV)(x,t)\|_{L^\infty(R)} \right\}$$

$$\leq \frac{1}{\nu} \sup_{0 \leq t \leq T} \left| \int_0^t d\zeta \int_{-\infty}^{+\infty} \left(\frac{e^{-|x-y|^2} \cdot 2|x-y| \cdot \zeta^\beta}{4 \cdot \zeta^{1/2} \cdot \zeta} \right) \right|$$

$$\times \|U^2(\cdot,t) - V^2(\cdot,t)\|_{L^\infty(R)}$$

$$\leq \frac{1}{2\nu} \sup_{0 \leq t \leq T} \left| \int_0^t d\zeta \zeta^{\beta - \frac{1}{2}} 2 \cdot c \|U(\cdot,t) - V(\cdot,t)\|_{L^\infty(R)} \right|$$

$$\times \overbrace{(\|U(\cdot,t)\|_{L^\infty(R)} + \|V(\cdot,t)\|_{L^\infty(R)})}^{\leq 2R}$$

$$\leq \frac{2c}{\nu} R \cdot \frac{T^{\beta + \frac{1}{2}}}{(\beta + \frac{1}{2})} \left(\sup_{0 \leq \zeta \leq T} \|U(\cdot,t) - V(\cdot,t)\|_{L^\infty(R)} \right) \tag{A.18}$$

So $\hat{F}$ will be a contractive application in $E_{(T,R)}$ if

$$\frac{4cRT^{\beta + \frac{1}{2}}}{\nu(2\beta + 1)} < 1 \Leftrightarrow T^{\beta + \frac{1}{2}} < \frac{\nu(2\beta + 1)}{4cR}. \tag{A.19}$$

We have thus the existence and unicity for the Burger equation in the intergral form (A.15) for global times satisfying Eq. (A.19).

Appendix B. Some Comments on the Support of Functional Measures in Hilbert Space

Let us comment further on the application of the Minlos theorem in Hilbert spaces. In this case one has a very simple proof which holds true in general Banach spaces $(E, ||\ ||)$.

Let us thus, give a cylindrical measure $d^\infty \mu(x)$ in the algebraic dual E^{alg} of a given Banach space E (Chap. 5).

Let us suppose either that the function $||x||$ belongs to $L^1(E^{\text{alg}}, d^\infty \mu(x))$. Then the support of this cylindrical measures will be the Banach space E.

The proof is the following:

Let A be a subset of the vectorial space E^{alg} (with the topology of pontual convergence), such that $A \subset E^c$ (so $||x| = +\infty$). Let the sets be $A_\mu = \{x \in E^{\text{alg}} \mid ||x|| \geq n\}$. Then we have the set inclusion $A \subset \bigcap_{n=0}^\infty A_n$, so its measure satisfies the estimates below:

$$
\begin{aligned}
\mu(A) &\leq \liminf_n \mu(A_n) \\
&= \liminf_n \mu\{x \in E^{\text{alg}} \mid ||x|| \geq n\} \\
&\leq \liminf_n \left\{ \frac{1}{n} \int_{E^{\text{alg}}} ||x|| d^\infty \mu(x) \right\} \\
&= \liminf_n \frac{||x||_{L^1(E^{\text{alg}}, d_\mu^\infty)}}{n} = 0.
\end{aligned}
\tag{B.1}
$$

Leads us to the Minlos theorem that the support of the cylindrical measure in E^{alg} is reduced to the own Banach space E.

Note that by the Minkowisky inequality for general integrals, we have that $||x||^2 \in L^1(E^{\text{alg}}, d^\infty \mu(x))$. Now it is elementary evaluation to see that if $A^{-1} \in \oint_1(\mathcal{H})$, when $E = \mathcal{H}$, a given Hilbert space, we have that

$$
\int_{\mathcal{H}^{\text{alg}}} d_A^\infty \mu(x) \cdot ||x||^2 = \text{Tr}_{\mathcal{H}}(A^{-1}) < \infty.
\tag{B.2}
$$

This result produces another criterium for supp $d_A^\infty \mu = \mathcal{H}$ (the Minlos Theorem), when $E = \mathcal{H}$ is a Hilbert space.

It is easy to see that if

$$
\int_{\mathcal{H}} ||x|| d^\infty \mu(x) < \infty
\tag{B.3}
$$

then the Fourier-transformed functional

$$Z(j) = \int_{\mathcal{H}} e^{i(j,x)_{\mathcal{H}}} d^{\infty}\mu(x) \tag{B.4}$$

is continuous in the norm topology of $\mathcal{H}$.

Otherwise, if $Z(j)$ is not continuous in the origin $0 \in \mathcal{H}$ (without loss of generality), then there is a sequence $\{j_n\} \in \mathcal{H}$ and $\delta > 0$, such that $||j_n|| \to 0$ with

$$\begin{aligned}
\delta \leq |Z(j_n) - 1| &\leq \int_{\mathcal{H}} |e^{i(j_n,x)_{\mathcal{H}}} - 1| d^{\infty}\mu(x) \\
&\leq \int_{\mathcal{H}} |(j_n,x)| d^{\infty}\mu(x) \\
&\leq ||j_n|| \left(\int_{\mathcal{H}} ||x|| d^{\infty}\mu(x) \right) \to 0, \tag{B.5}
\end{aligned}$$

a contradiction with $\delta > 0$.

Finally, let us consider an elliptic operator B (with inverse) from the Sobolev space $\mathcal{H}^{-2m}(\Omega)$ to $\mathcal{H}^{2m}(\Omega)$. Then by the criterium given by Eq. (B.2) if

$$\mathrm{Tr}_{L^2(\Omega)}[(I + \Delta)^{+\frac{m}{2}} B^{-1} (I + \Delta)^{+\frac{m}{2}}] < \infty, \tag{B.6}$$

we will have that the path-integral written below is well-defined for $x \in \mathcal{H}^{+2m}(\Omega)$ and $j \in \mathcal{H}^{-2m}(\Omega)$. Namely

$$\exp\left(-\frac{1}{2}(j, B^{-1}j)_{L^2(\Omega)} \right) = \int_{\mathcal{H}^{+2m}(\Omega)} d_B\mu(x) \exp(i(j,x)_{L^2(\Omega)}). \tag{B.7}$$

By the Sobolev theorem which means that the embeeded below is continuous (with $\Omega \subseteq R^{\nu}$ denoting a smooth domain), one can further reduce the measure support to the Hölder α continuous function in Ω if $2m - \frac{\nu}{2} > \alpha$. Namely, we have a easy proof of the famous Wiener theorem on sample continuity of certain path-integrals in Sobolev spaces

$$\mathcal{H}^{2m}(\Omega) \subset C^{\alpha}(\Omega) \tag{B.8a}$$

The above Wiener theorem is fundamental in order to construct nontrivial examples of mathematically rigorous Euclideans path-integrals in spaces R^{ν} of higher dimensionality, since it is a trivial consequence of the Lebesgue theorem that positive continuous functions $V(x)$ generate functionals integrable in $\{\mathcal{H}^{2m}(\Omega), d_B\mu(\varphi)\}$ of the form below

$$\exp\left\{-\int_\Omega V(\varphi(x))dx\right\} \in L^1(\mathcal{H}^{2m}(\Omega), d_B\mu(\varphi)). \qquad \text{(B.8b)}$$

As a last important remark on cylindrical measures in separable Hilbert spaces, let us point at to our reader that the support of such above measures is always a σ-compact set in the norm topology of $\mathcal{H}$. In order to see such result let us consider a given dense set of $\mathcal{H}$, namely $\{x_k\}_{k \in I^+}$. Let $\{\delta_k\}_{k \in I^+}$ be a given sequence of positive real numbers with $\delta_k \to 0$. Let $\{\varepsilon_n\}$ be another sequence of positive real numbers such that $\sum_{n=1}^\infty \varepsilon_n < \varepsilon$. Now, it is straightforward to see that $\mathcal{H} \subset \bigcup_{h=1}^\infty \overline{B(x_k, \delta_k)} \subset \mathcal{H}$ and thus $\limsup \mu\{\bigcup_{k=1}^n \overline{B(x_k, \delta_k)}\} = \mu(\mathcal{H}) = 1$. As a consequence, for each n, there is a k_n, such that $\mu(\bigcup_{k=1}^{k_n} \overline{B(x_k, \delta_k)}) \geq 1 - \varepsilon$.

Now, the sets $K_\mu = \bigcap_{n=1}^\infty [\bigcup_{k=1}^{k_n} \overline{B(x_k, \delta_k)}]$ are closed and totally bounded, so they are compact sets in $\mathcal{H}$ with $\mu(\mathcal{H}) \geq 1 - \varepsilon$. Let us now choose $\varepsilon = \frac{1}{n}$ and the associated compact sets $\{K_{n,\mu}\}$. Let us further consider the compact sets $\hat{K}_{n,\mu} = \bigcup_{\ell=1}^n K_{\ell,\mu}$. We have that $\hat{K}_{n,\mu} \subseteq \hat{K}_{n+1,\mu}$, for any n and $\limsup \mu(\hat{K}_{n,\mu}) = 1$. So, $\text{supp}\, d\mu = \bigcup_{n=1}^\infty \hat{K}_{n,\mu}$, a σ-compact set of $\mathcal{H}$.

We consider now an enumerable family of cylindrical measures $\{d\mu_n\}$ in $\mathcal{H}$ satisfying the chain inclusion relationship for any $n \in I^+$

$$\text{supp}\, d\mu_n \subseteq \text{supp}\, d\mu_{n+1}.$$

Now it is straightforward to see that the compact sets $\{\hat{K}_n^{(n)}\}$, where $\text{supp}\, d\mu_m = \bigcup_{n=1}^\infty \hat{K}_n^{(m)}$, is such that $\text{supp}\{d\mu_m\} \subseteq \bigcup_{n=1}^\infty \hat{K}_n^{(n)}$, for any $m \in I^+$.

Let us consider the family of functionals induced by the restriction of this sequence of measures in any compact $\hat{K}_n^{(n)}$. Namely

$$\mu_n \to L_n^{(n)}(f) = \int_{\hat{K}_n^{(n)}} f(x) \cdot d\mu_p(x). \qquad \text{(B.8c)}$$

Here $f \in C_b(\hat{K}_n^{(n)})$. Note that all the above functionals in $\bigcup_{n=1}^\infty C_b(\hat{K}_n^{(n)})$ are bounded by B.1. By the Alaoglu–Bourbaki theorem a compact set in the weak star topology of $\left(\bigcup_{n=1}^\infty C_b(\hat{K}_n^{(n)})\right)^*$ is formed, so there is a subsequence (or better the whole sequence) converging to a unique cylindrical measure $\bar{\mu}(x)$. Namely,

$$\lim_{n \to \infty} \int_\mathcal{H} f(x)d\mu_n(x) = \int_\mathcal{H} f(x)d\bar{\mu}(x) \qquad \text{(B.8d)}$$

for any $f \in \bigcup_{n=1}^\infty C_b(\hat{K}_n^{(n)})$.

Chapter 11

The Atiyah–Singer Index Theorem:
A Heat Kernel (PDE's) Proof*

Dirac-type operators in Riemannian manifolds are mathematical objects of cornerstone importance in differential geometry[1-3] and with a growing usefulness in quantum physics.[4,5] Unfortunately, many studies related to the theory of Dirac operators are very sophisticated for applied scientists due to the use of sophisticated and cumbersome methods of algebraic topology and bundles manifolds.[1]

In this chapter, we intend to present in a relatively simple way and based in the elementary aspects of the Seeley theory of pseudo-differential operators, one of the most celebrated differential topological theorem: The Atiyah–Singer index theorem which (in one of its "palatable" version) says roughly that the trace of the evolution operator associated to a certain class of Dirac operators defined in two-dimensional orientable compact Riemannian manifolds possess a manifold topological index — its Euler–Poincaré genus.

Let us thus start with A denoting an elliptic differential operator of second order acting in the space $C_c^\infty(R^2)_{q \times q}$ formed by all these functions infinitely differentiable with compact support in R^2 and taking values in $M_{q \times q}(C)$ (the vector space of the complex matrices $q \times q$)

$$A = \sum_{|\alpha| \leq 2} a_\alpha(x) D_x^\alpha, \qquad (11.1)$$

where $\alpha = (\alpha_1, \alpha_2)$ are non-negative integers associated with the basic self-adjoint differential operators

$$D_x^\alpha = \left(\frac{1}{i} \frac{\partial}{\partial x_1} \right)^{\alpha_1} \left(\frac{1}{i} \frac{\partial}{\partial x_2} \right)^{\alpha_2}, \qquad (11.2)$$

with $A_\alpha(x) \in C_c^\infty(R^2)_{q \times q}$.

*Author's original results.

Let us now consider the (contractive) semigroup generated by the operator A through the spectral calculus for the operators in Banach spaces

$$\exp(-tA) = \frac{1}{2\pi i} \oint_{\widehat{C}} d\lambda e^{-t\lambda} (\lambda \mathbf{1}_{\varepsilon \times \varepsilon} - A)^{-1}, \tag{11.3}$$

where $\widehat{C}$ is a given closed curve containing in its interior the spectrum of A: $\sigma(A)$ which is supposed to be in the semiline R^+.

According to Seeley,[1,2] let us consider the operational symbol of the operator $(\lambda \mathbf{1}_{q \times q} - A)$ which is defined by the relationship below:

$$\sigma(A - \lambda \mathbf{1})(\xi)^{\mathrm{def}} = e^{-ix\xi}(A - \lambda \mathbf{1})e^{ix\xi}$$

$$= \left(\sum_{|\alpha| \le 2} A_\alpha(x)(\xi_1)^{\alpha_1}(\xi_2)^{\alpha_2} \right) - \lambda \mathbf{1}_{q \times q}$$

$$= \sum_{|j|=0} A_j(x, \xi, \lambda), \tag{11.4}$$

where for $0 \le j < 2$,

$$A_j(x, \xi, \lambda) = \sum_{|\alpha|=\alpha_1+\alpha_2=j} A_\alpha(x)(\xi_1)^{\alpha_1}(\xi_2)^{\alpha_2}, \tag{11.5a}$$

and for $j = 2$,

$$A_2(x, \xi, \lambda) = -\lambda \mathbf{1}_{q \times q} + \left(\sum_{|\alpha|=\alpha_1+\alpha_2=2} A_\alpha(x)(\xi_1)^{\alpha_1}(\xi_2)^{\alpha_2} \right). \tag{11.5b}$$

Let us now consider the usual Fourier transforms in $L^1(R^2) \cap L^2(R^2)$ with its operational rule

$$\widehat{F}(\xi) = \frac{1}{\sqrt{2\pi}} \int_{R^2} d^2 x e^{i\xi x} f(x) \tag{11.6a}$$

$$(Af)(x) = \frac{1}{(2\pi)^2} \int_{R^2} d^2 \xi e^{ix\xi} [\sigma(A)(x, \xi)] \widehat{F}(\xi). \tag{11.6b}$$

At this point, it is of crucial importance to note that the functions $a_j(x, \xi, \lambda)$ are all homogeneous functions of degree j when considered as functions of the variable ξ and $\sqrt{\lambda}$:

$$a_j(x, c\xi, c^2 \lambda) = (c^j) a_j(x, \xi, \lambda). \tag{11.7}$$

Now it is an important technical theorem of Seeley that the Green function of the operator $(A - \lambda \mathbf{1})$ has a symbol which is given by a series of smooth functions $\{C_{-2-j}(x, \xi)\}_{j=0}$ in the Seeley topology

$$\sigma((A - \lambda \mathbf{1})^{-1}) = \sum_{j=0}^{\infty} C_{-2-j}(x, \xi). \tag{11.8}$$

As a consequence of the above formulae, we have the following exact formulae for the inverse of any elliptic inversible operator A

$$(A^{-1}f)(x) = \int_0^{\infty} dt (e^{-tA}f)(x)$$

$$= \sum_{j=0}^{\infty} \left\{ \frac{1}{2\pi i} \oint_C d^2\xi d^2x d^2y e^{i\xi(x-y)} \left[\oint_{\widehat{C}} \frac{C_{-2-j}(x, \xi, \lambda)}{\lambda} d\lambda \right] f(y) \right\}. \tag{11.9}$$

A crucial observation can be made at this point. One can introduce a noncommutative multiplicative operation among the Seeley symbols associated to elliptic operators A and B acting on the domain $C_c^{\infty}(R^2)_{q \times q}$ namely,

$$\sigma(A) \circ \sigma(B) \stackrel{\text{def}}{=} \sigma(A \circ B), \tag{11.10}$$

where $A \circ B$ is the operator composition. Explicitly, we have that

$$\sigma(A) \circ \sigma(B) = \sum_{|\alpha|=\alpha} [D_\xi^\alpha \sigma(A)(x, \xi)][i D_x^\alpha \sigma(B)(x, \xi)]/\alpha!. \tag{11.11}$$

Here

$$\alpha! = \alpha_1! \cdot \alpha_2!. \tag{11.12}$$

From the obvious relationship

$$(A - \lambda \mathbf{1}) \circ (A - \lambda \mathbf{1})^{-1} = 1, \tag{11.13}$$

we have that

$$\sigma(A - \lambda \mathbf{1}) \circ \sigma((A - \lambda \mathbf{1})^{-1}) = 1. \tag{11.14}$$

As a consequence it yields the result

$$\left\{ \frac{1}{\alpha!} \sum_{|\alpha| \leq 2} (D_\xi^\alpha \sigma(A - \lambda \mathbf{1})(x, \xi)) i D_x^\alpha \left(\sum_{j=0}^{\infty} C_{-2-j}(x, \xi) \right) \right\} = 1, \tag{11.15}$$

or in an equivalent way

$$\frac{1}{\alpha!} \sum_{|\alpha|\leq 2} \left\{ \sum_{j=0}^{\infty} [(D_\xi^\alpha \sigma(A - \lambda\mathbf{1})(x,\xi)) i D_\xi^\alpha (C_{-2-j}(x,\xi))] \right\} = 1. \qquad (11.16)$$

The above equations can be solved by introducing the scaled variables $(t > 0)$

$$\xi = t\xi'; \quad \lambda^{1/2} = t(\lambda')^{1/2}, \qquad (11.17)$$

and

$$C_{-2-j}(x, t\xi', ((t\lambda')^{\frac{1}{2}})^2) = t^{-(2+j)} C_{-2-j}(x, \xi', \lambda'). \qquad (11.18)$$

After substituting Eq. (11.18) into Eq. (11.16) and by comparing the resulting power series in the scale factor $1/t$, one obtains the famous Seeley recurrence equations which for $t \to 1$, give us explicitly expressions of all Seeley coefficients $\{C_{(-2+j)}(x,\xi)\}$ of the Green function of the operator A (note that they are elements of $M_{q \times q}(C)$)

$$C_{-2}(x,\xi) = (a_2(x,\xi))^{-1}, \qquad (11.19)$$

$$0 = a_2(x,\xi) C_{-2-j}(x,\xi)$$

$$+ \frac{1}{\alpha!} \left(\sum_{\substack{\ell < j \\ k-|\alpha|-2-\ell=-j}} (D_\xi^\alpha a_k(x,\xi))(iD_x^\alpha C_{-2-\ell}(x,\xi)) \right). \qquad (11.20)$$

For the Laplacian-like elliptic operator below,

$$A = -\left(g_{11}(x_1,x_2) \frac{\partial^2}{\partial x_1^2} + g_{22}(x_1,x_2) \frac{\partial^2}{\partial x_2^2} \right) \mathbf{1}_{q \times q}$$

$$+ (A_1(x_1,x_2))_{q \times q} \frac{\partial}{\partial x_1} + (A_2(x_1,x_2))_{q \times q} \frac{\partial}{\partial x_2} + (A_0(x,x_2))_{q \times q}; \qquad (11.21)$$

where $\{g_{11}(x), g_{22}(x); x \in R^2\}$ are positive definite functions in $C_c^\infty(R^2)_{q \times q}$, we can evaluate exactly the Seeley relationship (Eq. (11.20)).

Now, it is a tedious evaluation to see that (important exercise left to our readers) we have the result

$$a_2(x,\xi,\lambda) = (g_{11}(x)\xi_1^2 + g_{22}(x)\xi_2^2 - \lambda\mathbf{1}), \tag{11.22a}$$

$$a_1(x,\xi,\lambda) = -iA_1(x)\xi_1 - iA_2(x)\xi_2, \tag{11.22b}$$

$$a_0(x,\xi,\lambda) = -A_0(x), \tag{11.22c}$$

$$C_{-2}(x,\xi,\lambda) = (g_{11}(x)\xi_1^2 + g_{22}(x)\xi_2^2 - \lambda\mathbf{1})^{-1}, \tag{11.22d}$$

$$c_{-3}(x,\xi,\lambda) = i(A_1(x)\xi_1 + A_2(x)\xi_2)(C_{-2}(x,\xi,\lambda))^2$$

$$- 2ig_{11}(x)\xi_1\left[\left(\frac{\partial}{\partial x_1}g_{11}\right)(x)\xi_1^2 + \left(\frac{\partial}{\partial x_1}g_{22}(x)\right)\xi_2^2\right]$$

$$\times (C_{-2}(x,\xi,\lambda))^3$$

$$- 2ig_{22}\xi_2\left[\left(\frac{\partial}{\partial x_2}g_{11}(x)\right)(\xi_1)^2 + \left(\frac{\partial}{\partial x_2}g_{22}(x)\right)(\xi_2)^2\right]$$

$$\times (C_{-2}(x,\xi,\lambda))^3. \tag{11.22e}$$

We now analyze the symbolic–operational Seeley expansion for the evolution semigroup Eq. (11.3)

$$\mathrm{Tr}_{(C_c^\infty(R^2))_{q\times q}}[\exp(-tA)]$$

$$= \sum_{j=0}^{\infty}\left(\frac{1}{2\pi}\right)^2\int_{R^2}d^2x\,\sigma(\exp(-tA))(x,\xi)$$

$$= \sum_{j=0}^{\infty}\left(\frac{1}{2\pi}\right)^2\frac{1}{2\pi i}\int_{+\infty}^{-\infty}d(is)e^{ist}\int_{R^2\times\bar{R}^2}d^2x\,d^3x\,C_{-2-j}(x,\xi,-is)$$

$$= \sum_{j=0}^{\infty}\left\{\frac{1}{t}\frac{1}{(2\pi)^3}\int_{R^2}d^2\xi\int_{R^2}d^2x\int_{-\infty}^{+\infty}e^{is}C_{-2-j}\left(x,\xi,\frac{-is}{t}\right)ds\right\}$$

$$= -\left\{\sum_{j=0}^{\infty}\frac{1}{t}\frac{1}{(2\pi)^3}\int_{R^2\times R^2}d^2x\,d^2\xi\int_{-\infty}^{+\infty}e^{is}(t)^{\left(\frac{2+\xi}{2}\right)}C_{-2-j}(x,t^{1/2}\xi,-is)\right\}$$

$$= \sum_{j=0}^{\infty}\left\{t^{\frac{(\xi-2)}{2}}(2\pi)^3\int_{R^2}d^2\xi\int_{R^2}d^2x\int_{-\infty}^{+\infty}ds\,C_{-2-j}(x,\xi,-is)\right\}. \tag{11.23}$$

After substituting Eq. (11.22) (together with the term $C_{-4}(x, \xi, \lambda)$ not written here), we have the Seeley short-time expansion for the Heat–Kernel evolution operator associated with our Laplacean A given by Eq. (11.21):

$$\text{Tr}_{C_c^\infty(R^2)_{q\times q}}[\exp(-tA)]$$

$$\overset{t\to 0^+}{\sim} \frac{1}{4\pi t}\left(\int_{R^2} d^2x \left(\sqrt{g_{11}g_{22}}\right)(x)\right) 1_{q\times q}$$

$$+ \frac{1}{4\pi}\left(\int_{R^2} d^2x \sqrt{g_{11}g_{22}}\left(-\frac{1}{6}R(g)\right)\right) 1_{q\times q}$$

$$- \frac{1}{2}\times\frac{1}{\sqrt{g_{11}g_{22}}}\left[\left(\frac{\partial}{\partial x_1}(\sqrt{g_{11}g_{22}}\cdot A_1)\right) + \left(\frac{\partial}{\partial x_1}(\sqrt{g_{11}g_{22}}\cdot A_1)\right)\right](x)$$

$$- \frac{1}{4}\left[\tilde{g}_{11}(A_1)^2 + \tilde{g}_{33}(A_2)^2 + A_0\right](x) + O(t). \tag{11.24}$$

Here $R(g)$ is the curvature scalar associated with the metric tensor $ds^2 = g_{11}(dx_1)^2 + g_{22}(dx_2)^2$. The inverse metric is denoted by $\{\tilde{g}_{11}, \tilde{g}_{22}\}$.

Let us give a proof of the famous Atiyah–Singer index theorem in the context of the heat kernel PDE's techniques.

Let us thus consider complex coordinates in a given (bounded) open subset $W \subset R^2$, supposed to be a chart of a given Riemann surface $\mathcal{M}$:

$$z = x_1 + ix_2; \quad \frac{\partial}{\partial z} = \frac{\partial}{\partial x_1} - i\frac{\partial}{\partial x_2},$$

$$\bar{z} = x_1 - ix_2; \quad \frac{\partial}{\partial \bar{z}} = \frac{\partial}{\partial x_1} + i\frac{\partial}{\partial x_2}. \tag{11.25}$$

For each integer j, let us introduce Hilbert spaces $\mathcal{H}_1$ and $\bar{\mathcal{H}}_j$ defined as[5]

$$\mathcal{H}_j = \begin{cases} \text{vectorial set of all complex valued functions which under the} \\ \text{action of a complex coordinate transformation } z = z(w) \text{ of } W, \\ \text{has the tensorial like transformation law} \\ f(z, \bar{z}) = (\partial w/\partial z)^{-j}\tilde{f}(w, \bar{w}). \end{cases}$$

$$\tag{11.26}$$

We now introduce a Hilbertian structure in $\mathcal{H}_j$ by the following inner product:

$$(g, f)_{\mathcal{H}_j} = \int_{R^2} dz d\bar{z}(\rho(z, \bar{z})^{j+1} f(z, \bar{z})\overline{g(z, \bar{z})}, \tag{11.27}$$

where $\rho(z, \bar{z})$ denotes a real continuous function in W and of the provenance of a conformal metric structure in a given Riemann surface with local chart W

$$ds^2 = \rho(z, \bar{z})dz \wedge d\bar{z}. \tag{11.28}$$

The Hilbert space $\bar{\mathcal{H}}_j$ is defined in an analogous way to the definition Eq. (11.26), but with the dual tensor transformation law

$$f(z, \bar{z}) = ((\overline{\partial w/\partial z}))^{-j} \tilde{f}(w, \bar{w}). \tag{11.29}$$

As an exercise for our readers we have that the above inner products are invariant under the action of the conformal (complex) coordinate transformations of the open set W.

Let us now introduce the following self-adjoint operators in $L^2(\bar{W})$:

$$L_j : \mathcal{H}_j \longrightarrow \bar{\mathcal{H}}_{-(j+1)}$$
$$f \longrightarrow \rho^j \partial_{\bar{z}} f \tag{11.30}$$

and its adjoint operator

$$L_j^* : \bar{\mathcal{H}}_{-(j+1)} \longrightarrow \mathcal{H}_{(j}$$
$$f \longrightarrow \rho^{-(j+1)} \partial_z f. \tag{11.31}$$

We consider the further positive definite (inversible) self-adjoint operators

$$\mathcal{L}_j : L_j^* L_j : \mathcal{H}_j \longrightarrow \mathcal{H}_j, \tag{11.32a}$$

$$\mathcal{L}_j^* : L_j L_j^+ : \bar{\mathcal{H}}_{-(j+1)} \longrightarrow \bar{\mathcal{H}}_{-(j+1)}. \tag{11.32b}$$

Note the explicit expression:

$$\mathcal{L}_j \mathcal{L}_j^* = (-\rho^{-1} \partial_{\bar{z}} \partial_z) + ((j+1)\rho^{-1}(\partial_{\bar{z}}(\ell g \rho \partial_z))), \tag{11.33a}$$

$$\mathcal{L}_j^* \mathcal{L}_j = (-\rho^{-1} \partial_z \partial_{\bar{z}}) - (j\rho^{-1}(\partial_z \ell g \rho \partial_{\bar{z}})). \tag{11.33b}$$

By using now the Seeley asymptotic expansion in $C_c^\infty(R^2)$, we have the following $t \to 0^+$ expansions for the heat kernels associated with the Dirac operator Eq. (11.33)[5]:

$$\lim_{t \to 0^+} \mathrm{Tr}_{C_c^\infty(R^2)_{2\times2}}(\exp(-t\mathcal{L}_j \mathcal{L}_j^*))$$

$$= \int_W \left(\frac{dz \wedge d\bar{z}}{2i} \right) \left(\frac{\rho(z,\bar{z})}{2\pi t} - \frac{(1+3j)}{12\pi}(\Delta \ell g \rho)(z,\bar{z}) \right) + O(t) \quad (11.34a)$$

$$\lim_{t \to 0^+} \mathrm{Tr}_{C_c^\infty(R^2)_{2 \times 2}}(\exp(-t\mathcal{L}_j^* \mathcal{L}_j))$$

$$= \int_W \frac{dz \wedge d\bar{z}}{2i} \left(\frac{\rho(z,\bar{z})}{2\pi t} + \frac{(2+3j)}{12\pi}(\Delta \ell g \rho)(z,\bar{z}) \right) + O(t). \quad (11.34b)$$

Now the famous topological heat kernel index can be obtained easily through the Atiyah–Singer definition:

$$\mathrm{index}(\mathcal{L}_j) = \lim_{j \to 0^+} [\mathrm{Tr}_{C_c^\infty(R^2)_{2 \times 2}}[\exp(-t\mathcal{L}_j \mathcal{L}_j^*) - \exp(-t\mathcal{L}_j^* \mathcal{L}_j)]]$$

$$= \frac{(1+2j)}{4\pi} \left[\int_W \frac{dz \wedge d\bar{z}}{2\pi i} \rho(z,\bar{z}) \left(+\frac{1}{\rho(z,\bar{z})} \Delta \ell g \rho(z,\bar{z}) \right) \right]$$

$$= \left(\frac{1+2j}{4\pi} \right) \chi(\mathcal{M}). \quad (11.35)$$

Here, the curvature of the Riemann surface $\mathcal{M}$ with metric $ds^2 = \rho(z,\bar{z})dz \wedge d\bar{z}$ is given by

$$R(z,\bar{z}) = \left(\frac{1}{\rho} \Delta \ell g \rho \right)(z,\bar{z}), \quad (11.36)$$

which is related to the topological invariant of the Euler–Poincaré genus of $\mathcal{M}$ through the Gauss theorem

$$\int_{\mathcal{M}} \rho(z,\bar{z}) R(z,\bar{z}) = 2\pi(2 - 2g). \quad (11.37)$$

References

1. N. Berline, E. Getzler and M. Vergne, *Heat Kernels and Dirac Operators*, Vol. 298 (Springer-Verlag); M. F. Atiyah, R. Bott and V. K. Patodi, On the heat equation and the index theorem, *Invent. Math.* **19**, 279–330 (1973); Errata, *Invent. Math.* **28**, 277–280 (1975); M. F. Atiyah and I. M. Singer, Dirac operators coupled to vector potentials, *Proc. Mat. Acad. Sci. USA* **81**, 2597–2600 (1984).
2. P. B. Gilkey, *Invariance, the Heat Equation and the Atiayah-Singer Index Theorem* (Publish or Perish, Washington, 1984).
3. J. M. Bismut, The infinitesimal Lefschetz formulas: a heat equation proof, *J. Funct. Anal.* **62**, 435–457 (1985).
4. A. S. Schwartz, *Commun. Math. Phys.* **64**, 233 (1979).

5. B. Durhuus, P. Olesen and J. L. Petersen, Polyakov's quantized string with boundary terms, *Nucl. Phys. B* **198**, 157–188 (1982).

6. R. S. Hamilton, The Ricci flux on surfaces, *Contemp. Math.* **71**, 237–262 (1988); B. Osgood, R. Phillips and P. Sarnak, Extremals of determinants of Laplacians, *J. Funct. Anal.* **80**, 148–211 (1988); L. C. L. Botelho and M. A. R. Monteiro, Fermionic determinant for massless QCD, *Phys. Rev.* **30D**, 2242–2243 (1984); A. Kokotov and D. Korotkin, Normalized-Ricci flow on Riemann surfaces and determinant of Laplacian, *Lett. Math. Phys.* **71**, 241–242 (2005).

7. L. C. L. Botelho, Path integral Bosonization for the thirring model on a Riemann surace, *Eur. Phys. Lett.* **11**, 313–318 (1990); L. Alvarez-Guamé, G. Moore and G. Vafa, *Commun. Math. Phys.* **106**, 1–35 (1980); M. F. Atiyah and I. M. Singer, Dirac operators coupled to vector potentials, *Proc. Math. Acad. Sci., USA* **81**, 2597–2600 (1984).

8. V. N. Romanov and A. S. Schwarz, Anomalies and elliptic operators, *Teoreticheskaya, Matematicheskaya Fizika* **41**(2), 190–204 (1979); M. Nakahara, *Geometry, Topology and Physics*, Graduate Student Series in Physics (IOP Publishing, London, 1990); L. C. L. Botelho, Covariant functional diffusion equation for Polyakov's Bosonic String, *Phys. Rev. D* **40**, 660–665 (1989); D'Hoker and D. Phong, The geometry of string perturbation theory, *Rev. Mod. Phys.* **60**, 917 (1988).

Appendix A. Normalized Ricci Fluxes in Closed Riemann Surfaces and the Dirac Operator in the Presence of an Abelian Gauge Connection

In this appendix we wish to expose a new result of ours on the Riemannian geometry of Riemann surfaces $\mathcal{M}$. Let us consider the nonlinear parabolic PDE's equation governing the flux dynamics of a given metric $h_{\mu\nu}(x)$ in $\mathcal{M}$, namely,

$$\frac{\partial}{\partial t}h_{\mu\nu} = (R_0 - R)h_{\mu\nu}, \qquad (A.1)$$

where R is a scalar of curvature and R_0 its mean value.

An important result in this subject of Ricci fluxes is the famous theorem of Osgood–Phillips–Sarnak, which shows that in the class of all metric structures with a fixed conformal structure, the determinant of the Laplace operator takes its maximal value in the metric of constant curvature.[6] Let us present a generalization of such result for a Dirac operator in the presence of an Abelian gauge connection (with a fixed spin structure).

Theorem A.1. *The determinant of the Dirac operator in the presence of a gauge connection in a compact orientable Riemann manifold (equivalent to a complex curve = Riemann surface) has a monotonic grown under the Ricci flux.*

Proof. Let us consider a Dirac operator with a spin structure (v^i, u^i) in the presence of a $U(1)$ Abelian gauge connection which in the physicists tensor notation reads as[7]

$$\not{D}(A, \hat{h}) = i\gamma^a \hat{e}^\mu_a \left(\partial_\mu + \frac{1}{8} W_{\mu,ab}(\hat{e}) \varepsilon^{ab} \gamma_5 + A_\mu \right). \qquad (A.2)$$

The matrices $\gamma^\mu = \hat{e}^\mu_a \gamma_a$ are the (Euclidean) Dirac matrices associated with the metric Riemann surface structure $\hat{h}^{\mu\nu} = \hat{e}^\mu_a \hat{e}^\nu_a$. Note that $\hat{h}^{\mu\nu}$ can always be written in the canonical conformal form ($\zeta \in \mathcal{M}$)

$$\hat{h}^{\mu\nu}(\zeta) = \frac{1}{\rho(\zeta)} \tilde{h}^{\mu\nu}(\zeta), \qquad (A.3)$$

where $\tilde{h}^{\mu\nu}$ is the element representative of $\hat{h}^{\mu\nu}$ in the Teichmiller modulo space of $\mathcal{M}$. We also note that the Abelian gauge connection (the Hodge Abelian connection) has the usual decomposition[7]

$$A_\mu = -\frac{\varepsilon^{\mu\nu}}{\sqrt{h}} \partial_\mu \phi + A^H_\mu \qquad (A.4)$$

with

$$\nabla^\mu (A_\mu - A^H_\mu) \equiv 0. \qquad (A.5)$$

It is useful to remark that the effects of the nontrivial topology of the genus of Riemann surface $\mathcal{M}$ reflects itself in the Hodge harmonic term A^H_μ of the $U(1)$-Connection through the Abelian differential forms (and its respective complex conjugates)[7] $\qquad \square$

$$A^H_\mu = 2h \left(\sum_{\ell=1}^g (p_\ell \alpha^\ell_\mu + g_\ell \beta^\ell_\mu) \right), \qquad (A.6a)$$

$$\alpha^i_\mu = -\bar{\Omega}_{ik}(\Omega - \bar{\Omega})^{-1}_{ks} w^f_\mu + \text{c.c}, \qquad (A.6b)$$

$$\beta^i_\mu = (\Omega - \bar{\Omega})^{-1}_{ij} w^j_\mu + \text{c.c.} \qquad (A.6c)$$

The Riemann surface matrix period Ω is defined by the usual homological relationships (Abel integrals)

$$\int_{a^j} \alpha^i = \delta_{ij}; \qquad \int_{b^i} \beta^j = \Omega_{ij}, \tag{A.7}$$

where $\{a^i\}$ and $\{b^j\}$ are the canonical homological cycles of $\mathcal{M}$.

Let us now consider the variation of the (functional) determinant of the Dirac operator $\det^{1/2}(\not{D}\not{D}^*) = \det(\not{D})$ in relation to an infinitesimal variation of the metric with a fixed conformal class[8]

$$\frac{\partial}{\partial t} \ell g \left\{ \frac{\det(\not{D}(A, \hat{h}))}{\text{area}(\mathcal{M}) \det(\not{D}(A = 0, \tilde{h}))} \right\}$$

$$= \frac{1}{\pi} \int_{\mathcal{M}} \frac{dz \wedge d\bar{z}}{2i} [\tilde{h}^{z\bar{z}} (\partial_z \varphi \partial_{\bar{z}} \varphi)(z, \bar{z})]$$

$$+ \frac{1}{12\pi} \int_{\mathcal{M}} \frac{dz \wedge d\bar{z}}{2i} \tilde{h}^{z\bar{z}} [(\ell g \rho)_t (\ell g \rho)_{z\bar{z}}]. \tag{A.8}$$

Note that this metric variation is evaluated for the normalized Ricci flux below

$$\frac{\partial}{\partial t} \ell g \rho \equiv [\ell g \rho]_t = R_0 - R, \tag{A.9a}$$

$$R_0 = \frac{2\pi(2 - g)}{\text{area}(\mathcal{M})}, \tag{A.9b}$$

$$R = -\frac{1}{\rho} \partial_{z\bar{z}} (\ell g \rho). \tag{A.9c}$$

As a consequence we have the positivity of Eq. (A.8):

$$\frac{\partial}{\partial t} \ell g \left\{ \frac{\det \not{D}(A, \hat{h})}{\text{area}(\mathcal{M}) \det(\not{D}(A = 0, \tilde{h}))} \right\} \geq 0. \tag{A.10}$$

At this point we remark that Eq. (A.10) vanishes solely for all those metric of constant curvature, which at the asymptotic limit $t \to \infty$ leads us to the usual result $(R \to R_0)$

$$\int_{\mathcal{M}} (R_0 - R) R \cdot \sqrt{h} \, dz \wedge d\bar{z} \overset{t \to \infty}{\longrightarrow} 0,$$

since for $t \to \infty$ all (smooth-C^∞) metrics on $\mathcal{M}$ are attracted to the metric of constant curvature under the action of Ricci flux with a fixed area $(\int_{\mathcal{M}} \sqrt{\hat{h}} \frac{dz \wedge d\bar{z}}{2i} = \text{area}(\mathcal{M}))$.

Index